Research Reports ESPRIT

Project 2092 · ANNIE · Vol. 1

Edited in cooperation with
the Commission of the European Communities

I. F. Croall   J. P. Mason (Eds.)

Industrial Applications of Neural Networks

Project ANNIE Handbook

Springer-Verlag

Berlin Heidelberg New York London Paris
Tokyo Hong Kong Barcelona Budapest

Editors

Ian F. Croall
John P. Mason
AEA Technology, Harwell Laboratory
Didcot, Oxon OX11 0RA, UK

ESPRIT Project 2092 "Applications of Neural Networks for Industry in Europe (ANNIE)" belongs to the Subprogramme "Information Processing Systems and Software" of ESPRIT, the European Specific Programme for Research and Development in Information Technology supported by the Commission of the European Communities.

Project ANNIE was established to investigate the performance of neural networks in a selection of applications, and to compare them with conventional approaches in order to determine their real advantages (if any). Applications studied included: pattern recognition in inspection and non-destructive testing; control systems for collision avoidance and path planning for autonomous vehicles; optimisation of airline crew scheduling. Project ANNIE was led by AEA Technology Harwell, and drew together expertise from ten European industrial and research organisations, including specialists in the conventional methods of each of the applications studied. This project handbook summarises the aims and findings of this extensive programme, and gives an indication of its considerable scope.

CR Subject Classification (1991): J. 2, I.2.6, I.5.1, I.5.4, G.1.6, J. 7

ISBN-13:978-3-540-55875-0   e-ISBN-13:978-3-642-84837-7
DOI: 10.1007/978-3-642-84837-7

Publication No. EUR 14679 EN of the
Commission of the European Communities,
Scientific and Technical Communication Unit,
Directorate-General Telecommunications, Information Industries and Innovation,
Luxembourg

LEGAL NOTICE
Neither the Commission of the European Communities nor any person acting on behalf of the Commission is responsible for the use which might be made of the following information.

Typesetting: Camera ready by authors
45/3140 – 543210 – Printed on acid-free paper

Foreword

Neural networks mimic the basic structures used in the brain for information processing and have obvious applications in advanced robotics, pattern recognition and devices that 'learn' by experience. The technology is a rapidly developing area of advanced computing, offering solutions for many information processing tasks in industry.

The ESPRIT Project 2092, ANNIE (Applications of Neural Networks for Industry in Europe) was established in November 1988 to investigate the performance of neural networks in a selection of applications, and to compare them with conventional approaches. Applications studied included:

- pattern recognition in inspection and non-destructive testing
- control systems for collision avoidance and path planning, for autonomous vehicles
- optimisation of airline crew scheduling.

ANNIE drew together expertise from ten European industrial and research organisations, led by AEA Technology, Harwell.

This project handbook summarises the aims and findings of the programme, gives an indication of its considerable scope, and offers some pointers to future directions and conclusions which may be derived from the extensive experiences of the ANNIE project team.

On behalf of the consortium, I should like to acknowledge the contribution of the CEC, not only through its financial support, but also through the invaluable guidance provided by the Project Officer, Mr Jean-Jacques Lauture, and the Reviewers: Dr Simon Garth, Dr Robert Linggard and Dr Francoise Fogelman-Soulie.

Having stepped into the project part way through to take over from its original manager, Dr Andrew Chadwick, I am personally most grateful to all of the team for the spirit of cooperation which has marked ANNIE throughout, and which has been a major element in its achievements.

J C Collingwood, ANNIE Project Manager
AEA Technology
Harwell Laboratory
Oxon OX11 0RA, UK

Table of Contents

Chapter 1

Introduction

1.1   Purpose of the Handbook

This handbook has been compiled to disseminate the results of the ANNIE project among members of the European IT industry in a usable manual format, and is a major deliverable to ESPRIT DG XIII who have funded 50% of the cost of the programme. Over three years, the ANNIE team, whose organisations have invested the other 50% of the costs, has investigated the potential of different types of neural networks for industrial applications that have proved difficult to solve using conventional computing methods.

While the first year was devoted largely to generic studies, assessing the performance of various kinds of neural network in different situations, the latter two years of the programme have been geared towards solving actual industrial problems, in areas such as control, optimisation and pattern recognition. Applications tackled include adaptive control of robotic vehicles, where advances have been made in the use of sensor data for navigation; optimisation of air crew scheduling, where progress has been made with a recognised 'difficult problem'; automatic classification of solder joint defects in printed circuit boards, for which an actual working prototype system has been developed and tested and is about to go into factory service; and classification of welding defects on the basis of ultrasonic images. The principles explored by the project are of potential benefit to a wide range of IT applications across the spectrum of European industry.

The handbook sets out to summarise the findings of the programme, to give an indication of its considerable scope, and to offer some pointers to future directions and conclusions which may be derived from the extensive experiences of the ANNIE project team. A guide to its layout is given in section 1.7.

1.2   Origins of the ANNIE Project

Recognising the future importance of neurocomputing for the European IT industry, and aware of the extent of investment already being made in the US, the original seven proposers combined to form a consortium to seek ESPRIT funding for a collaborative R&D programme to explore the industrial potential of neural networks for Europe.

Further interested partners were added to the team to give a balance of theoretical understanding, familiarity with conventional artificial intelligence methods, R&D capability, and availability of suitable applications. The hands-on industrial experience and ultimately commercial motivation of potential user companies was a further asset in the development of techniques which were industrially viable and for which there was a genuine need.

The proposal was accepted by ESPRIT, and the inaugural meeting of the project was held in September 1988.

1.3   The ANNIE Team

The ANNIE team combined the expertise of seven industrial and research organisations in four European countries, drawing upon the further resources of two subcontracted universities and a software consultancy. The team comprised:

Partners:	AEA Technology Harwell (UK), (Prime contractor)
	Alpha SAI (Greece)
	British Aerospace (UK)
	CETIM (France)
	IBP Pietzsch (Germany)
	Siemens AG (Germany)
	Artificial Intelligence Ltd (AIL - UK)
	(AIL's role changed to that of consultant in 1990)
Associate Partner:	KPMG Peat Marwick (Germany)
Subcontractors:	National Technical University of Athens (Greece)
	Technical University of Darmstadt (Germany)

The project was originally managed by Andrew Chadwick of AEA Technology, who was succeeded in 1990 by John Collingwood. A full list of team members, and the addresses of participating organisations are listed in Appendix 1.

1.4   Overall Objectives of the ANNIE Project

The overall aim of the ANNIE project, as set out in the original proposal, was to advance the knowledge and application of neural network architectures in Europe. This was intended to meet an urgent need within the European IT community; to assess and then exploit the industrial problem areas, such as pattern recognition and sensor fusion, which were proving difficult for existing technology.

When the programme was first put forward, it was known that neural networks were being exploited by over 100 companies in the USA and Japan in a range of software and hardware simulations, but were not at that time making any significant impact in Europe. The project was designed to pave the way for their take up into

manufacturing industry in the European Community through an unbiased assessment
of their merit in truly industrial environments.

Neural network architectures are based on ideas derived from many years of
research into the way in which the human brain functions. The claim is that massively
parallel adaptive networks of simple processing units (nodes) can recall information
associatively from incomplete or noisy inputs, and can derive optimal mapping
operations solely by generalisation from examples. The network architectures which
are being studied presently are defined by the transfer functions of the nodes and their
pattern and strength of interconnection. It is posited that with this technology it is
possible to go directly from the mathematical description of a problem to a
specification of a network solution, bypassing lengthy stages of program development
and debugging.

The programme set out to apply existing neural network simulators directly to real
problems, and to investigate the behaviour of different network architectures so that a
picture of likely performance could be built up which would be independent of the
software and hardware details of the simulation. Quantitative results on network
performance were obtained, covering the network size required, error rates and speed
of learning. Software was developed as a high level interface between neural networks
and applications, and results from application to real industrial problems publicised
widely among European industry.

It was realised that neural networks could be used in several important classes of
problem in industrial automation. The applications of neural architectures include
pattern recognition, optimisation, parallel signal processing and on-line adaptation,
which are frequently needed for instrumentation and control using multiple sensors
and robotics. They thrive on continuous variables, and can work with partial data, all
of which suggests they should be powerful tools for use in many industrial problems.

The research programme, which was devised to investigate the likely potential for
neural networks in IT, included the following tasks:

(i) to identify the 'difficult' functions which different neural network architectures
     can carry out efficiently, and to derive quantitative performance figures
     independent of the hardware and software used for simulating the network
(ii) to characterise selected problem areas in IT applications important for the
     competitive future of European industry, such as inspection and quality
     technology, and to develop prototype neural network solutions
(iii) to demonstrate the use of neural network architectures in application
      examples chosen to allow easy transfer of general results into widespread
      industrial use.

The original example systems designated for application development included
pattern recognition in automatic inspection, and multi-sensor fusion/trajectory
planning for robotics, both of which have been pursued with significant success. Initial
studies into parallel signal processing for on-line condition monitoring were
abandoned in favour of the optimisation work where existing methods were less
satisfactory and it was felt that neural networks had a more significant role to play.

The project aimed to develop this enabling technology towards high value applications and provide information of use to a very wide range of potential users.

The problems which these areas present contain features in common with larger-scale problems in areas of high priority to ESPRIT and the IT industry in general. As in conventional IT, it has been vital to analyse and formulate problems properly so that efficient solutions can be implemented. For neural networks this involves considerations such as how constraints can be presented to the network, the amount of training data which is necessary, the level at which the network is required to generalise, and any justification and explanation of results required by the user. A primary objective has been to analyse example problems in the required depth and to demonstrate the required methods to potential users of networks in other application areas. Necessary software for user support has been developed amongst the partners as a step towards high-level interfaces, for transparent use in commercial neurocomputer simulations.

1.5   Applications Selected for Demonstration of Neural Network Capability

It has been suggested that neural network architectures can provide the most efficient solution in at least 10% of IT applications. The ANNIE project has gone some way towards bridging the gap in Europe between the existing theoretical knowledge and the practical requirements of potential industrial users, by identifying which networks perform best in certain situations, and then demonstrating their use in real applications. The project set out to discover which IT applications could most benefit from the neural network approach, and to demonstrate the power of neural networks to solve a number of important generic problems. Application areas which offered the best chance of progress within the timescale of the project, and which were of wide industrial interest were chosen.

The generic areas studied by Project ANNIE have yielded development work on the following applications:

(i) Pattern recognition
   - the detection and classification of defects in large welds from different ultrasonic scanning methods;
   - the characterisation of solder joint defects on printed circuit boards from the 3D shape profile (provided by an optical 3D scanner)
(ii) Control
   - motor sensor interaction, signal classification, sensor fusion and control for an automatic guided vehicle (Lernfahrzeug)
   - sensor fusion in a position finding system for moving robots
(iii) Optimisation
   - comparison of conventional and neural network methods using hybrid network techniques for solving the air crew scheduling problem.

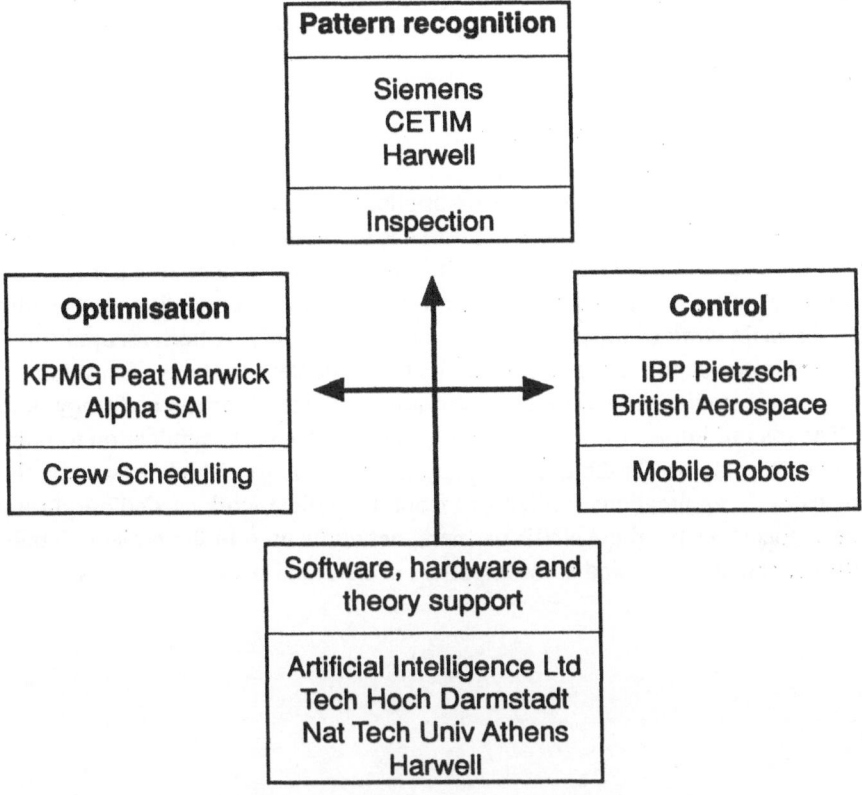

Fig 1.1 Applications of the ANNIE project

1.6 Relationship to ESPRIT Aims and Objectives

Project ANNIE was conceived as a programme which fitted well into the ESPRIT initiative, with objectives falling within its broad aims.

One of ESPRIT's objectives is to fund the provision of basic technology for the competitive requirements of the European IT Industry in the 1990s. Neural networks constitute an enabling technology with the clear potential to be used across a wide range of applications. Furthermore, if neural networks were applicable to only 10% of IT applications, their potential value to Europe has been estimated at 5 billion ECU per annum (1988 costs and market sizes). When Project ANNIE was first proposed, relatively little work had been done towards commercial application in Europe, and the programme has sought to promote understanding and trust of the new technology among potential users as well as developing networks for use in actual industrial situations.

The project has integrated user industries closely with software consultancy organisations, hardware specialists and multidisciplinary research organisations, thus furthering ESPRIT's aim of industrial cooperation in IT. While small organisations have made important contributions to the programme, the user companies have been of sufficient size to enable strong take-up of successful results in production.

1.7   Layout of the Handbook

The primary focus of this handbook is on applications, and the means of implementing them. A brief introduction to neural networks is given in Chapter 2, but the details are reserved for an appendix. The techniques for network implementation are reviewed in Chapter 3. Chapters 4, 5 and 6 deal in detail with the three application areas which formed the main work of the project. They are, respectively, pattern recognition, control and optimisation. Each of the applications chapters contains case studies as well as general studies. These are used to outline aspects of the methodology in these areas. The overall impact of neural networks and their general application to industrial processes are contained in the final chapter on methodology, which draws together lessons from the applications studied and from theoretical work carried out through the project. Appendices list the ANNIE partners, networks used in the project, details of ANNIE benchmark code, and some suppliers of neural network simulators.

Chapter 2

An Overview of Neural Networks

2.1   The Neural Network Model

We use the term *neural networks* (NNs) to denote a class of models which are
referenced in the literature under various names, including: *artificial neural systems*,
connectionist models and *parallel distributed processing models*. The name *neural
networks* was selected from the variety of currently used names because it is the most
popular and widely accepted. Moreover, it is historically justifiable and, although
perhaps less accurate than some of the other terms, has an intuitive appeal. These
names are used to denote mathematical models of brain function, which are intended
to express the massively parallel processing and distributed representation properties
of the brain (Arbib, 1964; Grossberg, 1988; Hestenes, 1986; Lippmann, 1987;
Rumelhart *et al*, 1986; Simpson, 1988).

This chapter presents a brief introduction to neural networks. The more detailed
reviews of the network architectures used in the project and some of the theoretical
studies are reported in Appendix 2.

The computational paradigm behind these networks is based on an idealised model
of a biological *neuron*. The main features of this object are that it takes signals from
other units and produces a signal which can be passed on to more units. Each unit
operates independently of every other unit. The structure of the weights which connect
the units determines the system's behaviour. Figure 2.1 shows the salient features of
the model neuron; the notation is explained in subsequent sections.

2.1.1   Connections

Units are connected to one another. *Connections* correspond to the edges of the
underlying directed graph. There is a real number associated with each connection,
which is called the *weight* of the connection. Other terms used for weight are *strength*
and the collection of strengths is referred to as the *long term memory*. We denote by
w_{ij} the weight of the connection from unit u_i to unit u_j. It is then convenient to
represent the pattern of connectivity in the network by a weight matrix W whose
elements are the weights w_{ij}. Two types of connection are usually distinguished:
excitatory and *inhibitory*. A positive weight represents an excitatory connection
whereas a negative weight represents an inhibitory connection. The pattern of

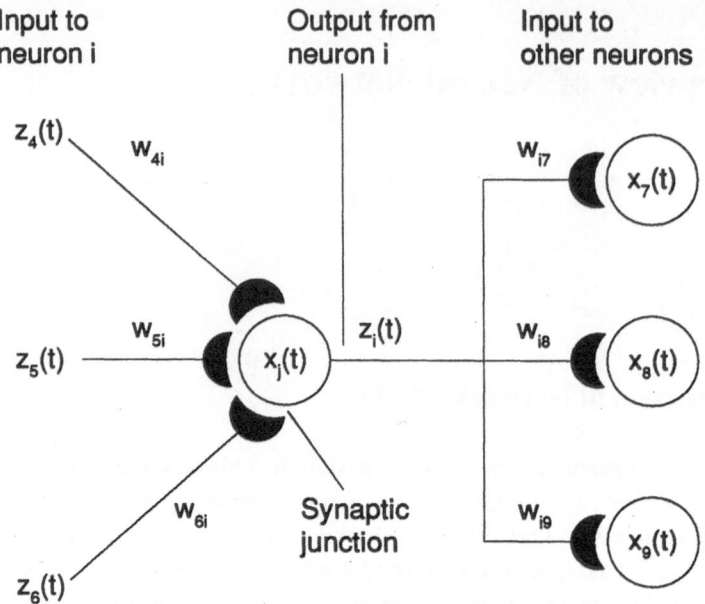

Fig 2.1 An idealised model of a biological neuron

connectivity characterises the *architecture* of the network (see section 2.2.1 and Figure 2.3).

2.1.2   States

At each point in time, there is a real number associated with each unit, which is called the *state* of the unit. The state is also referred to as the *activation* of the unit. We will denote by $x_j(t)$ the state of unit u_j at time t, and by $x(t)$ the state vector over the set of units. In various models, differing assumptions are made concerning the values taken on by the state of units. Thus, state values may be *continuous* or *discrete*. If they are continuous, they may be unbounded or bounded, eg take any real value in the interval $(0,1)$. If they are discrete, they may take on binary values or any of a set of values. In the case of binary values, which is the most frequently encountered, the values assumed are either 0 and 1, or -1 and 1 (the latter case usually being referred to as bipolar).

2.1.3   Unit Output

There is a signal associated with each unit, which is transmitted to its neighbours through connections. The value of the output signal depends directly upon the state of the unit. This dependence is expressed by the *output transfer function f_j* for unit u_j.

If we denote by $z_j(t)$ the output of unit u_j at time t then

$$z_j(t) = f_j(x_j(t)) \qquad (2.1)$$

or in vector notation

$$z(t) = f(x(t)) \qquad (2.2)$$

where $z(t)$ is the output vector over the set of units and f denotes application of the corresponding function to each individual component of the state vector. In some cases, the transfer function f_j is the identity function, hence, there is no distinction between state value and output value of a unit. Usually, transfer functions are *nonlinear threshold functions*, which are often bound to the range (0,1). Some of the most common threshold functions are the step threshold function, the ramp threshold function and the sigmoid function. Figure 2.2 shows the three commonly used threshold functions. Sometimes, the transfer function f is assumed to be a stochastic function, in the sense that the output of a unit depends upon its state in a probabilistic manner.

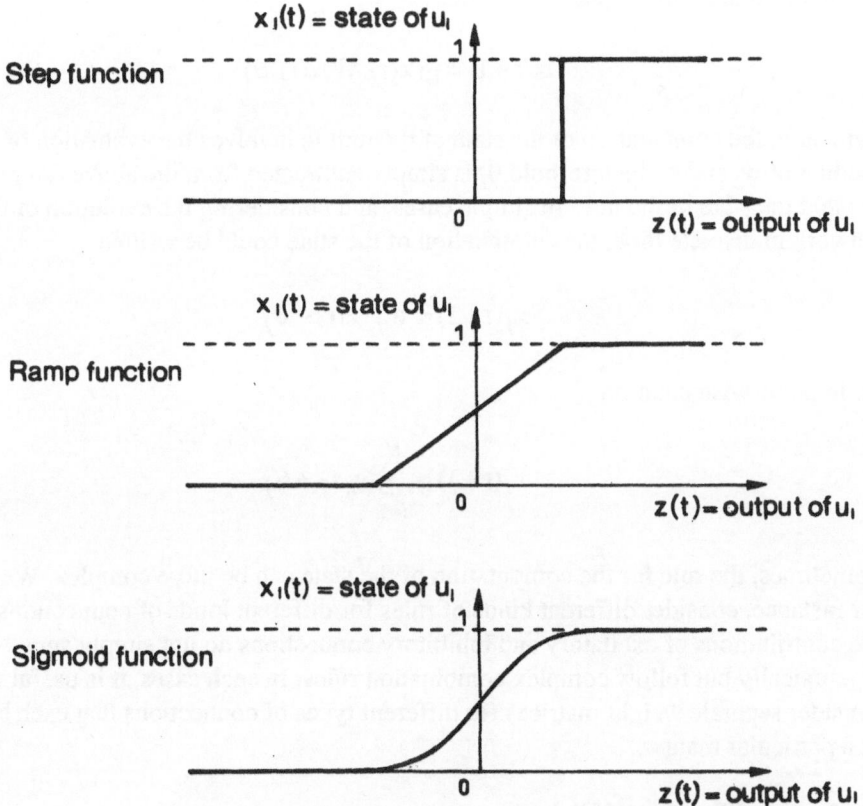

Fig 2.2　Three commonly used threshold functions

2.1.4 State Function

At any point in time, the state of a unit u_j is computed as a function g_j of its current state and the outputs of the other units to which it is connected. The function g_j, which we will call *state function*, is also referred to as *activation function*, and is often described in terms of a differential equation, which expresses the *dynamics* of the network.

In general, the computation of the new state of a unit, as time evolves, depends on the current state of activation of the unit, connection strengths of the unit to other units and the outputs of all the other units in the network. It is also common to associate a threshold θ_j with a unit u_j, even though this threshold could be formally assimilated into the activation function.

So we have in the general case, for discrete time steps, that the state of activation of a unit at time $t + 1$ is:

$$x_j(t+1) = g_j(x_j(t), \mathbf{w}_j, \mathbf{z}(t), \theta_j) \qquad (2.3)$$

where $\mathbf{w}_j$ is the column vector corresponding to the jth column of the weight matrix W. In vector notation this equation becomes:

$$\mathbf{x}(t+1) = \mathbf{g}(\mathbf{x}(t), \mathbf{W}, \mathbf{z}(t), \theta) \qquad (2.4)$$

Typically, the computation of the state of the unit u_j involves the evaluation of the dot product of $\mathbf{w}_j$ and $\mathbf{z}$. the threshold θ_j is simply subtracted from the above dot product to yield the state of the unit. In simple cases, and considering the evolution of the network in discrete time, the computation of the state could be written

$$x_j(t+1) = \mathbf{w}_j \cdot \mathbf{z}(t) - \theta_j \qquad (2.5)$$

or, in point-wise notation

$$x_j(t+1) = \sum_{i=1}^{N} w_{ij} z_j - \theta_j \qquad (2.6)$$

Sometimes, the rule for the computation of the state can be more complex. We may, for instance, consider different kinds of rules for different kinds of connections, so if the contributions of excitatory and inhibitory connections do not simply sum algebraically but follow complex combination rules. In such cases, it is useful to consider separate weight matrices for different types of connections that each behave in a particular manner.

2.1.5 Modification of Weights

Every neural network possesses knowledge which is contained in the values of the connection weights. Modifying the knowledge stored in the network as a function of experience implies a *learning rule* for changing the values of the weights (see section 2.2.2).

2.1.6   Environmental Inputs and Outputs

Neural networks can memorise information, which they subsequently use to respond to inputs from the environment. The latter can be generally viewed as a space of *input patterns*, over which is defined a time-varying probability distribution that can be history-sensitive. In the simple case, input patterns are *spatial*, ie there is a stationary probability distribution over the set of possible patterns, which can usually be enumerated. In the general case, input patterns are *spatiotemporal*, ie constitute a temporal sequence of spatial patterns. Most neural network models concern spatial input patterns, which can be considered as vectors of input values. In some models, the set of input patterns, is allowed to be arbitrary. In other models, it can be restricted to orthogonal or linearly independent vectors.

Input values can be either *analogue* (unrestricted or restricted within some given interval), or *discrete* (usually binary or bipolar). Sometimes, the inputs are assumed to be unit-interval valued, ie correspond to members of *fuzzy* sets. An input pattern is represented by the vector $a = (a_1, a_2, ..., a_n)$, where the dimension n is equal to the number of input units in the network. The possible input patterns to the network are usually numbered from 1 to M, so that the kth pattern is represented by the vector $a^k = (a_2^k, a_2^k, ..., a_n^k)$.[1] A similar notation is used for output patterns, which are composed of the values of output units. An output pattern is represented by the vector $b = (b_1, b_2, ... b_p)$ where p is the number of output units.

2.2   Principal Features

The previous section provided a brief description of the basic concepts involved in neural network models. In this section, we examine in more detail aspects, architecture and learning, which are of major importance for the characterisation of neural networks.

2.2.1   Architecture

It is possible to work with neural networks where there is little apparent structure to the connections between nodes. Figure 2.3 shows a general interconnection architecture. Some nodes are designated input nodes, another set output nodes; these two sets provide a means for the network to communicate with the outside world. The rest of the nodes are called hidden nodes. In general, input nodes may connect output nodes, nodes may mutually connect with each other and sometimes nodes may connect with themselves (if the connection is an inhibitory one, this is known as *reciprocal inhibition*).

---

[1] Sometimes the notation a+ and a- is used to distinguish the excitatory and inhibitory components of an input pattern (Grossberg)

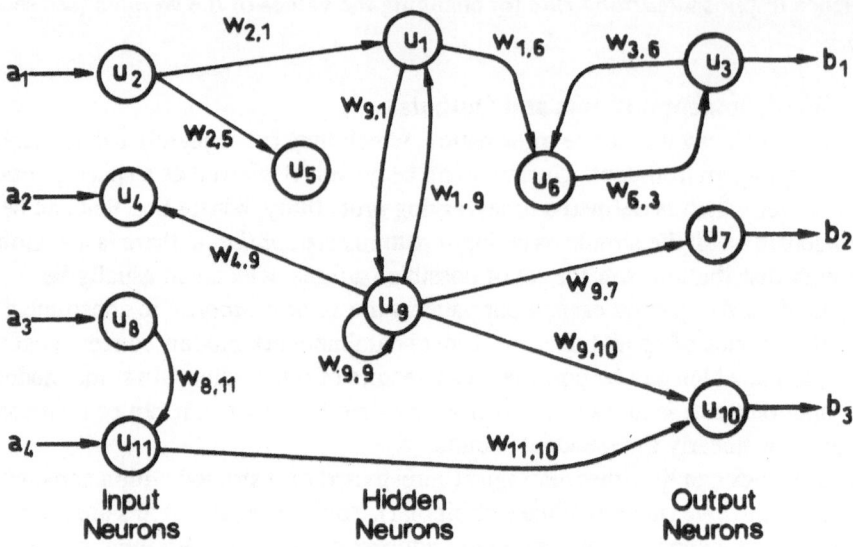

Fig 2.3 A general architecture for a first order neural network

However, most researchers impose a global structure on the organisation of the connections in the network: the *architecture* or *topology* of an *NN* is formed by organising the units into *layers* (also called *levels, fields* or *slabs*). Following this organisation, we distinguish between two interconnection schemes:

(i)   *intra-layer connections* are connections between units in the same layer, also referred to as lateral connections;

(ii) *inter-layer connections* are connections between units in different layers. Inter-layer connections allow signal propagation in one of two ways:
  • *feedforward*, ie in one direction. This type of organisation is also referred to as bottom-up or top-down
  • *feedback*, ie in either direction. In this case the flow of signals can also be recursive. Feedback architectures are also referred to as *interactive* or *recurrent*, although the latter term is often used to denote feedback connections in single-layer organisations.

Following the three types of units (input, output and hidden) we distinguish the *input layer* (that receives signals from the environment), the *output layer* (that emits signals to the environment) and *hidden layers* (that lie between the input and output layers).

We distinguish *single-layer* and *multi-layer* architectures. The single-layer organisation, in which all units are connected to one another, constitutes the most general case and is of more potential computational power than hierarchically structured multi-layer organisations. In multi-layer networks, units are often numbered by layer, instead of following a global numbering. In this case, we denote by u_j^k the jth unit of the kth layer and similarly use a superscript notation for all the quantities concerning the unit. Also, w_{ij}^k denotes the weight of theconnection from the ith unit of the $(k - 1)$th layer to the jth unit of the kth layer. In the case of two layers the superscript is usually left out.

2.2.2 Learning

The memorisation of patterns and the subsequent response of the network can be categorised into two general paradigms:

(i) *associative mapping* in which the network learns to produce a particular pattern on the set of output units whenever another particular pattern is applied on the set of input units. The associative mapping can generally be broken down into two mechanisms: *auto-association* and *hetero-association*. In the first case, an input pattern is associated with itself and the states of input and output units coincide. This is used to provide pattern completion, ie to produce a pattern whenever a portion of it or a distorted pattern is presented. In the second case, the network actually stores pairs of patterns building an association between two sets of patterns. Hetero-association is related to two recall mechanisms: *nearest-neighbour* recall, where the output pattern produced corresponds to the input pattern stored, which is closest to the pattern presented, and *interpolative* recall, where the output pattern is a similarity dependent interpolation of the patterns stored corresponding to the pattern presented. Yet another paradigm, which is a variant of associative mapping is *classification*, ie when there is a fixed set of categories into which the input patterns are to be classified

(ii) *regularity detection* in which units learn to respond to particular properties of the input patterns. Whereas in associative mapping the network stores the relationships among patterns, in regularity detection the response of each unit has a particular 'meaning'. This type of learning mechanism is essential for feature discovery and knowledge representation.

Information is stored in the weight matrix W of a neural network. *Learning* is the determination of the weights. Following the way learning is performed, we can distinguish two major categories of neural networks:

(i) *fixed networks* in which the weights cannot be changed, ie $dW/dt = 0$. In such networks, the weights are fixed *a priori* according to the problem to solve

(ii) *adaptive networks* which are able to change their weights, ie $dW / dt \neq 0$.

All learning methods used for adaptive neural networks can be classified into two major categories, *supervised learning* and *unsupervised learning*.

Supervised learning incorporates an external teacher, so that each output unit is told what its desired response to input signals ought to be. During the learning process global information may be required. Paradigms of supervised learning include *error-correction learning, reinforcement learning* and *stochastic learning*.

An important issue concerning supervised learning is the problem of error convergence, ie the minimisation of error between the desired and computed unit values. The aim is to determine a set of weights which minimises the error. One well-known method, which is common to many learning paradigms is the least mean square (LMS) convergence.

Unsupervised learning uses no external teacher and is based upon only local information. It is also referred to as *self-organisation*, in the sense that it self-organises data presented to the network and detects their emergent collective properties. Paradigms of unsupervised learning are *Hebbian learning* and *competitive learning*.

Another aspect of learning concerns the distinction or not of a separate phase, during which the network is trained, and a subsequent operation phase. We say that a neural network learns off-line if the learning phase and the operation phase are distinct. A neural network learns on-line if it learns and operates at the same time. Usually, supervised learning is performed off-line, whereas unsupervised learning is performed on-line.

We now describe briefly the principal learning rules or learning algorithms mentioned above.

Hebbian learning
All learning rules used in practice can be considered as variants of the Hebbian rule, named after Donald Hebb who formulated the concept of correlation learning, but not its mathematical formalisation (Hebb, 1949). The basic idea is that a weight should be adjusted according to the correlation of the values of the two units it connects. In a simple version this can be stated as

$$\Delta w_{ij} = \eta z_i x_j \tag{2.7}$$

Two important extensions of the simple Hebbian learning rule are *Signal Hebbian learning* and *Differential Hebbian learning* (Kosko, 1987).

Error-correction learning
The weight of the connection from unit u_i to unit u_j, where u_j is an output unit, is adjusted in proportion to the difference between the desired state value d_j and the actual state value x_j of unit u_j. This can be stated in discrete time as

$$\Delta w_{ij} = \eta z_i (d_j - x_j) \tag{2.8}$$

where η is a constant representing the learning rate. This rule is called the *Widrow-Hoff rule* (Widrow and Hoff, 1960) or *delta rule*, and constitutes a generalisation of the perceptron learning rule (Minsky *et al*, 1969).

Competitive learning

Competitive learning, introduced by Grossberg (1987), is a procedure based on neighbour inhibition (or lateral inhibition) within a layer . When a pattern is presented to a layer of the network, each unit competes with the others by sending excitatory signals to itself and inhibitory signals to all its neighbours. Eventually the unit with the highest state value will remain active while all the others become quiet. Another example of competitive learning is the self-organising feature map introduced by Kohonen (1984).

Reinforcement learning

This rule is similar to the error-correction rule in that weights are reinforced for well performed actions and punished for badly performed actions. The major difference is that, instead of using one error value for each output unit as does error-correction learning, reinforcement learning requires only one scalar value characterising the performance of the output layer, which is provided by the environment (Williams, 1987).

Stochastic learning

Stochastic learning requires the definition of an energy function depending upon the parameters of the neural network. At each step, a random weight change is performed and the resulting energy is determined. If the network energy is lower (improved) after the random weight change, then the change is accepted, otherwise acceptance or rejection of the change is decided according to a given probability distribution. This learning algorithm is based on the principle of *simulated annealing* (Kirkpatrick *et al*, 1983), which allows the avoidance of local energy minima.

2.3 Neural Networks used in ANNIE

A more detailed treatment of those networks used in ANNIE is to be found in Appendix 2 of this handbook. In the three years of the project there has been a very rapid development of the theoretical background, and hence some of the latest techniques would probably improve on the early results reported in later chapters. Where possible, references to the latest work are given.

References

Arbib (1964) Brains, machines and mathematics. McGraw-Hill, New York

Grossberg S (1988) Nonlinear neural networks: Principles, mechanisms and architectures. Neural Networks *1*, 17-61

Hebb D (1949) Organisation of behaviour. John Wiley, New York

Hestenes D (1986) How the brain works: The next great scientific revolution. In: C Smith (ed) Maximum entropy and Bayesian spectral analysis and estimation problems, Reidel Press, Boston, USA

Kirkpatrick S, Gelatt C and Vecchi M (1983) Optimisation by simulated annealing. Science *220*, 671

Kohonen T (1984) Self-organisation and associative memory. Springer-Verlag, Berlin, Germany

Kosko B (1987) Competitive adaptive bidirectional associative memories. IEEE First Int Conf on Neural Networks, San Diego CA, USA, June 1987

Lippmann (1987) An introduction to computing with neural networks. IEEE ASSP magazine *4*, 4-22

Minsky M and Papert S (1969) Perceptrons: An introduction to computational geometry. MIT Press, expanded edition

Rumelhart D E, McClelland J L and the PDP Research Group (1986) Parallel distributed processing: Explorations in the microstructure of cognition. Bradford Books *1 and 2*, MIT Press, Cambridge, Massachusetts, USA

Simpson P K (1988) A review of articifial neural systems: Foundations, paradigms, applications and implementations. Submitted to CRC critical Rreviews in Articifial Intelligence

Widrow B and Hoff M (1960) Adaptive sampled data systems - a statistical theory of adaptation. 1959 IRE Weston Convention Record *4*

Williams (1987) Reinforcement learning connectionist systems. Technical report NU-CCS-87-3, Northeastern University, College of Computer Science

Chapter 3

Implementations of Neural Networks

This chapter presents an overview of the methods currently in use for the implementation of neural networks. The simulators developed within ANNIE are also covered here. Possible hardware (VLSI) simulators are also discussed.

In recent years a great number of neural network simulators have been developed. Within ANNIE an attempt was made to compare available systems. Although the results are now somewhat dated, the methodology is still relevant. Appendix 5 to this Handbook lists some of the current (1991) suppliers of such systems.

Implementations are discussed with reference to the more popular algorithms. This is done for two reasons. First, it is these algorithms that are most likely to be implemented. Second, the previous chapter describes a number of common features in these algorithms and this description is sufficiently detailed to reason about coding strategies.

Whenever a neural network algorithm is mentioned in the chapter, it is considered to have the following characteristics:

- it fits the framework given in chapter 2. In particular, it simulates the evolution of a network of simple units, whose states change from step to step
- the algorithm simulates a network with only a few unit types
- the algorithm simulates a network with a regular connection arrangement
- only three kinds of operation need be simulated:
    - (i) calculation of unit states involving only local information
    - (ii) calculation of connection weight updates, using mainly local information
    - (iii) a limited amount of global synchronisation and calculation. The above two operations use local processing. However, these local operations are subject to netwise synchronisation. For example, learning usually occurs after the response to an input vector has been calculated, that is after a number of applications of the state rule.

Sections 3.1 and 3.2 consider the sequential implementation of such algorithms. This consists of representing the network state in main memory and then performing operations on it corresponding to the state and learning functions. It is seen that the regularity of the connection structure permits an efficient representation using arrays and matrices. This avoids the overhead of pointer following. Furthermore, since there are few unit types within the simulated network, only a few types of operations need be simulated. Thus the implementation of a neural network algorithm on a sequential computer involves a few array processing routines processing a large amount of data. This point is illustrated in section 3.2, where the sequential implementation of three network algorithms is described. Section 3.3 considers the parallel implementation of neural network algorithms. Two forms of parallelism are discussed, namely network parallelism and training parallelism. A brief summary of the salient features of network software implementations is given in section 3.4. Sections 3.5-3.9 deal with the issues of hardware implementations of neural networks. The problems of benchmarking neural network simulators is discussed in section 3.10, and the ANNIE philosophy on benchmarking is outlined. A brief overview of software simulators in section 3.11 is followed, in section 3.12, by a description of three illustrative software environments developed within ANNIE. Finally, section 3.13 is devoted to a discussion of dedicated neural network hardware.

3.1   Sequential Implementation

The sequential implementation of a network algorithm consists of a representation of the network state in main memory together with code to perform operations on it corresponding to the state and learning functions. It is therefore necessary to choose data structures to represent this state. The framework given earlier motivates particular data structure, namely arrays of unit states together with matrices for connection weights. However, there is a possibility that the weight matrix may be very large and sparse making this approach unattractive. Fortunately, this is not the case with the example algorithms of the previous chapter as it is always possible to order the units such that weight matrix is dense or regular. This being so, the task of calculating a unit's input can be performed by simple array processing routines.

3.1.1   Implementation of Regular Nets
This subsection describes the implementation of regular networks. It can be seen that the majority of the computational effort is spent executing a few matrix by vector, or vector by vector, operations. There are three things to note. First, because there are only a few basic operations, time may be spent in optimising their implementations. Second, some of the operations are matrix by vector operations which means that the vector data is used many times within an operation. An efficient implementation may take advantage of this by placing the vector data *near* to the processor such as in cache memory or registers. Such a scheme reduces the access time for this often used data and can free memory bandwidth if internal registers are used. The final value of constructing a neural network algorithm from a few matrix and vector operations is that such calculations are already commonplace in other application areas. Thus, much suitable hardware and software already exists. Hardware is discussed later where it is seen that a number of vector (pipelined) processors are marketed as neurocomputers.

In addition, scientific and signal processing software libraries contain implementations of many of the routines occurring in the section below. Using this already optimised code helps the quick and efficient coding of network algorithms.

3.1.2   Implementation of Irregular Nets

In general, the connection matrix for a neural network may be sparse and lack regular structure. One solution to this problem is the use of pointers, as in the data structures used in the Rochester Connectionist Simulator (RCS). The RCS was designed to allow the construction of nets with arbitrary connection arrangements. A net is represented by creating a collection of data structures, each representing a node. Hanging off each node data structure is a linked list of structures representing weights into the node. Each weight structure has a pointer to its source node. The RCS supports nets with more than one type of connection. This is done by interposing a structure called a site between a node and list of weights. Simulation using the RCS thus requires much pointer following.

3.2   Examples of Implementations of Neural Networks

This section describes the implementation of three common neural networks on a sequential computer. We consider one fixed network (Hopfield), one error correcting network (MLP with backpropagation learning), and one self-organising network (topology preserving map). For each network model, the network representation data structures are described, followed by a description of the code for the state and learning rules. A consistent notation is used to describe the data structures:

- with matrices, M stands for the entire matrix, $M[i]$ stands for the ith row of the matrix and $m[i,j]$ stands for an element of the matrix
- with arrays, a stands for an array and $a[i]$ for its ith element
- scalars are described by lower case letters.

The implementation of the learning and state rules are described in three steps. First the algorithm being implemented is given. Then the pseudocode that implements the algorithm is described. This pseudocode contains array operations. Finally, features of the pseudocode are described. The array operations forming the inner loops of the code are identified and any locality of data reference is noted. This occurs when an argument to an inner loop operation is not updated by the loop variable.

This section demonstrates two points made in the section above: first, each may be implemented using a few array processing routines, and second, the occurrence of matrix by vector operations allows a certain amount of data locality.

3.2.1   Hopfield Net

The Hopfield net is a fixed, asynchronous network. All units are of the same type and each is connected to every other. The data structures used to implement the network

are taken directly from the descriptive framework given in Appendix 2. The
implementation below shows that the vast majority of all the computation in the net
can be simulated by two array operations.

Data structures used
   W   matrix of weights
   out   array of outputs of units
   θ   array of unit thresholds
   x^k array representing the kth training vector
   p   number of training vector
   n   number of units in network.

State rule
Algorithm implemented. Since the Hopfield net is asynchronous, only one unit is
updated each iteration. Its new output is given by:

$$o_i(t+1) = bin\_step\left[\sum_{j=1}^{n} w_{ij}o_j(t) + \theta_i\right]$$ (3.1)

where *bin_step* represents the step function of Figure 2.2 in chapter 2.

Code. The above algorithm can be implemented very simply:
   randomly choose a unit, x;
   temp: = **W**[x].**out**;
   out[x]: = bin_step (temp + θ[x]).

Description of code. There are two things to notice about the above code. First, the
majority of the computation is performed by a dot product. Second, there is no locality
of data reference. Each item of network data is used only once. This occurs because
the net is an asynchronous net.

Learning rule
The Hopfield net is a fixed net and so connection weights are initialised with respect to
a set of training vectors, and are not adjusted afterwards.

*Algorithm implemented.*The learning procedure sets connection weights equal to the
sum of the outer products of the training vectors:

$$w_{ij} = \sum_{k=1}^{p}\left(2x_i^k - 1\right)\left(2x_j^k - 1\right)$$ (3.2)

Code. The implementation below moves the *j* loop inside the *k* loop. This allows the *j* loop to be performed by an array operation.

$$
\begin{aligned}
&\text{for } i = 1 \text{ to } n \\
&\quad \text{for } k = 1 \text{ to } p \\
&\qquad \mathbf{W}[i] = \mathbf{W}[i] + (4x^k[i] - 2).\mathbf{x}^k + (1-2x^k[i])
\end{aligned}
\tag{3.3}
$$

Description of code. The code above exhibits locality of reference. The $\mathbf{W}[i]$ array is used throughout the *k* loop. An efficient implementation would move this array into registers or equivalent at the start of each *i* loop. If this is done, it can be updated locally throughout the *k* loop.

3.2.2   MLP

The MLP has two main differences from the Hopfield network above. First, it has synchronous operation, so many units are updated at once. Second, its weights are not fixed, but can be changed by the backpropagation procedure. The implementation below involves the use of three array operations.

Data structures used
Rather than use a single connection matrix, a number of matrices are used. This is because the connection matrix given according to the descriptive framework is block structured.

$\mathbf{W}^l$ matrix of weights from layer *l*-1 to layer *l*.
out^l array of outputs of units in layer *l*
δ^l error of units in layer *l*
$width^l$ number of units in layer *l*.

State rule
Algorithm implemented. Unit activation feedforward in the MLP. This can be seen to be equivalent to a number of matrix vector multiplications sandwiching a number of nonlinear transfer functions. The propagation of signals from one layer to the next is given by:

$$
o_i^{m+1} = sig\left[\sum_{j=1}^{width(m)} w_{ij}^{m+1} o_j^m \right]
\tag{3.4}
$$

where *sig* represents the sigmoid function of Figure 2.2 in chapter 2.

Code. The above can be implemented as a collection of dot products:

```
for i = 1 to widthˡ
temp: = Wˡ[i].outˡ⁻¹
outˡ[i]:= sig (temp)
```

Description of code. Unlike the Hopfield network, there is locality of data reference when calculating the state function of the MLP. Thus, **out**$^{l-1}$ can be moved to local memory whilst each of its elements is accessed *width^l* times.

Learning rule

The learning rule described below is backpropagation learning. References to other possible learning rules for the MLP are given in Appendix 2.

Algorithms implemented. Backpropagation learning involves three stages. The first stage is the feeding forward of the input vector, as described above. The second stage involves the feeding back of errors, represented by δ. Finally, the third stage is the evaluation of the weight updates. The equation below defines the calculation of error in all layers except the output layer. It is based on the product of a transposed matrix with a vector:

$$\delta_i^m = \left[\sum_{j=1}^{width(m+1)} w_{ji}^{m+1} \delta_j^{m+1}(t) \right] . f'\left(a_i^m\right) \tag{3.5}$$

If the sigmoid function is used as the transfer function, f, then the functions derivative can be easily calculated:

$$f'\left(a_i^m\right) = f\left(a_i^m\right).\left(1 - f\left(a_i^m\right)\right) = o_i^m.\left(1 - o_i^m\right) \tag{3.6}$$

After the error has been fed back, the weights are updated:

$$w_{ij}^m := w_{ij}^m + \varepsilon.o_j^{m-1}.\delta_i^m \tag{3.7}$$

Code. There are two possible implementations of a product of a transpose matrix with a vector. The first is a collection of dot products where the matrix is accessed by columns rather than rows. The second is given here:

```
for i = 1 to width l
    δ^l-1: = δ^l-1 + δ^l[i] . W^l[i]
δ^l-1: = δ^l-1 * out^l-1 * (1 - out^l-1)
```
(3.8)

'*' is the component-wise product of vectors (Kronecker product). '1' is a vector of 1s. Weight update is accomplished by a superficially very similar piece of code:

```
for i = 1 to width l
    W^l[i]: = W^l[i] + (δ^l[i] ε) . out^l-1
```
(3.9)

Description of code. The error feedback code and the weight update code seem very similar both being based on:

$$a := a + \beta b$$

However, they are different as they exhibit different kinds of locality. In the error feedback code the first array argument, the one that is updated, can be moved to local storage. In the weight update code, however, it is the second argument that is local.

3.3   Parallel Implementation

This section describes the scope for the parallel implementation of current network algorithms. Because of the wide variety of parallel architectures, we discuss the characteristics of the algorithms that encourage parallel implementation, rather than particular implementations.

3.3.1   Network and Training Parallelism
The data dependencies of neural network algorithms permit two kinds of parallelism, namely network parallelism and training parallelism. Network parallelism occurs in the algorithms that implement a network's state or learning rules. This can be viewed as being caused either by the simultaneous operation of a number of units or by the parallelism inherent in the matrix vector operations occurring in network algorithms. Training parallelism is obtained by concurrently calculating the weight updates caused by different members of the training set. These independently calculated weight updates are then combined to update the copy or copies of network state. It can be seen that training parallelism can only provide speed up during training and cannot be used with networks that require that weight updates are implemented before the next weight updates are calculated.

3.3.2   Characteristics of both Network and Training Parallelism
An implementation can be based on either form of parallelism or can make use of both. Whichever approach is taken, the resulting parallel algorithm can be made to have the following four properties:

(i)   *Compile-time decomposition.* The decomposition of the algorithm into processes can be determined at compile time, as can the data flow between these processes. With network parallelism this occurs because networks have a fixed structure and a fixed method of operation. With training parallelism, each process corresponds to a portion of the training set and data flow corresponds to the exchange of weight update information.

Compile time decomposition allows the planning of data flows and processor loads. Planned data flow has two benefits. First, data can be sent to a processor without the processor needing to request it. This can halve the time and bandwidth required for interprocessor communication. Second, data contention can be prevented by planning that no two processors attempt to use the same datum at the same time. Planned processor loads means there is no overhead for runtime load balancing, or inefficiency caused by uneven partitioning.

(ii) *Regular decomposition.* The processes produced and the data flow is regular, in addition to being known at compile time. With network parallelism this occurs because networks have a regular connection structure and few unit types. With training parallelism, each process is doing the same work.

The regular decomposition of an algorithm greatly simplifies programming. It is far easier to coordinate a similar rather than dissimilar collection of processes. In addition, a regular decomposition may be modular and thus generalise to varying number of processors, depending on the number available.

(iii) *Possibility of purely local data flow.* In both forms of parallelism, the process decomposition can be arranged so each process only communicates with a number of local processes. This does not mean that global communication does not occur. For example, with training set parallelism, the processes can be arranged in a ring and the weight updates circulated. All processes access all other processes information, but only local communication is used.

Global data flow is difficult to implement on distributed memory systems and systems with only local interprocessor links. However on systems with broadcast communications or global fan-in, this requirement is less important.

(iv) *Reuse of data.* An *i/o bound* computation is one where each data item takes part in only one computation. A *compute bound* computation has data items that occur in many computations. Given a compute bound computation, it may be possible to increase execution rate whilst keeping the same i/o bandwidth by increasing the number of processors. With an i/o bound computation, it is not. In network parallelism, unit activations are shared, that is, the vector in a matrix calculation is reused. With training set parallelism, the network weights are shared between all processes. On a larger scale, there is much data reuse in both forms of parallelism, as network algorithms are iterative and many state variables are unchanged between iterations.

Data reuse prevents a computation from becoming i/o bound. That is, the limiting factor is not the rate at which data can be fed into the network simulation system, but the rate at which the system operates. Data reuse in parallel algorithms is equivalent to data locality in sequential algorithms. A sequential implementation that does not reflect data locality in its use of memory hierarchy can find memory bandwidth becoming a bottleneck.

It should also be noticed that the small number of operations within a network mean that there is only a small number of operations to parallelise. The global operations that occur in current network algorithms can be treated in two ways: they can be implemented using local only data flow as with the circulation of weight updates, or they can be regarded as short sequential hiccups occurring within largely parallel operation.

One review of parallel programming (Haynes *et al* 1982) concluded that there are five problems that can prevent the linear speed up of parallel algorithms with the addition of processors: the need for processor synchronisation; limited algorithmic parallelism; runtime system overhead; data contention; and i/o bottlenecks. These points are all addressed above.

To summarise, in theory the compile-time decomposition of network algorithms into a collection of regular processes with known data flow can help prevent many of the problems associated with parallel implementations. In practice, whether these problems are avoided depends on the implementation and so varies on a case to case basis.

3.4   Discussion

It should be remembered that this chapter only considers the implementation of network models that possess the features listed at the start of the chapter. That is, only algorithms that resemble a number of current algorithms are considered. With this rider, the following points can be made about the implementations of such algorithms:

- a large number of steps are required
- there are very few different operations performed within each algorithm, that is, each algorithm can be constructed out of calls to a small number of kernel routines. This is because there are a few unit types within each network, and each unit operates in a similar way
- if the network has a regular connection structure, the kernels can be coded in terms of simple array processing operations
- careful coding can allow some locality of reference between consecutive array processing operations. A second kernel call can use the same portion of one of its argument arrays as the first call. For example, consider calculating the net input of the second layer units in an MLP. Each unit has exactly the same set of source units for its connection, namely the first layer
- the data dependencies permit some parallel calculation
- network algorithms can be decomposed into processes at compile-time, and this decomposition features a regular pattern of data flow.

3.5   Hardware

Network simulation involves a large number of floating point calculations being performed on a large dataset. This requires that suitable hardware systems should have fast floating point performance and should have sufficient memory capacity to store the network state. In addition, a high rate of data flow through the floating point unit(s) is required. Each calculation involving a weight requires that a new weight be fetched, so the data path needs to have sufficient bandwidth to move these weights from wherever they are stored to the floating point units.

Due to the particular nature of network calculations, it is possible to use certain hardware specialisations to meet the above three requirements. The data dependencies within the simple array processing kernels permit a pipelined architecture to be used efficiently without stalling. Specialised vector processors, in addition to being pipelined, contain instructions that allow simple operations to be repeated along input arrays. Finally it should be remembered that it is possible to decompose the network algorithms to run on a number of processors.

Section 3.6 examines systems used for floating point calculation. The first subsection considers the architectures of currently available systems. Rather than attempt to describe all possible systems up to the performance and cost of a CRAY, it was decided to concentrate on affordable systems. Emphasis is therefore given to personal computers (PCs) and workstations, and to devices aimed at increasing their performance. The following four classes of system are considered: CISC processors together with coprocessors; RISC processors; pipelined floating point processors; and parallel systems. Neurocomputers are pipelined floating point processors marketed with network algorithm development software.

Section 3.7 describes recently announced processors: general purpose RISC and specialised maths processors are examined. These processors are still under development and so give a picture of the short term future. Two very encouraging trends are noticed. First, the floating point performance of general purpose processors continues to increase. This is partly because more attention is given to it and partly because new processors are designed to maintain high data throughput rates. Second, there are a number of new processors designed for numerical computing. The architecture of these processors mirrors the architectures of pipelined processor boards constructed from components.

Section 3.8 considers systolic computing. Systolic computing has been used for a number of years to perform array processing and recently a number of systolic implementations of neural network algorithms have been developed. The final section overviews the hardware variations seen within the chapter.

3.6   Floating Point Systems

PCs and workstations are often used as the starting point for neural network simulation. These are general purpose computing devices and often have no dedicated hardware support for floating point calculation. Simulation speed can be slow. Faster simulation can be achieved by adding special floating point hardware. This can be achieved in two ways. The simplest method is to add a tightly coupled coprocessor where the coprocessor appears to add floating point specific instructions to the main processor. This approach is discussed in the first subsection. The second approach is to add a second processing board to the system. This gives access to more advanced or specialised technologies. The following three subsections each examine one approach.

First, there are new RISC processors. These have better data throughput than older processors as the FPU is often more tightly integrated with the CPU. Second, there are pipelined floating point processors. Such boards are marketed as neurocomputers when bundled with network development software. Third, there are parallel systems.

3.6.1   Coprocessors

The floating point performance of a microprocessor can be improved by using a numeric coprocessor. As its name suggests, this coprocessor assists the operation of the main processor by speeding arithmetic calculations in hardware. The main processor executes the program and is the only processor allowed to access memory. When it needs to do a floating point calculation, rather than do it itself, it passes the operands and operator over to the coprocessor. The main processor is then free to continue until it needs its result which it obtains from the coprocessor. Concurrent execution can thus occur limited by the need for the two processors to synchronise to pass data between each other.

It is worth examining how the addition of a coprocessor can help a system meet the requirements set out at the start of the chapter. First, floating point performance is clearly enhanced both by dedicated hardware support and by allowing the calculations to happen in parallel, with the main processor accessing main memory for more operands. Memory for holding the new state is available with the normal PC workstation memory. However, there can be a problem with data throughput. Data retrieval consists of two parts, the memory to main processor transfer and the main processor to coprocessor transfer. The first transfer can happen in parallel with coprocessor calculations. All the coprocessors listed at the end of this section are sufficiently slow that this main memory access is not a bottleneck and occurs 'invisibly' during coprocessor computation. In contrast, the other transfer, that between processor and coprocessor, does affect total computation time. The transfer occupies both processors, so the time taken must be added to the rate determining thread.

There are three main coprocessor architectures. First there are *standard* coprocessors. A standard coprocessor receives both its operands and operator as data over the data bus from the main coprocessor. Secondly there are *memory mapped* coprocessors. These receive operands from the main processor via the data bus operator and other control information is sent over the data bus. This reduces the time spent in processor to coprocessor communication at the expense of slightly reducing the address space for main memory. The third type of coprocessor is the *integrated* coprocessor. Here the coprocessor is designed into the instruction and chip set of the main CPU. As a result interprocessor communication is much better: integrated coprocessors waste less time on communication than memory mapped coprocessors which in turn are better than standard coprocessors.

Coprocessors for the i386 and the M680x0

Intel and Motorola produce standard coprocessors for their 32 bit processors. Intel produce the i80387 coprocessors for the i80386. Motorola produce the M68881 and the newer M68882 for both the M68020 and the M68030. Weitek produce memory mapped alternatives to these coprocessors. The w1167 and the newer w3167 are for the i386, and the w3168 interfaces to an M68020 or an M68030.

3.6.2   RISC Processors

The use of a reduced instruction set for a processor is based on a collection of separate ideas. First, simple instructions speed up instruction fetch and decode. An instruction can be decoded in a single cycle without the need for complex time and space consuming microcoding. The single cycle decode allows instruction fetch to be pipelined. The instruction to be decoded on the next cycle can be fetched while the current one is decoding. Second, simple regular instructions make it simpler to construct an optimising compiler. Finally, RISC chips have a load/store architecture. This means that main memory may only be accessed by a *load* or *store* instruction and may not directly supply an operand to any other instruction such as an addition.

The development of a load/store architecture was motivated by the fact that instruction execution time became less than memory access time. If an instruction specifies that a memory location needs to be accessed directly, then the processor can do nothing while the memory is accessed. A load/store architecture allows a load request to be started on one cycle, like any other instruction. However, the processor can then continue leaving the memory interface hardware to access the memory and place the value in a register. In effect, the processor consists of two execution units, a memory interface and an instruction execution unit that may only access registers. These units may operate concurrently and communicate with each other via registers. All communication with outside memory occurs via load and store instructions executed by the memory interface.

The RISC approach has three potential advantages for neural network simulation. The simpler decode hardware and lack of microcode store, counter etc. frees chip space. This space can be used to speed selected functions in hardware, in particular floating point calculation. On the T800 this space has been used to provide 4 DMAs, an FPU and a process scheduling kernel. The second benefit of network simulation comes from the extensive use of pipelining. This provides the necessary data and instruction throughput. Finally, the load/store architecture involves the separation of memory interface from the instruction execution units. This suggests a processor architecture with multiple on-chip execution units communicating via registers. This provides one method of integrating FP unit(s) into the processor architecture.

RISC accelerator boards for the PC

RISC processors are used in the majority of new workstations. In addition, a number of RISC processor boards are available for the PC. For example: an AMD 29000/29027 based board is sold by Yarc Systems; a Cypress CY 7C608 SPARC CPU with TI 8847 FPU based board is sold by Definicon; and various vendors (eg Microway and Definicon) sell T800 based boards.

3.6.3   Pipelined Floating Point Processors

General purpose computers may not provide sufficient performance for simulating large networks. Pipelined floating point processors offer higher performance. These have been sold by a number of manufacturers, including HNC and SAIC who market them as neurocomputers. (The SAIC board has been used in ANNIE.)

The SAIC Delta 2 FPP has 8M of data memory and a peak sustainable performance of 22 Mflops. Its architecture attempts to meet the three criteria stated at the start of this chapter, namely fast floating point performance, large memory and sustainable data throughput. The heart of the Delta 2 board are two floating point units: a floating point multiplier (BIT 2110) and a floating point adder (BIT 2120). These can perform their respective operations in one of the processor's 100ns cycles. Furthermore they may be formed into a two step pipeline to allow them both to operate in the same cycle. The adder holds a running sum, and within one cycle the multiplier can multiply from the product of two new operands while the adder adds the result produced by the multiplier in the previous cycle to the running sum. This allows a *multiply and accumulate* (MAC) to be performed in one cycle.

The performance of the floating point units creates a great demand for new operands. The Delta board has architectural features to help provide the necessary throughput, namely a *Harvard architecture, dual memory banks, pipelining* and *static column DRAM*.

Systems with the Harvard architecture have separate data and instruction buses. This avoids a possible bottleneck caused by the limited bandwidth of a combined instruction and the data bus. Dual databanks simply means splitting the data memory into two separate parts, each with their own data bus. This allows two simultaneous data access, effectively halving access time, and two data buses, effectively doubling data bus bandwidth. The tradeoff is that performance is affected by data placement. If data is only fetched from one bank the performance benefits of the second bank are not felt. This is an example of a complexity/performance tradeoff, a trade-off that will be seen throughout this section.

The Delta 2 operates in a three-deep pipelined mode. The first stage is concerned with instruction fetch, the second stage with instruction decode and data fetch, and the third stage with a floating point operation. Also, when the multiply and addition units are chained together to produce the single cycle MAC operation, the processor has four rather than three pipeline stages. The nature of these pipeline stages shows that the processor is optimised to execute instructions consisting of a data fetch followed by a floating point operation.

Static column DRAM allows each 1M word memory bank to be viewed as consisting of 1,024 pages, each of 1,024 words. If a memory fetch is from the same page as the previous fetch, then the access time is less than one cycle so no wait states need to be inserted. Thus, if a large array is being accessed sequentially, 99.9% of access occurs without wait states. It should be noted that there is no mention of caching. The Delta board was designed to maintain performance on data with little locality of data reference.

Other pipelined floating point processors
In addition to the Delta 2, floating point processing boards from HNC, IO Systems, Myriad Solutions, Mercury, and Sonitech have been marketed. It is interesting to note that the newer boards use specialised maths processors rather than components. This seems to provide a better price to performance ratio.

3.6.4   Parallel Systems
Nearly every multiprocessor board for the PC has been based on the T800 transputer. This processor has an on-chip FPU giving it approximately the floating point performance as a 386/w1167 combination. In addition, each T800 has four i/o ports to allow direct communication with four other transputers. A complete transputer system consists of independently executing T800s, each with local memory, communicating via their links. It is the responsibility of the programmer to control information exchange between these processors.

3.7   New Processors and Components

This section examines the processors which are already in use. In addition, it is possible to speculate about the performance of systems to be released in the next year or two as they will probably be based on these processors.

The first subsection examines the variety in modern RISC architecture. Four separate approaches to integrating the FPU with the operation of the ILU are seen. The following three subsections cover hardware specifically aimed at numeric computing. In turn these cover the Intel i860, DSP processors and Intel's iWARP. All these processors have both fast floating point performance and exhibit a concern for data throughput.

3.7.1   RISC Processors
This subsection examines architectural features of new RISC processors, with emphasis on those features that improve floating point performance and data throughput. The M88000 examined below has multiple on-chip execution units, register scoreboarding and a Harvard architecture. The multiple execution design places the FPU on an equal footing to the IPU within the processor. This is one method of coordinating the operation of an FPU with the operation of the processor as a whole.

Motorola M88000
The M88000 has three on-chip execution units; a memory interface unit, an integer instruction execution unit and a floating point instruction execution unit. These three units communicate via a 32 element register file and all access this file using 332 bit internal buses. Motorola claim these buses provide necessary bandwidth for up to eight execution units to communicate satisfactorily. In particular, the buses can provide enough on-chip bandwidth to keep pace with the FP unit.

Multiple execution units raise the problem of synchronisation. This problem occurs in a simpler form with all load/store architecture processors. The contents of a register are not valid until a number of cycles after a load instruction has been started. A simple solution is to ask the compiler. It is required not to access a register until a safe number of cycles have elapsed since the 'load' instruction was issued so the register contents are valid. The 88000 provides a backup in hardware called *register scoreboarding*. When an instruction is executed that will alter the value of a register a number of cycles in the future, then the register is marked with a *dirty* bit. If an execution unit attempts to access a register with this bit set, then the unit suspends. When a register contents become valid the bit is cleared and instruction units suspended on its value are allowed to continue. Register scoreboarding thus provides hardware protection that prevents invalid registers from being accessed.

The M88000 has separate instruction and data buses, ie a Harvard architecture. This shows a concern for data throughput. The separate buses can each be attached to a cache and memory management unit (CMMU) chip. The memory ports of the CMMUs can be attached to the same bus, giving a processor with an internal Harvard architecture and an external von Neumann appearance. The single bus will cause a bottleneck if cache misses occur in both the instruction and data paths. This hybrid architecture will also be seen in the next two subsections.

Others

Other general purpose RISC chip sets include the MIPS R3000, the SPARC chipset and the Fairchild Clipper 3000. These each have a different approach to integration of the FPU. The MIPS processor consists of a dual chipset, but the FPU is an integral part of the set. Its instruction stream, integrated into the instruction pipeline, avoids the problems of processor to coprocessor communication. This approach seems to have paid off as the R3000/R3010 combination produces 7 Linpack Mflops (SP, HC). The SPARC architecture is a specification of an instruction set rather than a particular implementation. There are a variety of ways in which the floating point instructions can be performed. The Cypress chipset uses a TI8847 FPU whereas the Fujitsu ILU in a Sparc Station uses a Weitek w3170. This is a version of the w3167 mentioned in an earlier section. The Fairchild Clipper provides floating point performance on chip. Half of the die area is given over the FPU. Since floating point instructions can take longer to execute than integer instructions, there is a potential problem synchronisation with an architecture that executes both kinds of instructions. Like the M88000, the Clipper uses scoreboarding to safeguard data integrity rather than rely on the compiler alone.

3.7.2   Intel i860

Scientific computing can involve processing large arrays of numbers both in calculating results and displaying them in a graphic fashion. The i860 is targeted at this use and its fast pipelined arithmetic makes it very suitable for network simulation.

The 860 is a load/store RISC processor with an on-chip FPU. Floating point performance is enhanced by two major differences in design from the M88000. Firstly, the 860 has two instruction fetch modes, a regular mode and a dual fetch mode. In this second mode two 32 bit instructions are fetched in a single cycle. The first is passed to the IPU and the second to the FPU. This dual instruction fetch enables the IPU to be instructed to load an operand into a floating point register whilst the FPU is instructed to operate on floating point operands from other register locations. The dual instruction mode is designed to send instructions to the 860 as fast as the 860 can execute them. With the single fetch mode, two cycles are needed to instruct both processing units causing a bottleneck.

The second difference between the 860 and the M88000 in the design of their FP operation is in the actual construction of the FP units. The FP unit in the 860 is constructed from a separate FP multiplier and an FP adder. To take advantage of the fact that the FPU contains two execution units, the FPU supports two types of FP instruction. The first type are scalar instructions. When executing a scalar instruction, the FPU may only use one of its execution units, and the FPU is not available for executing another instruction until the scalar instruction has been completed and the result stored in a register. The second type of instruction is a vector instruction. Vector instructions allow the FPU to execute instructions using a three deep pipeline. The first stage takes two operands from the FP register and multiplies them. The second stage takes two operands and adds them. The final stage stores the result back to the FP register. When the FPU is executing vector instructions, the FP adder and FP multiplier may execute concurrently, doubling FP performance. This can allow a single cycle MAC like the pipelined processors mentioned in the section above. With a 40 MHz clock, this produces a theoretical peak performance of 80 Mflops. However, this is only sustainable in very short bursts due to data and instruction bandwidth limitations noted below.

There are two potential problems with the 860. Firstly, it can be seen that the 860 has a tremendous appetite for instructions and data. If the 860 is executing in dual instruction fetch mode, and the FPU is executing vector instructions, then the 860 has three instruction execution units processing concurrently. This can create a need for up to two 32 bit instructions and two 32 bit operands to be fetched each cycle. There are wide buses and large caches to help data and instruction throughput on chip. There are separate instruction and data paths and a 4k instruction cache and an 8k data cache. However, off chip the 860 has only a single 64bit bus. Thus it is similar to the 88000 when the CMMU are attached to the same bus, it has an internal Harvard architecture and an external von Neumann architecture. If there are many cache misses, forcing the 860 to make many off-chip fetches, then this single bus may cause a bottleneck. The problem of actual memory access time is reduced by support for static column DRAM and prefetching up to three instructions.

The second worry concerns the programmability of the 860. Its dual instruction fetch modes and dual RP instruction types give it a complicated instruction set. This poses problems for both assembly language programmers and compiler writers. There is the possibility that an inefficient compiler prevents the high level programmer from using the theoretical performance of the 860.

3.7.3   DSP Processors

Digital signal processing (DSP) is the use of digital hardware for processing sampled signals. DSP has much in common with scientific array processing as they both involve the processing of large arrays of numbers. This similarity extends to many operations being common in both domains, such as a dot product or FFT. As a result, the hardware originally designed for DSP is suitable for general purpose array processing and vice versa. However, DSP computation has three characteristic features: the algorithms are often simple; the computation often needs to be carried out in real time; and there is often a need for only limited accuracy. The combination of simple algorithms and the need to provide real time response has meant that many DSP systems have been purpose built from simple arithmetic components, algorithm specific components or application specific components. However, a number of processors designed to support DSP processing have been developed.

The early DSP processors reflected the specialised nature of DSP processing and so have a number of features that detract from their fast maths capability. An example is the TMS320C10. Its positive features include a single cycle MAC, a Harvard architecture an an on-chip DMA. However, it also has a number of less appealing features that restrict its suitability for network simulation work. It has an irregular instruction set, it only supports fixed point numbers and has a very limited (64k) address space. Nevertheless, two TMS320C10s were used by Kohonen to implement his speech recognition systems. The first performed the signal processing on the input speech and the second was used to implement the neural net component of the system.

Several more recent DSP processors have been designed with the possibility of being used for more general purpose numeric work in mind. These do not have the TMS320C10's drawbacks but keep its positive features. They have more regular instruction sets, larger address spaces and support 32 bit floating point arithmetic in addition to single cycle MAC units and multiple data channels. The DSP processors with these features including the following: AT&T DSP32C; Fujitsu MB86232; Motorola M96001 and M96002; and the Texas Instruments TMS320C30. Two, the TMS320C30 and the DSP32C, have C compilers.

AT&T DSP32C

AT&T's DSP32 was the first DSP processor that supported 32 bit floating point operation. However, it only has a 16k address space and so has limited suitability for neural networking. The new DSP32C has resolved the limited addressing problem and has a 16Mbyte address space. Its clock cycle is 80 ns which gives it a peak performance of 25 Mflops. Unlike the majority of DSP processors, the DSP32C has a von Neumann architecture. However, concern for data bandwidth is shown by three 512 word banks of on-chip RAM, parallel and sequential i/o ports and an on-chip DMA. Though the conventional bus architecture might produce a bottleneck, it probably helped in the production of AT&T's optimising compiler.

Texas Instruments TMS320C30

The TMS320C30 is similar to the DSP32C in that it possesses large on-chip memories; an on-chip DMA; a single cycle MAC unit; parallel and sequential i/o ports; and a C compiler. However, it has a slightly faster clock (60 ns) giving it a higher peak performance and a different internal bus architecture. On chip, the TMS320C30 has a Harvard architecture, but from the outside it appears as a von Neumann machine. This simplifies the package but risks a single bus bottleneck as both data and instructions are frequently fetched using this main bus.

3.7.4 iWARP

The iWARP component can be viewed as an improved T800. It is designed to implement parallel array processing computers (including systolic systems) but can also be used as a stand alone processor. Each iWARP processor consists of a communication agent and a computation agent. Data channels are provided by a 64 bit bus to local memory (managed by the computation agent) and four duplex data links that may be attached to other iWARP processors (managed by the communication agent). Local memory can be used to store data and instructions. In addition, there is a 256 word RAM and a 2k word ROM within the computation agent for instruction storage. This subsection describes the computation agent.

Like the i860 and the M88000, the iWARP computation agent consists of multiple independent execution units.These is an integer RISC core (the ILU) and two floating point units (an FPA and FPM). The processor is based on the load/store idea, with only the ILU allowed to access main memory by use of load-to-register or store-from-register instructions. The other execution units and ILU instructions access a central 128 slot register file. The ILU executes an instruction each 50 ns cycle and each floating point unit takes two cycles. There are two instruction fetch modes, single fetch and long instruction word (LIW). As its name suggests, single fetch involves the fetching of a single 32 bit instruction word. This instruction is passed to the execution unit that implements it. The second mode involves fetching three 32 bit instructions, one each for the ILU, the FPM and the FPA. The ILU operation specifies two load or store operations and the FPM and FPA each operate on operands from specified registers. This multiple instruction takes two machine cycles to execute.

The LIW mode allows the FPA and FPM to execute concurrently, giving a peak performance of 20 Mflops, without resorting to special vector instructions. The floating point units always operate as scalar units. Within a double cycle, each FPU can fetch a pair of operands, perform an operation and place the result in the register file. Intel hope that this scalar operation of the FP units will facilitate efficient compiler construction. It should be noted that it is possible to construct a MAC pipeline: the add instruction specified by the current LIW adds the product specified by the previous LIW to the running sum.

As with the other processors in this section, attention has been given to ensure good data and instruction throughput. If a MAC pipeline is implemented as above, then in each double cycle the processor may need to fetch three instruction words and two data words. Within this time, an iWARP processor can access four words from the local memory and receive four words via the interprocessor links. Thus there is sufficient bandwidth to support the peak computation rate of 20 Mflops.

3.8   Systolic Computation

'In general, systolic designs apply to any compute-bound problem that is regular - that is, one where repetitive calculations are performed on a large set of data.' This claim (Kung, 1982) explains much of the interest in the systolic implementation of network algorithms: they hold out the promise of the parallel implementation of regular compute bound algorithms, a class to which many network algorithms belong. (To recall, a *compute bound* problem is one where a number of data items take part in more than one calculation, and an *i/o bound* problem is where each datum is used only once).

A *systolic system* (as described in Kung, 1982) consists of a regular network of processing elements through which data flows in a rhythmic fashion. Figure 3.1 gives an example of a linear system. More precisely, a systolic system has four principal features:

(i) *Multiple use of each datum.* Systolic systems are used to implement compute bound problems where a number of items are used more than once. However, if each datum is returned to memory after each use then the problem can become i/o bound. This is demonstrated by the first case in Figure 3.1 where the PE cannot operate at more than 5 Mops because the memory bandwidth is only 10 Mwords per second. The alternative is to use the data many times before returning it to memory thus increasing computation rate without requiring more memory bandwidth. This is demonstrated by the second case in the figure.

Data reuse requires either that data is broadcast to a number of PEs or is circulated. Systolic computation uses the latter as the former is difficult to realise in hardware

(ii) *Extensive use of concurrency.* Systolic systems allow a number of PEs to be used at once. Figure 3.1 shows this concurrency in the form of pipelining. Horizontal parallelism is also possible. If the data is pumped through a mesh rather than a linear array of PEs then both pipelined and horizontally parallel concurrent operation occur. Furthermore, it is possible to have concurrency in a systolic system at two other levels, apart from the PE level. Each PE itself can operate in a pipelined fashion giving rise to two level pipelining. In addition, a number of systolic systems can be chained together (eg Gentleman *et al*, 1981)

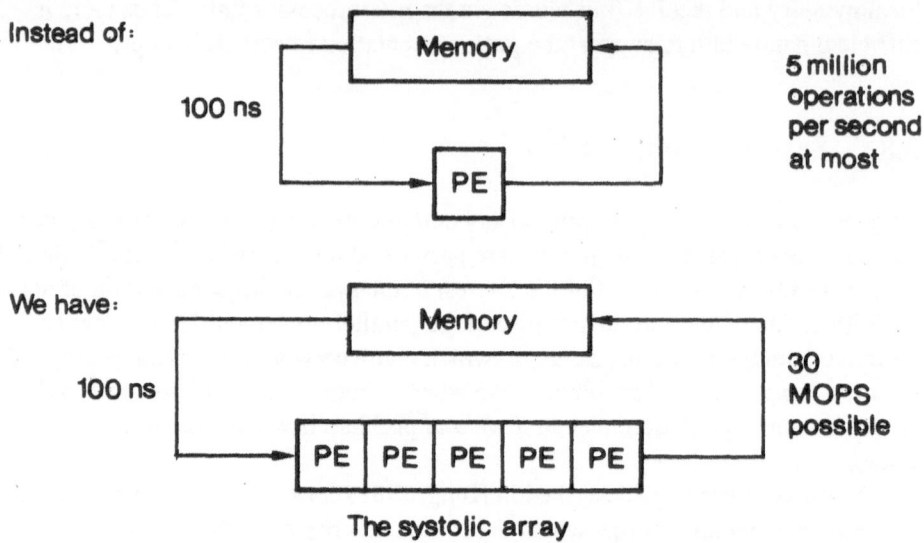

Fig 3.1 Basic principle of a systolic system (from Kung, 1982)

(iii) *There are only a few types of simple cells.* Systolic systems perform regular
     computations and thus make use of a few types of simple cells. This simplicity
     has two benefits. First, if the system is being implemented in
     hardware then simple repeated cells are easy to lay out. Second, in a general
     purpose system with programmable PEs, simple regular cells help
     programming. It is far easier to coordinate the operation of simple repeated
     units than a complex heterogeneous collection
(iv) *Data and control flows are simple and regular.* Systolic systems avoid non-
     local or irregular communication patterns. Non-local communication can be
     difficult to implement efficiently in hardware especially if there is an irregular
     pattern. In addition, the design of any irregular system is more complex than a
     regular one.

Kung (1982) states that these features give three very desirable properties, namely:
simple (a consequence of (iii) and (iv)); *expandable* (from (iv)); and *high performance*
((i),(ii) and (iv)).

Haynes *et al* (1982) state that the five principal problems for a parallel hardware
system are *PE synchronisation, algorithm data dependencies, runtime overhead,
contention for data* and *input/output.* It is instructive to consider how a systolic system
addresses these problems. PE synchronisation, runtime overhead and contention for
data are all handled by the planned nature of the computation data flow. The flow of
the data through the network means that one datum arrives at a PE rather than the PE
requesting the datum. Thus there is little need for synchronisation. There is no load
balancing or message passing to cause runtime overhead. Data contention does not

occur because data is planned to arrive at its destination rather than being accessed by an agent. Systolic systems are only used for computations whose data dependencies allow such a system to be derived and so the problem of insufficient data dependencies does not occur. Data reuse in a systolic system helps reduce the need for a high i/o bandwidth. In addition, data may be input and output on any boundary cells of a systolic array. Thus all i/o need not be channelled through a single PE.

There are a number of systolic implementations of neural network algorithms in the literature (eg Dowla *et al*, 1989; Kung *et al*, 1988; Millan *et al* 1989; Pomerleau *et al*, 1988 and Richards *et al*, 1988).

3.9   Summary of Architectural Features

The sections above contain descriptions of a variety of architectural features and operating modes designed to enhance floating point performance and throughput. This section contains a systematic description of these features and aims to indicate which areas of machine performance they affect. The first three subsections describe hardware subsystems, and the final subsection considers how their operation may be coordinated.

3.9.1   Floating Point Units
The performance of the floating point unit in a given computer defines the upper bound on possible floating point performance, though by no means the lower bound. The FPUs seen above differ from each other with respect to their *basic implementation*, their *integration* with the rest of the system and their *execution modes*.

The basic implementation of an FPU describes the hardware used to implement it. An FPU can consist of a single unit, a separate multiplier and adder or a single unit capable of performing two operations concurrently, such as a MAC unit. The basic performance of each of these units is determined by their cycle time.

The second axis of difference between floating point units is their integration with the rest of the architecture. Subsection 3.6.1 described the three types of coprocessors, namely standard, memory mapped or integrated.

The third difference is in mode of operation. The units can either operate in pipelined or scalar mode. This point is further considered in section 3.9.4 on control.

3.9.2   Memory Structure
Main memory is used to hold both instructions and data. The primary requirement on memory is that it is sufficiently large. Given a large enough memory, the next concern is access bandwidth. The FPUs and control and addressing logic have a great demand for instructions and data so the memory must be able to satisfy this need without inserting an excessive number of wait states.

Some systems above have memory grouped into banks whilst others have a particular organisation of memory within a given bank. A von Neumann computer has a single memory bank holding both data and instructions. A number of systems, for instance the SAIC and HNC processor boards, have separate data and instruction banks, the Harvard architecture. In addition, both split the data memory into two banks but in different ways. The Delta board has two explicitly separate databanks and data buses, with data stored explicitly to one or the other. This allows two data items to be accessed concurrently, provided they are in separate banks. The HNC Anza+ board uses two memory banks to implement an organisation called *interleaving*. The data memory logically appears as a single memory, but adjacent addresses are implemented in different banks. This organisation allows the effective access time of memory access to be halved, if each bank is accessed in turn, say by sequential access.

Within a given bank it is possible to have differing arrangements of memory. The most common is caching. This speeds up the memory access when there is locality of reference. An alternative to caching is to provide regions of fast memory whose addresses are known. Explicit control of the use of fast memory in effect allows programmer controlled caching, but at the cost of increased programmer work. TMS320C30 and transputer based systems have fast memory at known addresses, in the form of on-chip RAM.

3.9.3   Data and Instruction Paths

This subsection describes the combinations of buses and caches that are used to provide a path between the memory and floating point units and control logic. The most common approach in general purpose computing is to use the same path for both data and instructions. This approach opens the possibility that the limited bandwidth of this single path causes a bottleneck. The T800 transputer uses a single path architecture, but its designers believe that its use of a short instruction word might prevent this path from causing a bottleneck. The instruction word is only 8 bits long whereas the data/instruction bus is 32 bits wide. Therefore a new instruction fetch need be made only once every four instructions, assuming no jumps are made.

A common alternative to the single data/instruction path is to separate the data from instruction memory and have separate paths. This is called the Harvard architecture and is mentioned in the subsection above. A number of systems combine these two approaches to produce a combined Harvard and von Neumann architecture. This leads to a scheme where there are separate instruction paths within the processor but a single combined path outside. Such systems are easier to build, and provide the performance advantages of a Harvard architecture provided that data and instructions are not both fetched from outside the processor very often. The success of such a scheme therefore depends on the effectiveness of the usage of within-processor memory, either cache or user controlled. With the TMS320C30, the off processor bus is used if either an instruction is not found in the 64 word cache or if data is not found in the 2k of on processor RAM. The i860 has a single external bus but separate internal data and instruction data paths supported by an 8k data cache and a 4k instruction cache.

The purpose of data fetch is usually to move the data to the FPU. The ease with which this may be accomplished depends on how the FPU is integrated with the rest of the system. This is discussed above.

3.9.4   Control

This final subsection describes the different approaches to coordinating the operation of the above subsystems. All systems described in this chapter possess multiple execution units. A simple PC with a coprocessor has two, an i860 based system has four, a memory interface agent, an IPU core, an FPM and FPA. There are two approaches to getting these agents to operate together, either separate instructions for each agent or by *pipelined* mode of operation.

If the agents are to be separately instructed, then there is the issue of the instruction stream. With a PC and a coprocessor, there is a single instruction stream. Instructions arrive sequentially and are then executed by the main processor or the coprocessor. By way of contrast, the iWARP has multiple instruction streams. In LIW format, a separate instruction is fetched each double cycle for the IPU, the FPM and the FPA.

Pipelining allows multiple execution units to be instructed with a single instruction. However, there is a trade-off. Pipelined instruction modes are only suitable when the instruction sequence is repetitive. In addition, a pipelined instruction is special purpose in that it specifies only a certain combination of agent instructions. This causes two problems. First, a pipelined instruction may not exist for the desired combination of events. Second, special purpose instructions make it difficult to write efficient compilers. Despite the worries about pipelining, it is very prevalent in the systems described above. The neurocomputers boards are pipelined, as are the DSP processors and the i860.

3.10   Benchmarking

Measuring the performance of a computer ultimately reduces to calculating or timing the execution rates of particular instruction sequences and datasets, and then inferring execution rates in other situations given this data. This view can be seen to encompass both the use of benchmarks and calculations based on theoretical peak instruction rates. Performance measurement carried out in this fashion requires that three principal decisions are made, namely the *complexity of the instruction sequences and datasets*, the *particular examples of instruction sequences and datasets chosen* given a complexity level, and finally the *method of inferring other execution times*.

For example, a peak megaflop figure corresponds to the execution of the fastest floating point instruction on ideal data. This figure will only give a reliable indication of actual performance in an application program if the program involves the repeated execution of the fastest floating instruction on ideal data. This is an unlikely situation and so a more sophisticated approach such as benchmarking is usually taken. With benchmarking it can be seen that there are still three choices to be made, namely complexity of the benchmarks, choice of the benchmarks, and method of inferring performance in other situations.

The three choices to be made are interrelated. In general, using benchmarks of lesser complexity provides a number of advantages. The benchmark routines are easier to port, they are easier to implement consistently and are easier to interpret. However, the routines must be chosen so that it is possible to infer the performance on the computer when performing real applications from these simpler benchmarks. There are three dangers. First, the time taken to execute two simple instruction sequences may not equal the time taken to execute the single sequence made from their concatenation. In many modern processors, the time taken for a floating point operation, together with a load operation on a separate piece of data, is less than the sum of the times of the two operations taken separately. This is due to on-chip parallelism. Second, the execution time of simple routines may not scale simply with the size of dataset or instruction sequence length. This is often due to memory system effects such as caching. Finally, there is the problem of determining what mix ('workload') of simpler benchmark routines taken together provide an adequate abstraction of the work to be performed. If the chosen benchmark routines are not representative of the work performed by an application, or a non-representative workload is chosen, then the inferred figures will be meaningless.

3.10.1 Constraints on ANNIE Benchmarking

Having discussed benchmarking in general above, consideration is now given to more specific aspects encountered during the definition of ANNIE benchmarks. Deriving a set of benchmarks for use in ANNIE faced four difficult problems:

(i) *Sensitive hardware*. The performance of many hardware systems is *sensitive* to the code and data: that is, the performance can vary greatly according to the nature of the code and data executed. In general, this manifests itself in three ways. Firstly, there is sensitivity to the type of operation. Hardware possessing multiply and accumulate (MAC) units will perform calculations involving a MAC more quickly than other similar operations. Vector units will execute a regular instruction stream more quickly than an irregular one. The second sensitivity is to the degree of code optimisation. More performance can often be obtained by hand coding routines, and with some systems simple optimisations can produce a dramatic increase in performance. A common example is that of a compiler failing to vectorise code, and for which hand vectorisation produces a dramatic increase in performance. The third sensitivity is to data placement. Many benchmark operations involve a low calculation per datum ratio with corresponding emphasis placed on data throughput. Many of the systems have heterogeneous memory arrangements. Data placement can therefore have a great effect on execution rate.

(ii) *Network algorithms with few kernels*. Current network algorithms can be constructed from only a few kernels. Thus variations in kernel execution rates can be reflected in the variation of execution rates between different network algorithms.

(iii) *Unsure workload*. Neural network techniques are developing rapidly. It is very difficult to foresee which algorithms will be used in the future, what modifications will be made to existing algorithms, and what new algorithms will be developed.

(iv) *Third party testing*. Benchmarking tends to be performed by hardware vendors themselves. Much hardware is very sensitive to hand optimisation of the code so there is the problem of producing consistently optimised code on unfamiliar systems. The most consistent approach is to ask each vendor to provide as efficient code as possible in the hand coded part of the tests.

The above four constraints had a very great effect on the approach to benchmarking used in ANNIE. In particular, they encouraged the use of kernel routines rather than full network algorithms.

More detail about the benchmark code developed within ANNIE can be found in Appendix 3.

3.10.2 Use of Kernels

The selection of a benchmark code is not simple. It is necessary to derive a set of benchmarks to represent the execution of an under-defined workload whilst dealing with sensitive hardware. The approach taken is the use of a collection of kernels which are implemented at up to three levels of optimisation, and a core set repeated with a different locality of reference with the dataset. From the execution of these kernels in various forms a pool of figures is obtained, which can then be used to deduce facts about the hardware and its performance with complete algorithms.

From this pool three kinds of inference can be made. First, it is possible to get a general impression of the hardware. The range of times and their distribution gives a good idea of the raw power of a system and its sensitivity. Second, by using kernels that are similar to each other, and repeating kernel execution with differing degrees of optimisation and locality, the three facets of hardware sensitivity mentioned above can be investigated. Finally, given a particular algorithm, kernels can be selected from the pool in the correct proportions to deduce the approximate execution rate of the algorithm. This threefold interpretation is discussed more fully in Appendix 3.

Using this pool-of-kernel times idea has a number of advantages over simply using timed complete algorithms, especially when the four constraints worked under are considered. First, because each kernel is not algorithm specific, there is a better chance of predicting the execution time of new algorithms. Second, using kernels provides better value of information for implementation effort. Each kernel may be shared by a number of algorithms, and so the volume of benchmark code is reduced. This is a very important consideration when the benchmarking is being performed by an unpaid third party. Third, being simpler, kernels have allowed a greater possibility for specifying the use of *plain vanilla* code. There is a mismatch between the desire for consistent benchmark code, and the natural inclination to code each benchmark as efficiently as possible. By having a routine with a hopefully natural implementation, this problem is much reduced. Fourth, simple routines allow an easier specification of allowable optimisations with hand coding and data locality. Finally, these simple routines are understandable. If different execution rates are observed between simple routines, it is easier to infer the reason for this than if the difference was between the execution rates of complex routines. This makes it easier to understand the sensitivities of the system under test.

It is worth noting that, in general, the larger the number of kernels used, the more information can be derived. However, the choice of third party testing provides an opposite force. The size of set used was determined by a reconciliation of these opposing pulls.

3.11   Software

The many software simulators produced over the last few years fall into three classes:

- template
- non-template
- network specific.

3.11.1   Template Simulators
A template simulator allows ease of use at the expense of flexibility. Such a simulator allows the rapid construction of a net from a given collection and the easy modification of model parameters. a good user interface is provided. Defined nets can usually be called from C allowing nets to be incorporated into layer application.

3.11.2   Non-template Simulators
Non-template simulators allow the user to define new ANN models or implement major modification to existing models. A feature of all these simulators (with the exception of ANSpec) is the provision of a well-developed graphical interface. This allows unexpected features of the runtime behaviour of a new net to be observed and understood.

3.11.3   Network Specific Simulators
They are specific to a particular neural network type, allowing the user a limited choice of the network's parameters - number of units per layer. The application area is also limited due to the unique network type provided.

Rather than describe the many commercially available software packages here, we shall limit outselves to a discussion, in section 3.12, of the three principal simulation packages developed within ANNIE. For a list of some of the suppliers of network simulators, see Appendix 5.

3.12   Environments Developed within ANNIE

Each of the application areas within ANNIE has developed a network simulation
system for itself. The following subsections describe illustrative examples of these
briefly. Details of their use will be found in the application chapters of this handbook.

3.12.1   PROFAN (Prototype for Applications of Neural Networks)

PROFAN is a PC-based system for applying a variety of neural networks and
conventional methods to the problems of pattern recognition encountered in the
analysis of image-type data. The system is designed to be as flexible as possible so
that a variety of different tools can be integrated together, and users may easily add
any routines of their own as and when required.

Structure and functionality of the system

The basis of the system is a simple menu-based interface which links together various
preprocessors and algorithms for classification and/or image analysis (see Figure 3.2).
As standard formats are defined for the data and for the input and output of the
processing routines, it is possible to apply successive analyses in any order and to
explore the role of different preprocessing techniques. It is also possible to use the
same system to analyse data from very different types of problem as were encountered
within the ANNIE project. Thus it was possible to analyse images of ultrasonic
datasets and of solder joint scanning with the same set of algorithms. An important
feature of the system is that many of the routines can be used to fuse together
information from several different images simultaneously ie the system can effectively
cope with 3-D and 4-D *images*. This aspect has been found to be vital in obtaining the
best results in the pattern recognition problems studied within the project.

The user interface

The user interface is an easily modifiable and expandable menu, which enables the
user to call executable modules, to edit or view any PROFAN configuration files or
application-specific ASCII files, and to list the currently available PROFAN data files.
The list of current files to be processed is maintained by the system within a suitable
configuration file. Each menu page is described by one ASCII file, so it is easy to
create, delete or modify menu pages by simply editing ASCII files. The user is also
provided with display utility functions which enable images of the raw data to be
viewed on the screen. It is also possible to review and edit the raw data within the
image files.

Preprocessors

The preprocessors implemented within the PROFAN package included:

- median filtering
- smoothing filter
- Laplace filter

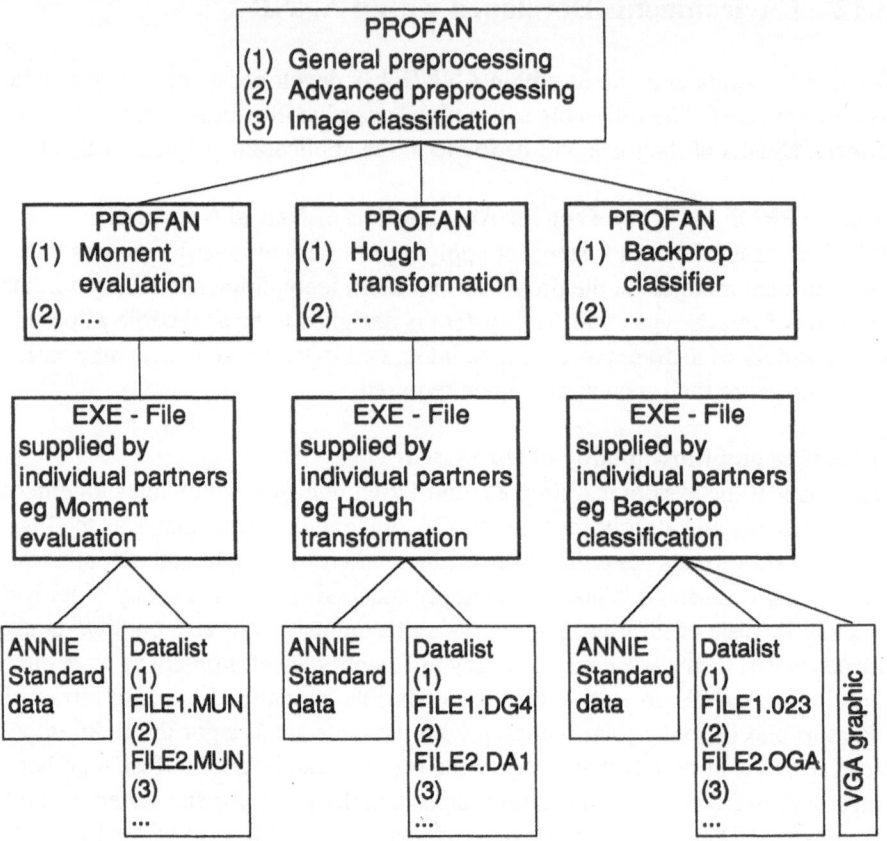

Fig 3.2 PROFAN menu interface screen

- scaling and normalisation
- thresholding Sobel operator filtering
- FFT
- low pass filtering
- highpass filtering.

Neural network algorithms and image processors
The following algorithms and processing routines have been installed within the
PROFAN system. All were developed within the project to help classify image data
directly rather than after the application of feature extraction:

- higher order network
- shared weight error backpropagation routine with up to 4-d receptive fields
- receptive field learning algorithm
- an adaptive field processor.

In addition, a number of other processing routines are included which facilitate the action of the learning algorithms. These cover generation of suitable files to define training and test examples within datasets, and the concatenation of datasets.

Documentation
Full documentation covered not only the use of the individual preprocessors and algorithms within the package, but also specified the standard data formats which were used for the image files and for the configuration files used by the PROFAN menu-driven system.

3.12.2 NetVision
NetVision uses an efficient dimensionality-reduction technique for visualising multivariate datasets in 2 dimensions. It involves unsupervised training of a neural network reversible mapping between the original data space and a 2D space, thereby allowing both data points and arbitrary locations to be mapped in either direction. These features make it possible to sample the characteristics of any associated data model on the same 2D map as the original data points. Using this technique, it is possible to generate 2D and 3D displays consisting of data points, dependent variables or classifications, and model regression surfaces or decision boundaries. Such displays may be interrogated interactively, and will provide a valuable insight into the accuracy, validity and scope of any model which is applied to the data.

A Turbo C program incorporating the ReNDeR technique was produced. The program was compiled for use on standard PCs and the Myriad DASH!860 card (an i860-based PC accelerator card for DOS and UNIX environments). The Myriad card gave a speed improvement of around 20 times over a 16 MHz 386/387, and increased the available memory space (for datasets and networks) from 400 Kbytes to over 5 Mbytes.

Applications
The reversible nature of the ReNDeR mapping makes it ideal for the analysis of relationships in multivariate data. Applications have been identified in the following areas :

- pharmaceutical industry (quantitative structure-activity relationships (QSAR) in drug design)
- agrochemical industry (QSAR for pesticide design)
- food industry (product design and quality control)
- manufacturing industry (robotics and sensor fusion)
- neural network research (real-time model analysis).

A description of the algorithm was presented at the Eurographics '90 conference (Montreux, September 1990), and appears in the Proceedings of the Graphics and Interaction in Esprit Sessions, under the title 'ReNDeR: Reversible non-linear dimensionality reduction by neural networks'.

3.12.3 NeVIS
NeVIS III

The intention of the NeVIS III simulation is to show the abilities of neural networks trained by error-backpropagation in a control task, namely the control of vehicle movements in a free definable environment. The software runs on two IBM-PC/ATs or 386 machines. The user interface is exceptionally friendly; all menu choices can be done by mouse. Every simulation step offers an online help facility.

The complete simulation run is done in five steps:

(i)   creation of a proving ground (PC1)
(ii) generation of a learning file by driving the vehicle inside the proving ground (PC2)
(iii) creation of a network with suitable input/output layers (PC2)
(iv) training of the network with the learning file (PC3)
(v)   use of the network to control the vehicle inside the learned proving ground or inside an unlearned proving ground.

The creation of a proving ground is done by moving various objects like circles, rectangles and blocks on a test area. Out of these objects the user is able to define an environment for training of the vehicle. The proving ground is saved in binary format and can be loaded for further changings.

The generation of a learning file is done by moving the vehicle through this proving ground. The learning file contains the coded sensor information and the associated actuator data. The learning files are saved in ASCII containing all information coded in either binary, decimal or analogue format.

The training of the network takes about 1-10 hours depending on the hardware speed and size of the network learning file. The simulation is limited to 30 nodes per layer. During learning a convergence file gets created which contains various network statistics and serves for graphics of the global error curves. The software is usable for training of analogue and binary input datasets. The algorithm is the original version of backpropagation as developed by Rumelhart *et al* (1986). The only difference to the learning formula is the use of a momentum term which increases learning speed.

The recall of the network can be done in two ways. The data file recall allows testing of the network with some test data files. The differences between desired output and actual output of the network are written into a file. This file can be investigated by use of a spreadsheet program, eg Microsoft-EXCEL. The differences between the curves allow evaluation of the performance of the trained network. This is appropriate for analogue trained networks only.

For recall with the NeVIS application both PCs were needed. One shows the neural network while the other shows the vehicle driving around in a known or unknown environment. The vehicle is able to drive in eight directions at each position. After each step, the new sensor information was detected and sent to the network. Some milliseconds later the actual driving reference is available at the output nodes.

They are decoded and transmitted back to the vehicle which drives in the corresponding direction.

NeVIS IV

The intention of this advanced simulation was to show the network's performance applied to a more realistic application. A vehicle with a free driving behaviour, ie acceleration and continuous turns, is trained by a backpropagation network to avoid obstacles. The neural network runs either on a PC, on a MicroVax/VAX station, or on a Silicon Graphics workstation. Thus, the principal change to the network software is in terms of graphical output. The application part is modelled by use of a 3D robot simulation package (KISMET: Kinematic simulation, monitoring and off-line programming environment for tele-robotics) running on a silicon Graphics Personal Iris (SG).

The complete simulation run is done in six steps:

(i)   creation of a proving ground (SG)
(ii) generation of teachfiles representing the way the vehicle has to learn (SG)
(iii) generation of training data by driving according to the teachfile in the proving ground (SG)
(iv) creation of a suitable network architecture (PC/VAX/SG)
(v)   training of the network (PV/VAX/SG)
(vi) use of network to control the vehicle inside the learned proving ground or inside an unlearned proving ground.

The proving grounds can be created in two ways. The easier way is to draw the obstacles by use of a CAD package and transferring them to the KISMET station. The other way is to create them by use of a limited functionality inside the KISMET environment.

For generation of learning data the vehicle has to be moved by the user, using mouse operations, inside an existing proving ground. Single points of this trace get saved in so-called KISMET - *Teachfiles* in IRDATA-Code. This file now describes the driving route of the vehicle in the corresponding proving ground.

In the next step these *Teachfiles* were driven in *repeat mode* with sensors switched on. In parallel the sensor information and the movements were written into *learning files*.

After creation of a suitable network the training phases start. Corresponding to the hardware used the time for learning varies between 1 and 10 hours.

The last step is the recall phase. The quality of the driving behaviour depends on the network quality. The network quality itself depends on the training time learning file and network parameters.

NeVIS reinforcement

The intention of this simulation is to show the abilities of neural networks trained by reinforcement learning in a control task, namely control of vehicle movements in a freely definable environment. The software is similar to the NeVIS III software and varies only by use of a different algorithm.

The complete simulation run is done in three steps:

(i)   creation of a proving ground (PC 1)
(ii) generation of a suitable network architecture
(iii) use of a network to control the vehicle inside the learned proving ground or inside an unlearned proving ground. The network is trained in parallel to the application run.

The creation of a proving ground is done by moving various objects like circles, rectangles and blocks on a test area. Out of these objects the user is able to define an environment for training of the vehicle. The proving ground is saved in binary format and can be loaded for further changes.

After each step the vehicle made the new sensor information was detected and sent to the network. Included in this sensor information there is an evaluation of the last step. These data are coded and fed into the network, which changes its weights according to the evaluation, *ie* reward or penalty. Some milliseconds later the actual driving reference is available at the output nodes. They are decoded and transmitted back to the vehicle which drives in the corresponding direction.

The aim is to increase the number of rewards and to minimise the number of penalties. This is ensured by the use of the reinforcement learning algorithm. Depending on the complexity of the proving ground a good driving behaviour is reached after 1-24 hours.

3.13   Dedicated Neural Network Hardware

During the last few years a large variety of very different approaches to neural network hardware have been developed. These approaches range from simple von Neumann machines with and without coprocessor and acceleration boards to parallel digital machines as multidigital signal processor solutions, transputer networks, SIMD arrays and systolic arrays, through dedicated analogue hardware (special neuro chips) to optical systems. When reporting on such a vast area of different solutions, it is first necessary to find criteria for a categorisation. These criteria may be:

•   type of network to be realised
•   biological evidence of realisation
•   implementation technology
•   cascadability
•   mapping of network onto processing elements
•   flexibility.

3.13.1 Biological Evidence

One possible way to categorise neural network hardware is using its biological evidence as a categorisation criterion as done by Przytula (1988). He distinguishes between:

- neural networks mimicking biological neural systems
- neural networks on a somewhat higher level, eg early vision functions such as edge detection and others (these approaches also use existing understanding of biological systems)
- neural networks inspired by biological evidence to a lesser degree.

Thereby, the task a network is doing and its organisation are considered rather than its physical implementation and its basic building blocks. Pulse stream implementations as done by Murray *et al* (1988) for example, do not belong to networks mimicking biological systems or performing biological functions on a higher level. Of course, pulse stream chips work with current and voltage pulses, which seem to be similar to biological cells. However, this is done to implement asynchronous analogue circuits by means of a digital CMOS technology to combine their respective advantages (Murray *et al*, 1989). Thus, pulse stream architectures are neither mimicking biological networks nor performing biological functions on a somewhat higher level.

3.13.2 Implementation Technology

Neural networks consist of very simple processing elements (neurons) and connecting elements (synapses). In general, the task of a processing element is to sum up arriving signals and to perform a nonlinear threshold operation on the sum, whereas a connecting element has to weight an arriving signal, which involves a simple multiplication. However, summations, multiplications, and threshold operations may be performed with many, very different technical elements. For example, by benefiting from Ohm's law, one may perform a weighting operation or multiplication with a simple resistor. The voltage at the resistor is proportional to branch current I through the resistor branch with resistance R as a proportionality factor. Using a potentiometer one may realise an adaptive synaptic weight. Amplifiers may serve as nonlinearities. One of the first (perhaps the first of all) learning machines built of artificial elements in 1951 was designed by Edmonds and Minsky. The machine consisted of nearly 300 tubes and a number of motors. Memory of the machine was stored distributed in adaptive weights. To be able to test learning theories (Hebb learning rule) the positions of 40 control knobs could be changed by the machine itself. This job was done by automatic electric clutches. The whole machine was working even if some tubes or wire connections were wrong, showing one advantage of neural networks over many classical methods for signal processing, namely their inherent fault tolerance. However, a machine constructed from potentiometers, motors, amplifiers, and clutches would never be able to simulate or copy a network suitable for technical applications. Networks of reasonable size comprise at least several tens or hundreds of neurons together with thousands of synapses. The large number of synapses and connections especially cause implementation problems. Two promising technology areas for the implementation of rather large networks are VLSI technology and optical technology. Currently, implementation with VLSI circuits is state of the art, whereas usable optical implementations are not expected within the next few years. A very attractive feature of optical implementations is the fact that an extremely dense storage of weights may be achieved. Furthermore, the interconnection problem may easily be solved by means of light beams, which do not interact with each other. But still, many problems have to be solved before a first competitive optical implementation of a neural network will be available.

3.13.3 VLSI Technology

Very large scale integration (VLSI) has been a mature technology for many years. It allows a large number of electrical elements to be implemented on a small area of silicon. State-of-the-art mask technology (optical lithography) allows the fabrication of structure of the size of less than 1μm permitting the implementation of up to several million transistors per chip. With further decrease of feature size by means of X-ray and e-beam lithography (feature size down to 0.1 μm) ULSI chips (ultra large scale integration) will be available. This trend is sketched in Figure 3.3 (McClean 1991a,1991b). The principal problem which arises from this trend, from the point of view of the manufacturers, is the high probability of a fault occurring in a large circuit. This again is an aspect that makes neural networks a promising future technology due to their inherent fault tolerance. Unfortunately, silicon technology forces the designer to implement circuits on a two-dimensional plane with few available layers. However, in a brain, neurons (processing elements) are arranged in a three-dimensional space, offering many possibilities for interconnection. On a chip, neurons and synaptic interconnections have to be arranged on a two-dimensional area and this will be the most severe limitation for the implementation of large networks. Another strong limitation is the number of available pins to get information off the chip. Even if fully cascadable architectures are used, this will be a limitation. Principally two different styles of VLSI design, digital design and analogue design, are possible and these will be discussed in the subsequent sections.

Digital VLSI design

Today, for technical products, digital design is the most important design style. Currently, CMOS technology dominates. Its main features are small structures, extremely low power consumption, and high signal to noise ratio (Weste *et al*, 1985). In brief, important positive characteristics of digital VLSI design styles for neural network design are:

- simple to design
- high signal to noise ratio
- cascadability easy to achieve
- high flexibility

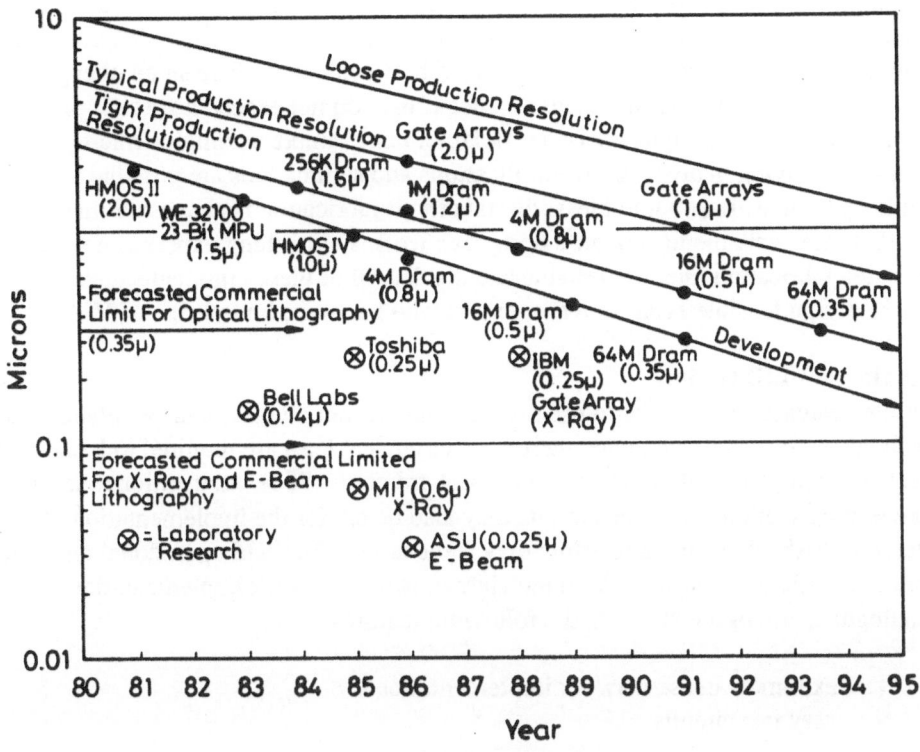

Fig 3.3 IC feature size trends

- cheap in fabrication
- rather small design experience necessary.

Simplicity of design is a very important point for the designer. As a result of years of practical experience with many commercial digital designs, a large variety of basic elements is available (Annaratone, 1986; Geiger *et al*, 1990).

Basic elements are circuits for multipliers, adders, ALUs (arithmic log units), more or less complex processing elements. An enormous advantage of digital circuits over analogue designs is the high signal to noise ratio. Whereas voltages or currents are directly used to carry information in analogue circuits, ie the value of voltage or current directly represents information which may be at least partly lost if noise is added to the signals, this is not the case in digital circuits. Voltages and currents are either on or off leaving a wide range for a signal to be interpreted as on or off. This range may be used to increase the signal to noise ratio. If the range is not exceeded by added noise it will not be harmful for circuit performance. Therefore, noise does not influence performance of digital circuitry. Furthermore, compared with analogue design, cascadability is easier to achieve. Due to parameter variations between different chips, absolute values of signals generated from equal analogue circuits on different chips may differ from each other. In addition, it is possible that the same signals may be interpreted differently from corresponding circuits on different chips. Digital signals, however, may be exchanged between chips, since small changes in an absolute signal value are of no significance if they do not exceed a specific range. Most of the digital designs are rather flexible. For example a complex microprocessor can serve solving many tasks. Multiplications and summations are possible as well as logic operations. Last but not least digital circuit fabrication is cheap and many design systems are available to support a designer's work. Due to long experience with digital circuits, fabrication lines are reliable and stable. All of these arguments make digital VLSI design feasible even for a non-expert who has some knowledge of VLSI.

Analogue VLSI Design

While designers now prefer to use digital methods, not all tasks can be solved by such techniques. In reality analogue interfacing circuits at least are required to be able to realise a complete system on one chip (Haskard *et al*, 1988). As mentioned in the preceeding section, analogue circuits may also be apt for the implementation of neural network basic element, eg resistors may serve as synapses and operational amplifiers (nonlinearities) as neurons. From the view of neural network implementations analogue methods are showing the following features:

- extremely dense network implementations
- very fast circuits.

By exploiting physical laws one may implement extremely small *computational entities* (eg resistor used as multiplier). Comparing the size of digital and analogue solutions, however, does not lead to a very clear statement. On the one hand, analogue circuits for neural operations may be extremely small which cannot be achieved by digital methods but, on the other, one has also to keep in mind calculation accuracy which is rather poor in analogue circuits, especially in simple and therefore small circuits. If high precision in mathematical operation is required, specially designed analogue circuits will become larger than digital circuits. A computational accuracy of 6-8 bit is probably the limit for analogue computations. If more is required, digital circuits will be more efficient. A feature which will never be reached by digital circuits, however, is the enormous computational speed which may be achieved by analogue methods. Since thousands of computations in a feedback network are done by a short relaxation phase (no clock scheme involved) which needs nanoseconds or a few microseconds, such circuits are of large computational power (Jackel *et al*, 1987). Unfortunately, analogue design has some severe disadvantages which are:

- tricky to design
- low signal to noise ratio
- influence of fabrication on performance (parameter variations)
- fabrication may be expensive (depending on implementation technology)
- cascadability difficult or impossible.

Everyone who has carried out an analogue design knows how difficult this may be. Extensive simulations on the level of physical behaviour have to be done (SPIC simulations) which is much more time consuming than a simple logic simulation of a digital circuit. In addition there are many parameters to adjust (width and length of transistors) if a specific behaviour of parts of the circuit is required. Furthermore, the signal to noise ratio is low, since even small disturbances of signals directly change calculated outputs (see section on digital VLSI design). Another common problem in analogue VLSI design is caused by the impact of fabrication parameters on final circuit performance since parameters cannot be held constant during a fabrication process. For example, changes in doping influence conductivity of transistors. Therefore, a difficult task for the designer is to take into account parameter variations and to minimise their influence on circuit performance. One effect of parameter changes which may vary significantly between different chips is that signal values of separated chips are too unreliable for direct exchange, limiting the cascadability of such designs. In recent years some designs were made using rather 'exotic' technology as Bismut resistors of NMOS technology. Many of these implementation technologies are mature for fabrication but only few manufacturers offer these technologies which are rather expensive compared with a widely spread standard CMOS process.

References

Annaratone M (1986) Digital CMOS circuit design. Kluwer Academic Publishers

Dowla F U, DeGroot A J, Parker S R and Vermuri V R (1989) Bacpropagation neural networks: Systolic implementation for seismic signal filtering. Int J Neural Networks, Research and Applications *1* 1, 138-153

Geiger R L, Allen P E and Strader N R (1990) VLSI design techniques for analog and digital circuits. McGraw-Hill

Gentleman W M and Kung H T (1982) Matrix triangularisation by systolic arrays. Proc SPIE *298*, Realtime signal processing IV, Society of photo-optical instrumentation engineers

Haynes LS, Lau R L, Siewiorek D P and Mizell D W (1982) A survey of highly parallel computing. IEEE Computer *15*, 1, 9-25

Jackel L D, Howard R E, Denker J S, Hubbard W and Sola S A (1987) Building a hierarchy with neural networks: An example-image vector quantization. Applied Optics *26*, 23, 5081-5084

Kung H T (1982) Why systolic computing? IEEE Computer*15*, 1

Kung S Y and Hwang J N (1988) Simulated annealing: Theory and applications. D Reidel Publishing

McLean W J (ed) (1991a) ASIC OUTLOOK 1991: An application specific IC report and directory. Integrated Circuit Engineering Corporation, Scottsdale, Arizona, USA

McLean W J (ed (1991b) STATUS 1991: A report on the integrated circuit industry. Integrated Circuit Engineering Corporation, Scottsdale, Arizona, USA

Millan J del R and Bofill P (1989) Learning by backpropagation: A systolic algorithm and its transputer implementation. Int J Neural Networks, Research and Applications *1*, 3, 119-137

Murray A F and Smith A V W (198) Asynchronous VLSI neural networks using pulse stream arithmetic. IEEE Journal of Solid-state circuits and systems *23*, 3, 688-697

Murray A F Hamilton A and Tarassenko L (1989) Analogue VLSI neural networks (pulse stream implementations). Journée d'Électronique 1989, 265-277

Pomerleau D A, Gusciora G L, Touretzky D S and Kung H T (1988) Neural network simulation at warp speed: How we got 17 million connections per second. Proc Int Conf on Neural Networks *II*, 143-150

Przytula K W (1988) A survey of VLSI iplementations of artificial neural networks. In R W
     Brodersen and H S Moscovitz (ed) VLSI Signal Processing III, IEEE Press, 221-231
Richards G and Smieja F (1988) Layered neural net simulator Edinburgh concurrent supercomputer
     newsletter, No 4, March, 6-12
Rumelhart D E, McClelland J L and the PDP Research Group (1986) Parallel distributed processing:
     Explorations in the microstructure of cognition *1 and 2*, Bradford Books, MIT Press, Cambridge,
     Massachusetts, USA
Weste N and Eshraghian K (1985) Principles of CMOS VLSI Design. Addison-Wesley

Chapter 4

Pattern Recognition

4.1 Introduction

4.1.1 Scope

Most tasks currently being tackled by neural networks might be viewed as pattern recognition tasks. For example, a control problem might be expressed as the production of a set response when a certain pattern is observed on some input sensors. The scope of this chapter has been limited to classification and clustering problems, where it is possible to set clear criteria for success, enabling ready comparison of neural network and conventional methods. These two problems are described below, in terms of supervised and unsupervised learning, and their application to a number of real problems will be described more fully later.

Although actual patterns from the applications of the partners will be described, and in some cases included in the comparisons of neural net and conventional methods, the initial thrust of the work on pattern recognition concerned the results obtained using generic representations of the patterns encountered in the applications. A generic representation can be defined as a pattern generated by computer, using parameters whose range covers the values which are present in the actual patterns. By concentrating initially on generic patterns, we were able to generalise our conclusions, outside of the particular applications of interest, to perform sensitivity analyses across ranges wider than that of any single dataset. At the same time, the discipline of basing the generic representations on patterns observed in real applications ensured that the conclusions were based on representative situations, and important details such as the size and quality of many training sets, or the problem of scaling, were not ignored.

The practical limitation in the number of available training and test patterns for many real applications is a good example of the need for generic representations. For instance, in the ultrasonic defect classification problem described below, a dataset of only 84 images was available. Each corresponded to an actual weld in which a defect of known type had been artificially introduced. It was not practical to obtain more data, so tests in which the number of training points was varied beyond 84 had to be performed on the computer-generated generic patterns.

Pattern recognition using conventional methods is a mature field, and several books have been written describing the standard methods. Any descriptions in this chapter are restricted to those best suited conventional methods which have been compared with neural network methods.

4.1.2   Image Representation

In many applications, raw data are obtained in the form of sets of examples, each of which is typically composed of hundreds of pixels arranged to form some image. Figure 4.1 shows a schematic representation of the image formed from the reflected intensity of a defect by an ultrasonic probe as a function of position. (Ultrasonic scanning is an important application for a number of the partners in ANNIE and is discussed extensively throughout this chapter. Background to the methods is given in section 4.8.) Note that formally we may consider such an image as being a (continuous) surface in a three-dimensional space. In some applications each pixel may have more than just a single intensity value associated with it. For example, in an ultrasonic scan each pixel may have probe angle and phase attributes as well as an intensity. However, the essence of data in the form of images is that they consist of pixels which are referenced by coordinates and are topologically related to each other.

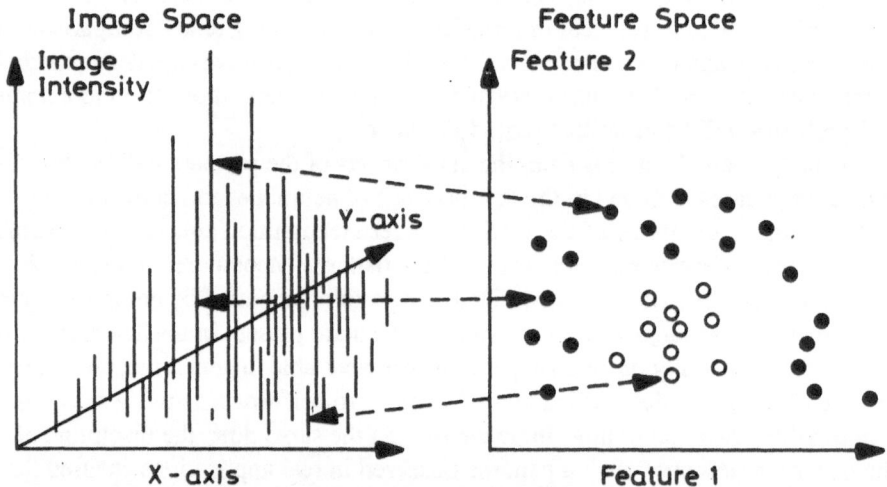

Fig 4.1 Image space compared with feature space. The image on the left represents a scan of defects in a material measured by an ultrasonic probe. The vertical axis measures the intensity of the ultrasonic reflection, and the two horizontal axes correspond to coordinates of the surface of the material. Neighbouring pixels in this image correspond to points close together in real space, and a single defect extends over a number of pixels. By comparison, on the right is a feature space distribution for ultrasonic defects. Here each point corresponds to a single defect, and points close together in feature space are likely to belong to the same class.

In some direct approaches to image analysis, each pixel is taken separately, and an image is considered no longer as a surface in a low-dimensional space, but as a point in a high dimensional space (dimensionally equal to the number of pixels). The drawback of such approaches derives from the fact that the topology of the surface is then ignored. For example, we know that when classifying an object within an image, moving the entire object just one pixel to the right should have no effect on its classification; it is the same object but in a different position. However, if we had chosen to express the image as a single point in a space with dimensionality equal to the number of pixels, movement of the object by a single pixel would change the axes completely, and the Euclidean distance of the new image from the old one would in general be vast. To be successful, ie to achieve object classification in a way invariant to such transformations as translation (invariant pattern recognition), methods which treat images as single points in high dimensional space must build in the underlying knowledge which relates to the proximity of various pixels in real space. This problem of invariant pattern recognition is one of the most difficult to be found in image analysis, and will occur in our discussions throughout the chapter.

The most common approach to processing image data is to compute a set of appropriate functions of the pixel values which encapsulate the information of interest. Such feature parameters reduce the dimensionality of the problem from hundreds, corresponding to the number of pixels, to a small number, typically less than 10, corresponding to the number of feature parameters. Each example in the dataset is therefore transformed from a large number of points in the image space to a single point in a multidimensional feature space (see Figure 4.1). An example of a standard image analysis method is to extract functions of the moments of the patterns as feature parameters. Classification and clustering can then be performed by looking at the distribution of examples in this feature space. Such procedures can be carried out by both conventional and neural methods, and comparisons of different methods are described in section 4.4. Because of the problem of invariance mentioned above, conventional techniques for processing the image directly are usually inappropriate. However, neural networks do offer the opportunity of direct image analysis, and these are explored in the specific application examples covered in sections 4.6 onwards.

4.2   Learning Mechanisms and Evaluation Criteria

In many applications, each example pattern is accompanied by a definition of the class to which it belongs. For example, the ultrasonic images may correspond to different known types of defect. In such cases, it is possible to define a classification process in which the input examples, whether images or feature space points, are each associated with a given class. For example, in the above, each example may correspond to an ultrasonic image or feature space point known to arise from a crack or from porosity. Figure 4.2(i) shows this in feature space with the classes labelled by different symbols. In general, the classes may correspond to distinct clusters in feature space, or the

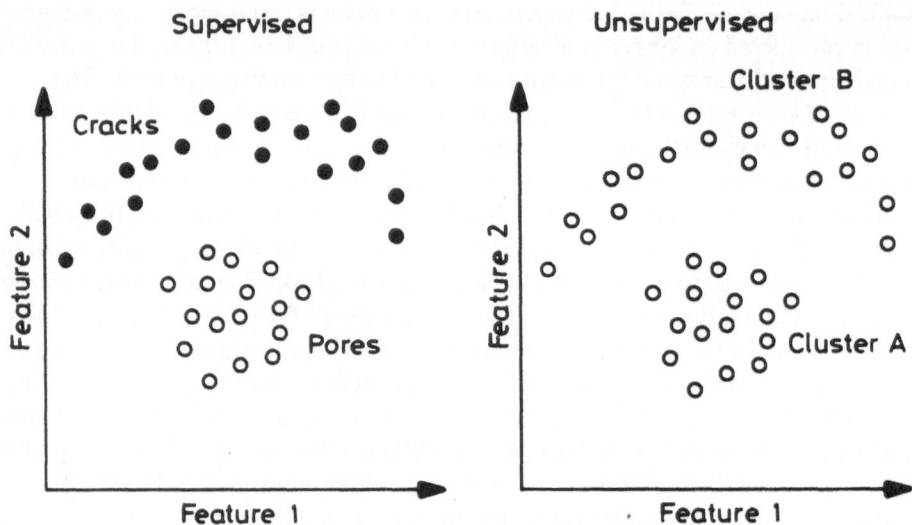

Fig 4.2 Supervised and unsupervised learning. (i) In supervised learning (left) points in a feature space are labelled according to their class. (ii) In unsupervised learning (right) the points have no label. Clusters may be seen and used to define likely classes within the system.

example points may overlap so that the definition of the class boundary is not straightforward.

In other applications the examples will not be associated with any known classes, and the objective is then to find the set of classes which best describes the data. If the points in feature space form distinct clusters, then these form the natural classes in a clustering process. Alternatively, they may form overlapping distributions which could be described as either one single cluster or two overlapping clusters. The process of defining the best clusters is the process of unsupervised learning. Ultrasonic defects may again be used to illustrate the process. In this case the dataset may have been collected without any prior knowledge of the defect types present. The objective would then be to define how many distinct types of defect are suggested by the data. Figure 4.2(ii) illustrates this in feature space. The defect examples are unlabelled and yet the clusters may be readily distinguished by the eye. The natural assumption is to ascribe separate classes to the clusters, although this may not always be valid.

In both supervised and unsupervised learning, the dataset of available examples should be divided into training and test datasets. For most classification and clustering methods it is possible to define a training process in which only the training dataset, whether in the form of images or feature space points, are presented. The performance should always be measured in a test process subsequent to the training, which uses examples from the independent test dataset. Sufficient test data points should be used such that the error in the estimate of classifier performance is much less than the variation of this performance with the experimental parameters, see Figure 4.3. When

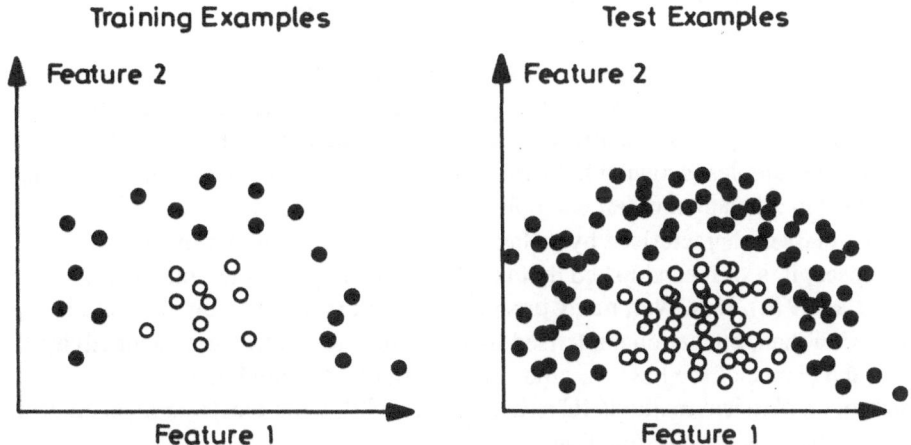

Fig 4.3 Training and test examples. To test classifiers it is necessary to divide the available examples into a fraction for training and to use the remainder for testing. The decision boundary derived during training will reflect the particular distribution of the training points, and will depend on the number of training points. The test examples are used to define a percentage success in classification. If there are enough test examples, all points in the space are covered, and the result becomes independent of the number of test examples.

artificially generated generic datasets are available, the number of test points can be made as large as necessary. The performance of classification methods can then be expressed by a single number: the percentage success in classification on a large test dataset.

4.3 Generic Problems Identified by the Partners

The following constitutes a set of problems of importance to the ANNIE partners which were considered to be suitable for possible solution using neural networks:

(i)   the detection and classification of defects in large welds from ultrasonic diffraction and reflection

(ii) the assessment of composite and other materials using acoustic emission

(iii) the characterisation of solder joint defects on printed circuit boards from their 3D shape profile.

Examples of the raw data and of the processed feature parameter data on which the conventional analysis methods are often based led the partners to believe that they had underlying similarities and might be a useful test of neural network methods. The form of data from these applications has been used to decide on the generic images and feature space distributions used in sections 4.4 and 4.5.

4.4   Supervised Learning on Generic Datasets

4.4.1   Supervised Learning of Classes in Feature Space

There are many practical classification problems in which each class is characterised by a clustering of points in some n-dimensional feature space. If we assume that a set of *training* examples is available in which the actual class of each point is known, the problem is to allocate new *test* examples to the correct class with the highest success rate. For classes characterised by a Gaussian probability density function in hyperspace, it is well established that the Bayes theory gives the optimal classification (Tou *et al*, 1974). However, many practical probability density functions are by no means Gaussian, but are characterised by extension, curvature, and connectivity. This makes the choice of the best classifier more difficult. Although there is an extensive literature on classical methods (Chatfield *et al*, 1980), there exists no ideal algorithm for all probability density functions.

Neural network methods have the potential advantage that, unlike linear classification methods, they are able to model complex decision boundaries between classes, whilst at the same time, unlike non-parametric methods, they are able in principle to make use of fully parallel computation to achieve classification in a few tens of computer cycles irrespective of the dimensionality of the problem. Even using a serial processing simulator, neural network methods have the useful properties that their classification speed is independent of the size of the training set, and classification may be rapidly adapted when new examples become available. In addition, neural network methods can work with clusters of arbitrary shape and connectivity; a property shared by several conventional classifiers.

The purpose of the work described below was to compare the actual performance of the neural network methods against several conventional classifiers for a series of model probability density functions of varying degrees of complexity. The models were created from analytic probability density functions so that the effects of the size of the training and testing sets could be investigated in a systematic fashion. The probability density functions were sampled repeatedly so that the effects due to using a limited size of training set could be evaluated statistically.

There have been several previous studies in which certain aspects of this comparison have been made. Huang and Lippmann (1987) have compared Bayesian and k-nearest neighbour algorithms with back-propagation for several two-dimensional probability density functions. Kohonen, Barna and Chrisley (1989) have compared back-propagation, learning vector quantisation and Boltzmann methods with a Bayes classifier for both slightly and strongly overlapping hyperspheres in 2-8 dimensions. The present work is aimed at being truly representative of applications found in the real world, with the model probability density functions based on observations by the ANNIE partners.

4.4.2 The choice of model feature parameter probability density functions

The basis of the choice of model feature parameter probability density functions was the types of function found in the data of applications considered later in this chapter. These are illustrated in Figure 4.4 and may be summarised as

(i) *Hypersphere:* a probability density function centred on a given point, with a Gaussian spread;

(ii) *Hyper-rectangle:* a uniform spread over a given volume, with an optional Gaussian spread;

(iii) *Banana:* a linear probability density function along a singly curved line, with an optional Gaussian spread perpendicular to the line;

(iv) *Umbrella:* a planar probability density function, with curvatures about two axes, and optionally with a Gaussian spread added perpendicular to the axes.

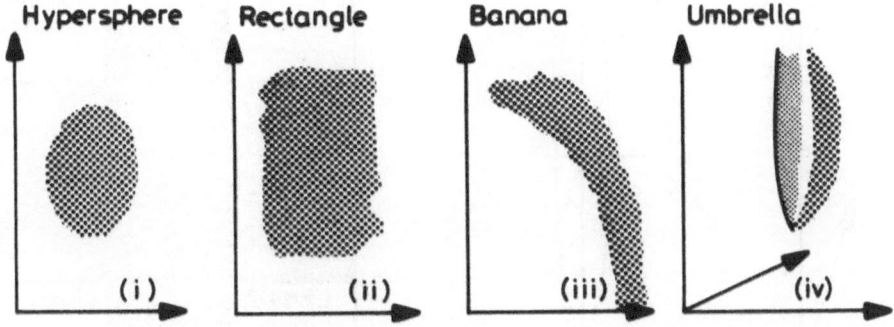

Fig 4.4 Some of the component shapes from which the generic probability density functions can be generated.

The method used was to construct model probability density functions from combinations of the above components. Each component may correspond to a separate class, or else disjoint classes may be constructed where a single class contains more than one component. The models described below have been chosen to enable the properties of the probability density functions, illustrated in Figure 4.5, to be explored in a systematic way.

4.4.3 The Algorithms used

Further details of the conventional classifiers described below can be found in Tou *et al* (1974), Duda *et al* (1973) and Fukunaga (1972).

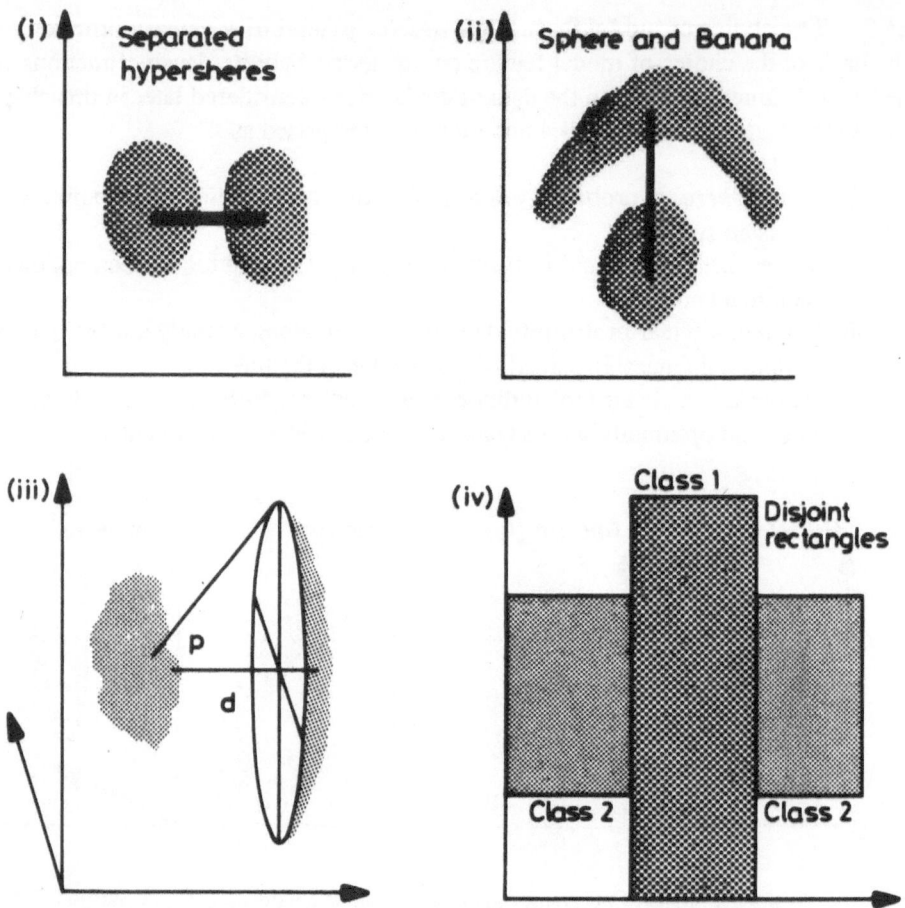

Fig 4.5 Model probability density functions which were considered. (i) Two spheres, separated by a variable distance as indicated by the heavy line. the spheres can be of any dimensionality from 2-8. (ii) A hypersphere separated from the curved banana may also be varied. (iii) The sphere and umbrella model. The variable parameters are the distance d between the sphere and the umbrella and the angle p of the umbrella. (iv) The disjoint rectangles model. The central class 1 is spanned by two regions of class 2. In this case the boundaries of the regions are quite sharp.

Minimum distance

In the minimum distance classifier, the n-dimensional coordinates of the class centres $(m_1, m_2, ...)$ are first found from the means of the feature values of the training examples within each class. For each test example, the distance from the point to each of the class centres is then derived using the Euclidean distance measure, ie

$$d^2 = (x_1 - m_1)^2 + (x_2 - m_2)^2 + ... \tag{4.1}$$

The test example is classified into the class giving the minimum value for this distance. For a two-class problem, it is clear that this method gives a decision surface which is the hyperplane that forms the perpendicular bisector to the line joining the two class centres.

Weighted minimum distance

The weighted minimum distance classifier is a slightly more sophisticated variant of the minimum distance classifier, designed to cope with hyper-elliptical Gaussian probability density functions, provided that the axes of the hyper-ellipses are aligned parallel to those of the feature-space. This is achieved by using a distance measure in which each component of the distance of a test example from the class centre is scaled by the measured standard deviation of the class in that direction, ie

$$d^2 = (x_1 - m_1)^2 / \sigma^2(x_1) + (x_2 - m_2)^2 / \sigma^2(x_2) + ... \tag{4.2}$$

where the notation is obvious.

Principal components

The principal components classifier used in this work is an extension of the principles used in the weighted minimum distance classifier, designed to handle the case of classes with hyper-elliptical Gaussian probability density functions, the axes of which may be at any angle to those of the feature space. This is achieved by use of a more elaborate metric known as the Mahalanobis distance, defined as

$$d^2 = (\mathbf{X} - \mathbf{M})^T \mathbf{S}^{-1} (\mathbf{X} - \mathbf{M}) \tag{4.3}$$

where $\mathbf{X}$ is an n-dimensional vector representing the test example, $\mathbf{M}$ is a vector representing the class centre and $\mathbf{S}$ is the covariance matrix of the particular class.

Bayes

The Bayes classifier seeks to find that decision boundary where the total probability of classification errors is minimised. In this work we used a form of Bayes classifier which assumes Gaussian probability density functions. This classifier corresponds to minimising the distance measure obtained by adding the logarithm of the determinant of the class covariance matrix to the Mahalonobis distance

$$d^2 = (\mathbf{X} - \mathbf{M})^T \mathbf{S}^{-1} (\mathbf{X} - \mathbf{M}) + \log|\mathbf{S}| \tag{4.4}$$

Fisher linear discriminant

There are many techniques, both parametric and non-parametric for designing linear discriminant functions. In the Fisher method, the discriminant function is chosen so as to maximise the ratio of the separation between class centres divided by the scatter within the classes. This is essentially a non-parametric method, but the decision surfaces are constrained to be hyperplanes.

K-nearest neighbour

The k-nearest neighbour classifier is the simplest non-parametric method used in this work, designed to handle problems in which the points do not fall in hyper-ellipsoidal clusters having Gaussian probability density functions. In this method, all of the training set of feature values are stored and the class of a test example is determined by the majority vote of its k-nearest neighbours in the feature space. In this context, *nearness* may be measured with any suitable metric, but Euclidean distance is used here for simplicity.

The k-nearest neighbour method makes no assumptions about the class probability density functions. Its main disadvantage is the amount of storage needed and the time taken to make a decision when the number of training examples is high.

Parzen window

The Parzen method is generally used as a non-parametric technique for estimating probability density functions. The method can also be used as a classifier by finding which class has the highest probability density function at the position of the test example in feature space. The Parzen estimate of a probability density function is based on a summation of kernels centred on each of the training examples in that class. Many forms of this kernel are possible, but in our method a Gaussian of suitable width was adopted for simplicity.

Potential function

The potential function classifier implemented in this work produces a decision function for each class consisting of a linear combination of Gaussians:

$$d(x) = \sum \exp\left[-(x_1 - a_{1i})^2 - (x_2 - a_{2i})^2 - ...\right] \tag{4.5}$$

The method has some similarities with the Parzen method, and is in principle capable of handling classification problems requiring highly complex decision surfaces for accurate results. Further details can be found in Tou and Gonzalez (1974).

The Rumelhart error back-propagation method

There are now many improvements possible to the method of error back-propagation described in *Parallel Distributed Processing* by Rumelhart, Hinton and Williams (1986), particularly for speeding up the training process, and for reducing its tendency to *over-fit* data (for example, by *pruning* the number of hidden units). However, the object of the present study was to compare performance rather than speed, so the method was used in its original form.

A fully-connected feedforward network of three layers was used in all cases with the activation function of individual neurons being modelled by a sigmoid function. The input layer was presented directly with the feature space variables, and the output layer had a number of units equal to the number of classes (usually two with targets of the form (1,0) for class 1 and (0,1) for class 2). In the testing phase, the largest output variable defined the chosen class. Thus there was no *uncommitted* class where targets were not reached. It was felt that this method, although not the usual practice, was most appropriate for comparison with the above conventional classifiers which all necessarily assign test examples to one class or the other.

The number of hidden units was varied to find the best performance. Experience dictated the number to use in other cases.

The learning vector quantisation method

In the learning vector quantisation method, developed by Kohonen (1989), a set of learning vectors m_i with n components comes to be scattered throughout the n-dimensional hyperspace of the feature parameters in such a way that the vectors represent the shape of the class probability density functions. Unknown test points are placed in the class of the nearest learning vector.

The learning scheme for adjusting the vectors is iterative as a function of time t. Example training vectors x_i are compared with the closest learning vector m_c. If this learning vector is of the correct class (i.e. is the same class as the example training vector), the learning vector is moved towards the training vector. If the class is incorrect, the vector is pushed away. Thus

$$m_c(t+1) = m_c(t) \pm a(t)[x_i - m_c(t)] \qquad (4.6)$$

where $a(t)$ is gradually reduced to a small value, and the + and - signs are used when the class is correct and incorrect respectively. The parameters of the method are $a(0)$, which was given the value 0.25, the number of learning vectors, which was chosen as 8, and the number of iterations, which was generally set at 5000. A graphical display of the movement of the learning vectors was used to decide if convergence was adequate.

4.4.4   Results

A wealth of results were produced on the different generic datasets. A typical example is shown in Figure 4.6. This shows the results for the banana and separated disk model. Each set was run as a function of the separation between the centre of the banana and that of the sphere.

The poorest performances are obtained from methods like the minimum distance, and Fisher linear discriminants, which are forced by their nature to have linear decision boundaries inappropriate to this shape of probability density function. Among the best classifiers are the nearest-neighbour methods, which are able to trace around the boundary points in the probability density function, whatever their shape.

Although the nearest-neighbour methods give the best results, the two neural network methods occupy two of the next five places. Examination of the decision boundaries produced by the Rumelhart method shows how the network is able to draw a boundary that successfully separates all the training points for the two classes. The boundary for a network with 8 hidden units also adopts a shape which ensures that all of the training points are correctly classified, but it is more complex than is necessary

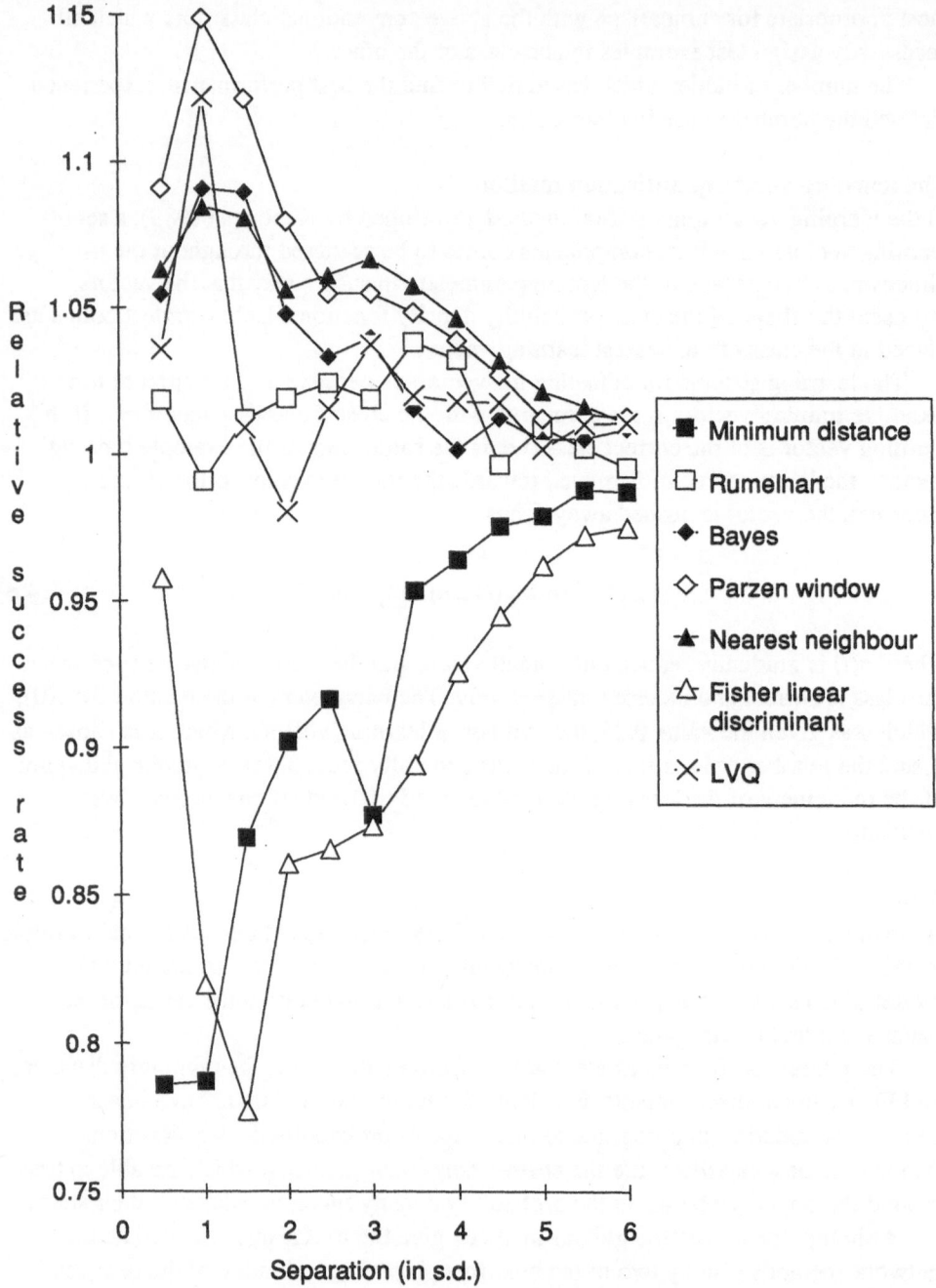

Fig 4.6 Success rates obtained with a curvature of 108°. At each separation distance the performance of each classifier is plotted as a ratio of the average performance of all the classifiers

or appropriate. Considerations like this are valuable when deciding on the number of hidden units which is most appropriate to avoid overfitting.

4.4.5   Conclusions

Results like those described above allowed a number of conclusions to be reached regarding the utility of neural network methods in preference to classical techniques.

The hope that neural network methods might prove to be superior in classification of extracted features has not been demonstrated in any convincing way, although in some cases clear advantages over conventional classifiers have been seen (speed of recall, applicability to a wide range of different feature parameter distributions).

The back-propagation and learning vector quantisation methods are both powerful classifiers which contain parameters enabling different levels of complexity of classification to be dealt with - the number of hidden units in back-propagation directly affects the topology and curvature range permitted in the decision boundary between classes. Both methods gave generally excellent results in classifying the model class distributions chosen as representative of real distributions from application areas. Both were generally fast in classifying test data, although back-propagation can be slow in training.

However, as classifiers they were not unique in having these advantages. The simple minimum distance classifier performed better than the neural classifiers (and other parametric classifiers) for certain well-suited model problems, like separated hyperspherical distributions, where the planar decision boundary given by the method matches the true planar decision boundary.

The k-nearest neighbour classifier is comparatively slow, but it can tackle class distributions of arbitrary complexity in a robust way. The learning vector quantisation method proved particularly reliable in our studies. However, classical methods such as the Parzen window behave in a very similar way and give similar results. Both methods describe training data by a defined number of suitably spaced class vectors or kernels.

A distinct advantage for neural network methods in classifying feature parameter data does become apparent when consideration is given to speed of operation as well as accuracy. Even without exploiting the opportunities of parallel and vector processing given by the neural network methods, these methods are quicker on recall than the only other method - nearest neighbour approaches - which performed with equal success over the entire range of distributions examined.

We can conclude that the choice of classifier most appropriate for a particular problem is dependent on the precise nature of the data to be analysed - in particular the form of the classification decision boundary. There is no unique *best* classifier which can be used at all times. If the problem has a linear decision boundary, then a linear classifier should be used. However, there are some instances where neural network approaches could be recommended on the grounds of ease of implementation and speed of recall:

(i)   when the classification is complex, with unknown decision boundaries and/or disjoint classes

(ii)   where speed of recall is at a premium (where its advantage over k-nearest neighbour becomes most evident)

(iii) where the data is of a less processed nature, ie the features extracted are not well defined or when it is difficult to know what features should be included

(iv) when the problem is subject to small changes requiring adaptability and robust retraining

(v)   when you want a rapid first look at an unknown feature space problem.

These considerations led the pattern recognition work in ANNIE to concentrate primarily on real problems for which classification could be attempted from 'raw' data such as direct input of images in pixel-by-pixel fashion. Before discussing this work we will, however, look at the related problem of unsupervised learning.

4.5   Unsupervised Learning

Unsupervised learning may be defined as the process of defining clusters of *similar* examples from an unlabelled training set. Clustering algorithms thus seek to attach a label denoting cluster membership to each individual example. When a labelled dataset is available, supervised learning is generally used but this is not always so, and in this case there may be several reasons for using unsupervised learning procedures.

(i)   the collection and labelling of a large set of example patterns can be surprisingly costly and time consuming. If a classifier can be crudely designed on a small labelled dataset of examples and allowed to run without supervision on a large unlabelled dataset, much time and trouble can be saved

(ii)   in many applications, the characteristics of patterns in a given class can change slowly with time. If these changes can be tracked by a classifier running in an unsupervised mode, improved performance can often be achieved, for example, in condition monitoring

(iii) in the early stages of an investigation a clustering analysis may give a valuable insight into the nature and structure of the data. The discovery of distinct subclasses or major departures from expected characteristics may significantly alter the approach taken to designing a suitable classifier, for example, in acoustic emission investigations.

In supervised learning, it is straightforward to measure the success in classification because the classes are known and statistically well defined, and each example within the dataset falls within a known class. On the contrary, in unsupervised learning, no prior knowledge is available on the structure of the data set and the success rate cannot generally be defined.

The method used here to define the success rates in unsupervised learning was to adopt the same datasets as in the supervised learning case. The actual class membership of each example is then known, and it is possible to define the percentage agreement between the clustering classes(see below) and the actual classes. The model distributions used were:

(i)   separated hyperspheres in 2-8 dimensions
(ii) a 2-dimensional distribution containing a disk and a curved banana.

Six conventional algorithms were used: dynamic clustering, hierarchical clustering, k-means, centres of aggregation, mutual nearest neighbour and an algorithm using the theory of fuzzy sets. If there are many neural networks that could be applied to the problem of supervised learning of clusters in an n-dimensional feature space, very few methods can be used directly for unsupervised clustering. In this work two neural network methods that can perform unsupervised learning were compared with conventional clustering techniques: back propagation and Kohonen mapping.

In order to perform unsupervised clustering with error-backpropagation we used a multilayer perceptron in auto-associative mode; a symmetrical network was constructed with the outputs constrained to be similar to the inputs. Clustering information was elicited by inspection of the levels of activation of the central layers of the network.

The Kohonen mapping was used as a pre-processor to the use of a conventional clustering technique. Figure 4.7 shows the procedure used.

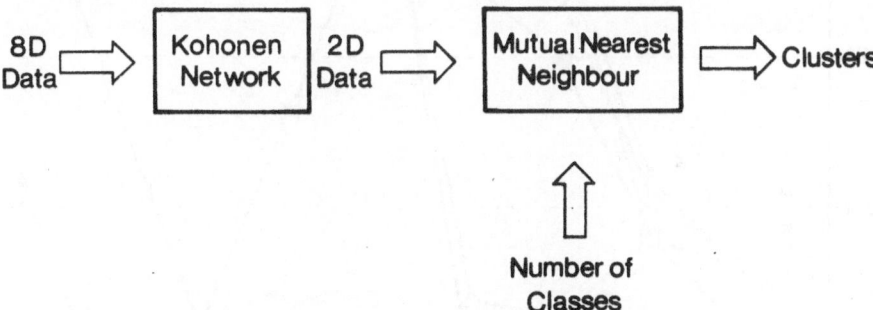

Fig 4.7 The use of the Kohonen self-organising mapping as a preprocessor to a conventional clustering algorithm. The mapping preserves the topology of the input space, hence makes the job of the clusterer easier

Results for separated hyperspheres

For separated hyperspheres the same data sets were used as in the investigation of the supervised learning algorithms. Each sample was labelled according to the class to which it belonged. This initial labelling was later used for evaluating the success of clustering. This quality criterion, whilst not perfect, is simple and allows the results to be compared with the supervised learning procedure.

The implementations of conventional clustering methods which were used allowed many parameters to be varied. However, only the best values in a particular class were chosen to illustrate a particular method.

The results for this section are summarised in the graph of Figure 4.8. In this particular case, the Kohonen mapping performed badly, although this was not generally true. For the disk and banana dataset rates of success were always greater than 88% and than 95% for $d > 0.5$. This result was better than the best conventional clustering algorithms.

The conclusion is that processing with a Kohonen network was useful in eliminating points which establish bridges between clusters, favouring fast and efficient subsequent clustering.

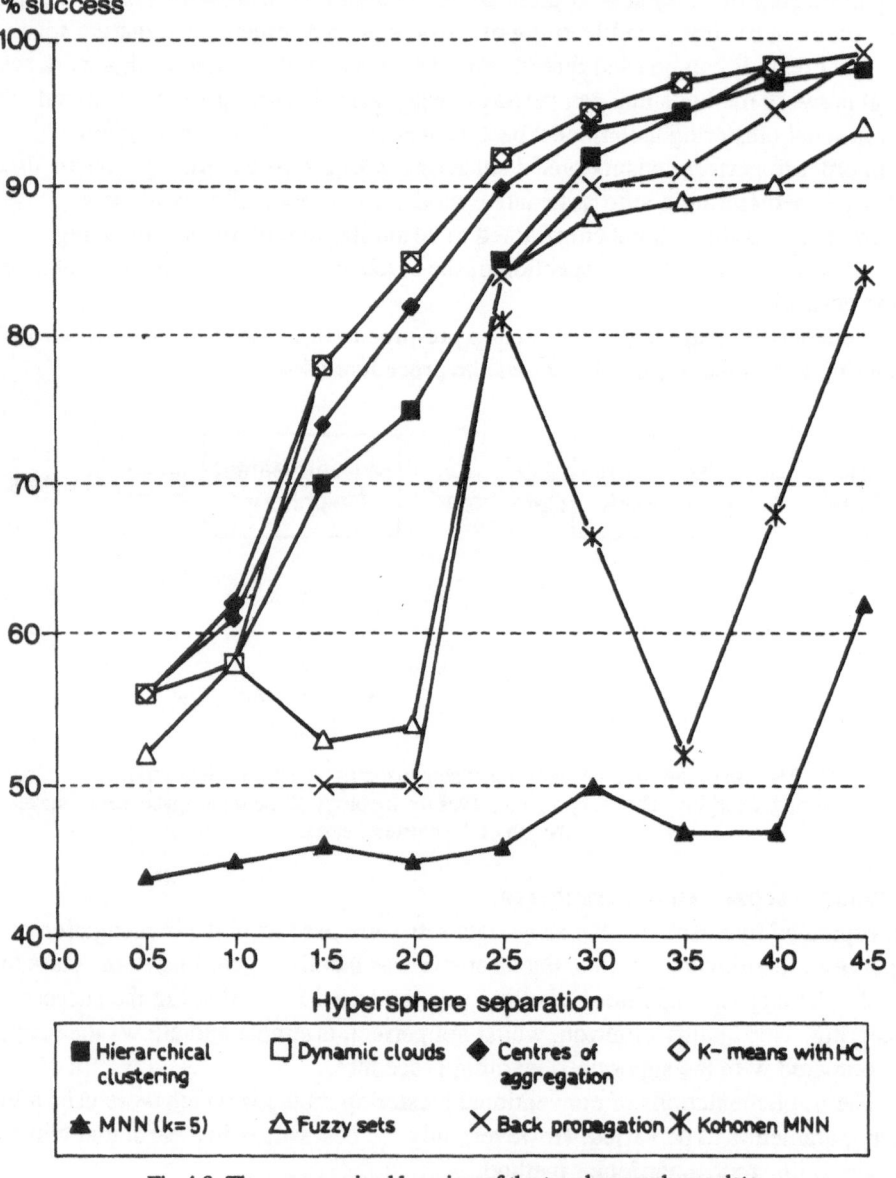

Fig 4.8 The unsupervised learning of the two hyperspheres data.
Example methods from each category are chosen.

Conclusions
Clustering analysis is more complex than supervised classification and it generally requires experience and intelligent input from the user. It is very difficult to make it automatic. Even with conventional algorithms, the best solution is not evident and it does not seem particularly well suited to neural network approaches, for they are generally a first step in the cognition process. However, examination of clusters can be important in complex classification problems for which neural networks may be a best 'first' attempted solution (see above). One such example was the case of pneumatic testing of pressure vessels, described below in section 4.7.

4.6   Applications of Neural Networks to Pattern Recognition in Acoustic Emission

4.6.1   Analysis of Acoustic Emission from Wear in Composite Components
Composite materials are now being considered for use in cars in applications such as springs and suspension parts where the safety of the component is critical. For such parts it is necessary to develop inspection techniques capable of demonstrating the component's quality and its suitability for the task. Acoustic emission during a stress sequence is currently being considered as one technique. The aim is to characterise the frequency distribution of the sound in terms of the possible types of defects. This is a generic problem in that the same techniques may be considered for tool monitoring surveillance and many other applications where a human operator uses his ears to obtain information.

The problem of acoustic emission from composite materials has been examined. The signals from a flexed bar of composite material were analysed to give the acoustic energies in nine equally spaced frequency bands. A statistical analysis showed that much of the useful information was in fact carried in the first three low frequency bands. In this three-dimensional space the clusters are observed as elongated distributions of the *banana* form, one of the generic forms studied within the project, and are shown in Figure 4.9. The results and conclusions of section 4.4 are therefore relevant to the problem.

4.6.2   Source Location Map
A rather different problem arises when acoustic emission is used to produce a so-called source-location map which is a real image rather than a feature space map. Figure 4.10 gives some examples of raw location maps obtained during acoustic emission (AE) inspection or surveillance of structures. The crosses in (i) denote the positions of transducers placed on the structure and each dot corresponds to an acoustic emission event which has been located by triangulation from the timing of the responses in the transducers. The centre line in (ii) shows the path of the weld as it is made. Clusters of dots correspond to defects located in the materials, and some known defects are marked and labelled on the figure. However, significant AE sources are lost in background points due, for example, to production processes. For example in welding monitoring, slag cracking during cooling generates an intense acoustic emission, sources of which are randomly located along the weld head.

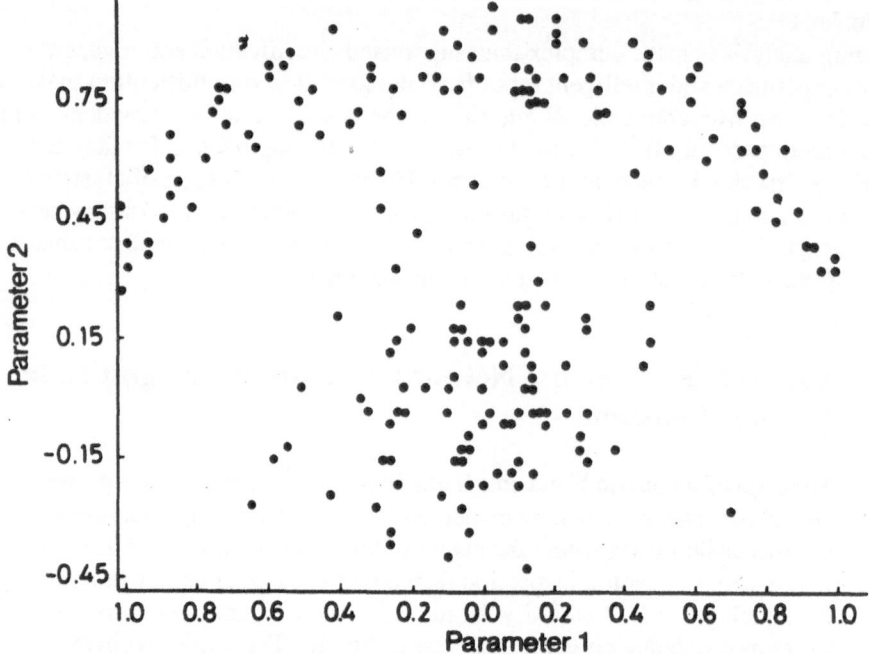

Fig 4.9   Artificial data generated to model the significant feature parameter data derived from the acoustic emission signals produced by a flexed bar of composite material

To aid the automation of the analysis of such maps, there are a number of algorithms, akin to image processing, which are generally very efficient and very fast. Thresholding on the density of sources by unit area can be used as input to a general algorithm such as mutual nearest neighbour. However, specific algorithms which use location and activity delay after the welding arc has passed have proved to be better and more robust. The generic results on unsupervised learning would suggest that the possible utilisation of neural networks in this area does not seem to be very clear except possibly in replacing the last types of algorithm which are time consuming and not very easy to use.

4.6.3   Characterisation of Materials

The characterisation of materials is a typical and classical problem in AE. The user gets a lot of signals during a test and wishes to classify them in order to assign a specific set of signals with a given source mechanism. In the case of AE it is very difficult to isolate one source mechanism as several of them are simultaneously activated. Therefore the problem is to create clusters of signals and to assign clusters to source mechanisms. It should be noted that this is necessarily an investigative process, where the decision belongs to the user. The solution can only be a proposal to be confirmed. The use of conventional algorithms usually allows a specialist to understand exactly what assumptions are being made about the clusters and what type of clusters are favoured. The 'blindness' of a neural network is therefore a distinct disadvantage in this application.

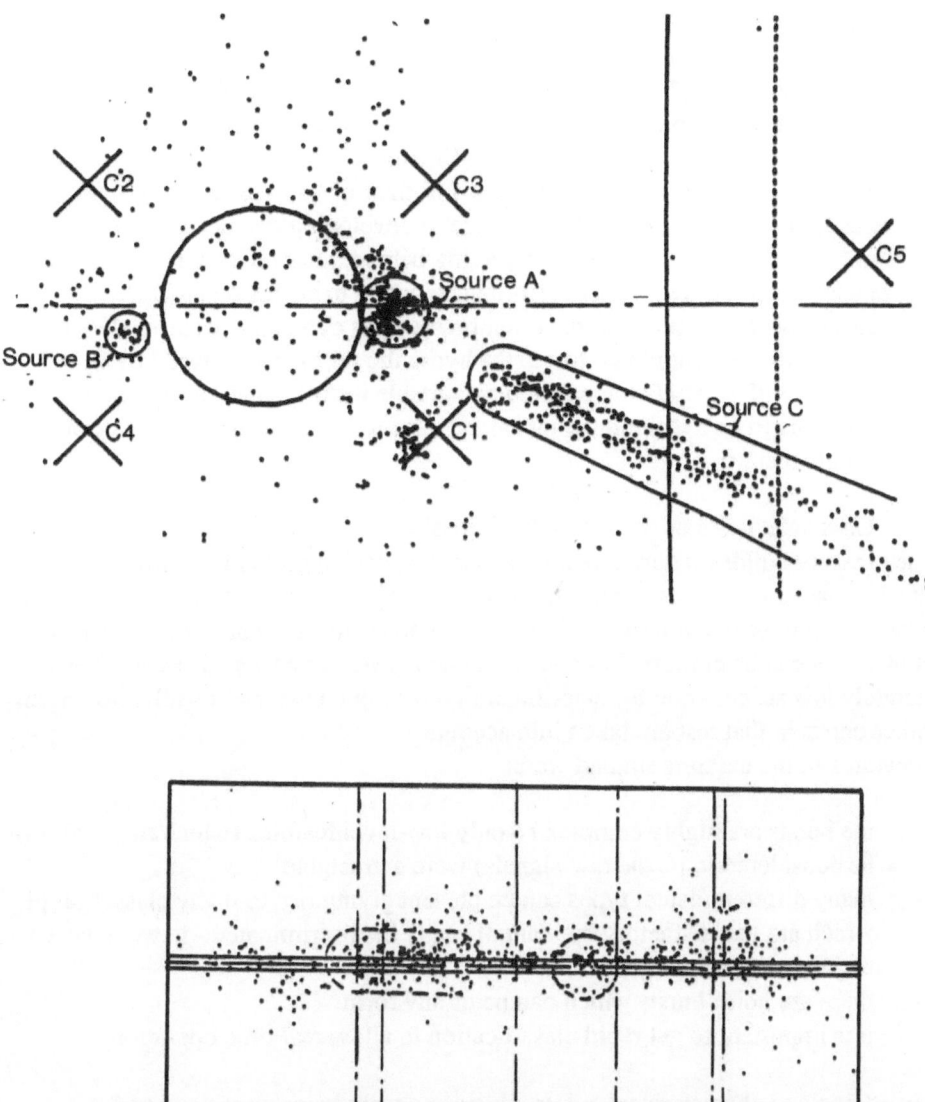

Fig 4.10 (i) The source location map obtained during monitoring of an offshore rig (top).
(ii) The source location map obtained during in-process weld monitoring (bottom)

The above (text) suggests that the most beneficial area for exploring neural networks in AE is in analysis of feature extracted data where there are some requirements, such as speed, which would make them favoured over conventional algorithms. An important such application, the proof testing of pressure vessels, is considered in the next section.

4.7   Proof Testing of Pressure Vessels

The hydraulic proof test of pressure vessels, which has been used for very many years, has the advantage of demonstrating that the vessel under test is capable of withstanding the maximum service pressure. However, at least for vessels intended to contain gases, it is expensive and offers poor information as to the quality of the structure and, in particular, about the presence of developing defects. It can even prove to be harmful because it runs the risk of leaving behind, after the test, traces of moisture which could affect vessel performance owing to corrosion and/or ice.

For these reasons manufacturers of compressed gas cylinders, accumulators or large pressure vessels, would like to see the hydraulic proof test replaced by a pneumatic test, with safety being ensured by suitable instrumentation. Acoustic emission appears to be a technique that can be used for this purpose, as has been demonstrated recently.

4.7.1   Characteristics of the Acoustic Emission

As the pressure builds up during testing acoustic emission occurs in a number of bursts. For each burst a number of low-level features are extracted: rise time, counts, energy, duration, peak amplitude. There may be 40 or 50 such bursts during a test, each of which can be classified as due to a defect or not, or as a *noise* event. Thus a moderately low success rate for classification of any one burst can be tolerated when all other bursts in that test are taken into account.

Features of the datasets studied were:

- the bursts are highly complex so only low-level features (which might almost be considered to be the raw signals) were extractable
- many different defect types can be present giving rise to many classes all of which are unsatisfactory but cannot easily be discriminated - ie we have a disjoint class structure
- there are noise bursts which can be of any form
- it is important to get rapid classification to allow real time operation.

In attempts to characterise the data, clustering techniques were used and visualisation was attempted using principal components analysis and the NetVision package. No large separation of the 7-parameter dataset was possible when reduced to 2D, confirming the assumed complex nature of defect-containing and sound specimen data.

4.7.2   Results

A first attempt at supervised learning was therefore undertaken using an error back-propagation network. Results are shown in Figure 4.11. This figure shows the marked difference during tests on sound data and defect pieces. The success rate in this case is about 60%. (Higher success rates during training gave worse test results. This corresponded to observations that overfitting was occurring where the decision boundaries as seen with NetVision began to fit around individual datapoints.) The reason that better success rates were unachievable was felt to be due to the nature of the data where up to 50% of the bursts arose from spurious (background noise) events. A way of deducing whether an event is real or noise is to combine the signals from more than one transducer, and use the timing information to deduce if the source is internal or external to the region of interest. In a revised approach two networks were therefore used: the first was trained to distinguish real from noise events; a second was trained to distinguish only those events classified as being real. Using this filtering approach, a success rate of about 83% for classifying individual bursts was achieved. It was concluded that this success rate on individual bursts was adequate (given 100-300 bursts for each classification) and allowed sufficient generalisation on the many defect responses. This work is very promising and will be pursued in the future as a real application.

Fig 4.11 The results of a trained back-propagation network at classifying acoustic emission signal bursts as arising from defect-containing or sound pieces. Each burst is classified as 1 for sound or -1 for defect. The x-axis represents time. As the pressure increases, successive bursts are emitted, each of which passes through the network for classification. Despite errors it is clear that the majority of bursts in the defect piece are classified as defects, whereas for the sound piece most bursts give outputs near to +1

4.8 Detection and Characterisation of Defects in Welds from Ultrasonic Testing

In recent years a great deal of effort has been placed on developing improved ultrasonic systems for non-destructive testing (NDT). Ultrasonic testing (UT), one of the main NDT techniques, is used to inspect pressure vessels, pipes and other components during construction and throughout their lifetime. Application areas include nuclear power plants, aerospace, offshore structures, chemical plant, oil refineries, and bridges. The precise techniques to be used will depend on the particular objectives of the applications such as the measurement of wall thicknesses, and the detection of cracks, pores or slags in ferritic welds or coarse-grained steel.

The ultrasonic inspection of large welded components, such as pressure vessels or pipes, may entail the collection of data from hundreds of metres of weld, with only a small probability of encountering a significant defect. When the inspection is done manually, the operator is relied upon to log defect signals when they occur and to investigate them sufficiently thoroughly to enable the location, character and significance of the defect to be estimated. It has long been recognised, however, that even well trained and diligent manual operators are subject to fatigue and distraction and may fail to perform the inspection exactly according to the required specification. The difficulty of working in muddy, cold water in the North Sea is easily imagined. This has led to a general view that, where very high reliability is required, inspections must be performed by automated equipment.

Such automated systems, when properly designed, can ensure that the inspection adequately covers the volume where defects might lie and can perform a crude assessment of the significance of signals, based purely on amplitude. However, amplitude, by itself, is an inadequate discriminator and the recording threshold consequently has to be set at a level where the majority of indications will subsequently be judged of no significance.

The sorting of significant indications from insignificant ones, defect detection, has been done in the past by direct application of human judgement to some sort of displayed view of the recorded indications. Very recently there have been attempts to automate this process using various filters and arc detection methods based on conventional algorithms. However, it has proved difficult to encapsulate within such methods the skill which enables an experienced human analyst to grasp the significance of a display 'at a glance', i.e. by absorbing all the information in the display in parallel. There appears to be an opportunity here for application of pattern recognition techniques based on neural network algorithms.

The more difficult and related tasks of defect characterisation, determining not only the presence of a defect but also classifying it (as for example a crack or porosity), is described in section 4.8.3. First we describe some general aspects of inspection using ultrasound which are important for both defect detection and defect characterisation.

4.8.1 Ultrasonic Inspection Techniques

Ultrasonic testing (UT) relies on observations of the behaviour of sound waves as they spread through the test medium. Variations, caused by reflection, refraction, diffraction and scattering, are measured by a number of different devices and are evaluated by various methods. Typical variables which can be measured are the intensity, transit time and phase of the reflected sound. Some systems rely purely on measurements of the total intensity of the transmitted or reflected continuous sound beam at each point on the specimen surface. These are often used to test plate-shaped objects. Such intensity systems can supply only poor information concerning the shape and location of a flaw.

Holographic systems utilise interference phenomena, incorporating additional phase information. These systems need complex processing methods and serve the purpose of showing a known flaw in detail rather than automatically detecting one in a large intact component.

Using the information provided by intensity and transit time leads to systems like the pulse-echo and the time-of-flight diffraction method. These methods accomplish a common approach for detection and sizing of general defects and rely on transducers emitting and receiving pulses of ultrasound. The characteristics of the ultrasonic signals obtained by these different methods are all similar, and will now be described in detail.

4.8.2   The Characteristics of Ultrasonic Signals
The A-scan
The basic signal unit in ultrasonic inspection is the A-scan: the signal detected at a receiving transducer some time interval after transmission of an ultrasound pulse from a transmitter. The A-scan is therefore a function of sound intensity versus transit time. The signal will partly consist of noise from various sources uncorrelated with the transmission but, since there are well-established methods of reducing those contributions to insignificant levels, we shall not consider them further. The remaining contribution to the signal will arise directly from the transmitted pulse, energy from which will reach the receiving transducer by reflection, refraction, diffraction and scattering.

In a fine-grained material, with no preferred grain orientation, at normal inspection frequencies, refraction and scattering from the grain can be ignored, so there will be no dispersion and no attenuation, other than that arising from geometrical beam spread. If we further stipulate a single block of material of uniform properties, there will be no other sources of refraction. Hence, reflection and diffraction will be the mechanisms by which signals arise. In simple terms each contribution to the signal can be imagined to arise from taking a replica of the transmitted pulse, delaying it by an amount determined from its velocity and path length, attenuating it by an amount determined from its pathlength, and applying such phase changes as are implied by its history of reflection and diffraction processes. Note that the signal can suffer conversion between compression and shear modes at a point of reflection or diffraction and this changes its velocity but not its frequency.

Transducer response
We must, however, allow for the properties of the ultrasonic transducers. In order to combine adequate range resolution with acceptable sensitivity, transducers are constructed to generate a pulse consisting of a few cycles of their resonant frequency, with an approximately Gaussian envelope (though the rise of the envelope is faster than the fall). The pulse therefore contains a band of frequencies of width typically up to half the resonant frequency. As long as we consider only points far from either transducer, the transmitter can be thought of as emitting ultrasound within a narrow cone around the beam centreline, and the receiver as being sensitive to ultrasound striking it from within a similar cone about its beam centreline. The amplitude of a signal will fall off as its path departs from the beam centrelines. However, the cone angle is dependent on the ratio of the transducer diameter to the ultrasound

wavelength, and so will be wider for lower frequencies. Hence, a signal generated by reflection near the beam edges will have relatively more low frequency and less high frequency content, depending on the bandwidth of the transducer.

At points near to a transducer, interference between waves from different parts of the transducer face makes the pulse envelope more complicated, and these effects again depend on frequency. Finally, the signal from an extended defect will be an integration of the signals from each of the defect's elements, and so may differ markedly in appearance from the prototype pulse. If the material is coarse-grained and the grains have anisotropic elastic properties, then each grain boundary may give rise to a small signal and there may be a noise-like background which is the residue of all these small signals after they have interfered with each other. Once more, this will be frequency dependent.

Data compression
The above remarks apply to the signal at the receiving transducer (often called the RF signal). However, in many inspection systems, the signal is rectified and smoothed before display or data acquisition, so that only an approximation to the envelope of the received signal is available for analysis. This rectification and smoothing removes any information about the phase of the signal but has the advantage of reducing the storage requirements.

4.8.3   Defect Characterisation
Ultrasonic defect characterisation is usually undertaken with reference to a B-scan of the material under inspection. An ultrasonic B-scan is constructed from a series of A-scans and can be thought of as a surface plot of amplitude as a function of transducer location and signal time. If we compress the amplitude information into a single bit by applying a binary threshold, we can talk of the shape of the signals in the scan-distance/transit time plane. The shape reflects the change in transit time of the signal as the transducers are moved and, since it can be predicted precisely, is a valuable means of discriminating between signals of different origin.

The synthetic aperture focusing technique (SAFT) makes use of this predictable shape to emphasise those signals which exhibit the particular shapes being sought. The SAFT algorithm works by computing the line integral along all possible arcs of appropriate shape in the B-scan (for a perfect, noiseless system a single point defect will give rise to an arc in the B-scan image). For RF data, the output amplitude is very sharply dependent on how well the actual signal fits the calculated arcs. This makes SAFT a powerful means of picking out wanted signals but also makes its vulnerable to factors which disturb the arc shape. For instance, the surface of the workpiece may not be perfectly flat and the transducers may be coupled through a water layer, giving a noticeable variation in signal timing as the path length in the water varies. For SAFT to work well in these circumstances, a *flattening* process (shifting individual signals to line up the surface *lateral* wave or back wall echo) may need to be applied, prior to application of SAFT, adding to the already considerable processing load of SAFT itself. The sensitivity to timing disturbance is reduced if the processing is applied to the data once it has been rectified and smoothed, but then the ability to discriminate against unwanted signals is also reduced.

By contrast, the human observer can often assess by eye the validity of signals in the presence of timing variations which would badly affect the performance of SAFT. If the B-scan contains considerable grain noise, SAFT will still enhance the signal-to-noise ratio of signals with appropriate arc shape, but it will not entirely suppress other features. The human observer, however, can usually make a judgement about the presence or absence of arcs of appropriate form with near certainty. Hence, the processing of the B-scan image which takes place in the brain has both a superior absolute performance to the SAFT algorithm, and also operates very much faster than the algorithm when implemented on typical computers. We have not tried so far to implement for this problem neural methods which have the brain's distortion tolerance built into them, but have concentrated on methods which, like SAFT, assume that the spatial relationship of image features is fixed.

The detection of cracks in welds is a problem of great importance. Pulses of ultrasound are reflected by cracks, but also by less deleterious defects, like regions of porosity or lumps of slag. The problem is to be able to distinguish the class of the defect from the form of the ultrasonic reflection. An extensive series of measurements has been made on some 84 defects of known class artificially introduced into test welds. In previous work (Kohonen, 1988) these data were transformed into up to 7 feature parameters which were analysed by a variety of conventional classifiers.

Figure 4.12 shows the 4 examples, displayed as points in a two-dimensional feature space of pulse amplitude against kurtosis. The classes of pores and slag are seen to be roughly represented by spherical distributions centred at different points in the scatter plot. This gives rise to the hypersphere distribution which was used in the generic studies described above. A hypersphere is centred at a given vector in feature space, then all points are given a statistical spread with a given standard deviation along all axes. The class of cracks is seen to be both extended and curved. Such *banana* distributions are simulated by a uniform distribution along a curved arc, with all points then being given a statistical spread in all dimensions. Figure 4.12 also shows the model distribution simulating these data composed of two spherical and one bent banana component. Using the model distribution it was possible to investigate such important factors as the number of training examples necessary to give a classification error below a prescribed value.

4.8.4 Defects in Welds: Direct Classification from Ultrasonic Images

This section reports on work which investigated the new opportunity opened up by neural net algorithms - to classify defects directly from ultrasonic images. It has important advantages over the current state of the art - feature extraction followed by classification. The direct image classification is rapid enough that on-line defect classification is practicable. Also the judgement involved in finding the best feature parameters is eliminated, since the neural networks find suitable features directly from the training set. It was concluded that to distinguish the classes of defects considered earlier, it was necessary to use three or four dimensional images, where the third dimension is the depth of the defect beneath the plane of the image, and the fourth

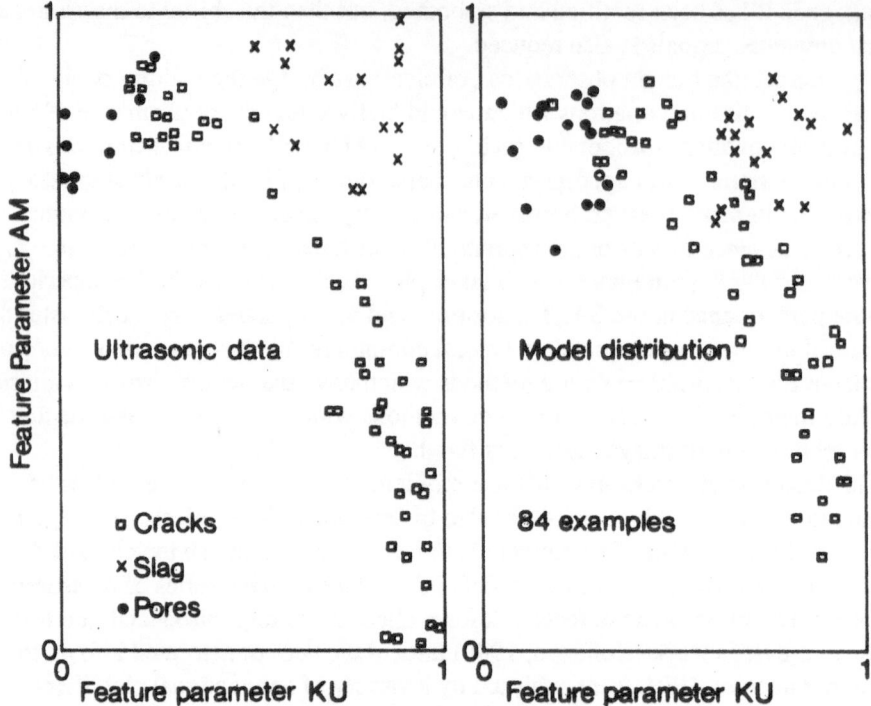

Fig 4.12 The left-hand graph shows a scatter plot of two of the feature parameters derived from ultrasonic interrogation of 84 examples of known defects. On the right is data produced to model the defect feature parameters data, composed of two spherical and one bent banana component

dimension is the angle of the ultrasound. Initial work was carried out on artificial datasets and was then applied to actual ultrasonic datasets from real defects, reported here.

The methods used were:

(i)   a conventional pattern matching method using prototypes for each class

(ii) an adaptive field method in which receptive fields centred on peaks in the image are first extracted and averaged together to give a prototype for each class of the image

(iii) direct input to a conventional back-propagation network

(iv) a receptive field input in which any field centred on a peak in the image becomes an example input to a back-propagation net

(v)   direct input to the shared weights adaptation of the back-propagation network.

The adaptive field method (ii), the neural net receptive field back propagation method (iv), and the shared weights method (v) all gave good results comparable to those obtained in conventional feature extraction studies. The results are industrially significant as far as that the speed of the neural network classification would allow on-line classification of defects.

The problem of the classification of defects in welds from data measured by ultrasonics has been a central theme of the ANNIE project. The importance of the problem stems from the fact that crack defects of a given size are more dangerous than the more benign slag inclusions and regions of porosity. The conventional state of the art for this problem is to extract feature parameters from the raw data and then apply a conventional statistical classifier. Feature parameters are functions of the raw image data which describe in a single number some physical effect, such as the amplitude fall-off with angle of ultrasound, or the kurtosis of the wave form. Human expertise was used over a period of many months to determine the best feature parameters for discriminating between classes. Between 3 and 8 feature parameters were used. Each example defect could therefore be represented by a point in a feature space of 3-8 dimensions and the feature space classification procedure consisted of assigning each point to a class. In most of the work four classes were defined, consisting of porosity, slag, and smooth and rough cracks.

At the start of the ANNIE project the Hopfield and back-propagation neural net methods had been applied to feature parameter data. The more general studies described earlier showed that the neural net methods gave good results but were not particularly better in performance than a suitably chosen parametric conventional classifier. In terms of computational speed, some of the best neural methods were excellent, but this advantage is nullified by the fact that the feature extraction process itself takes much more time than the classification. The conclusion is that methods in which the image is analysed directly should be explored, and the direct presentation of the ultrasonic pattern to the network was therefore considered. This process avoids the computation involved in feature extraction, and more fundamentally, avoids the expert work involved in the selection of optimal feature parameters for any given problem.

4.8.5 The Form of the Ultrasonic Images from Defects in Welds

The data used in this study was collected by Burch *et al* (1986) and used in their feature parameter studies. The defects were introduced artificially into test pieces, and so are of known class. The data took the form of three-dimensional scans of the ultrasonic reflected intensity as a function of position x, and y and depth z in the material. In addition each dataset was collected at two different angles, usually 0° and 20°. It was believed that data of this complexity was needed for good classification purposes. Each image contained a single defect of known class, which had been extracted by eye from a more extended image. The resolutions of the data were 2 mm in the x and y directions and 0.5 mm in the z direction. Each dataset was a different size, designed to cover the observed defect adequately. Typical sizes were 24x22x33x2 pixels in the x,y,z and angular axes respectively.

In previous publications the three-dimensional arrays were often presented through their three elevations. However, in this study the full three-dimensional data are used, together with two ultrasound angles. In Figure 4.13 elevations of such datasets are shown from four defects, each measured at two different angles of ultrasound, 0° and 20° from the vertical. Figure 4.14 shows these same four defects presented, at rather lower resolution, in the full four dimensions. All the pixels in the image are shown, in contrast to the elevations which show only integrals over the dataset. The level of shading shows the reflected intensity as sets of sections at constant depth z, of the sections x,y. The two angles of ultrasound are shown one above the other.

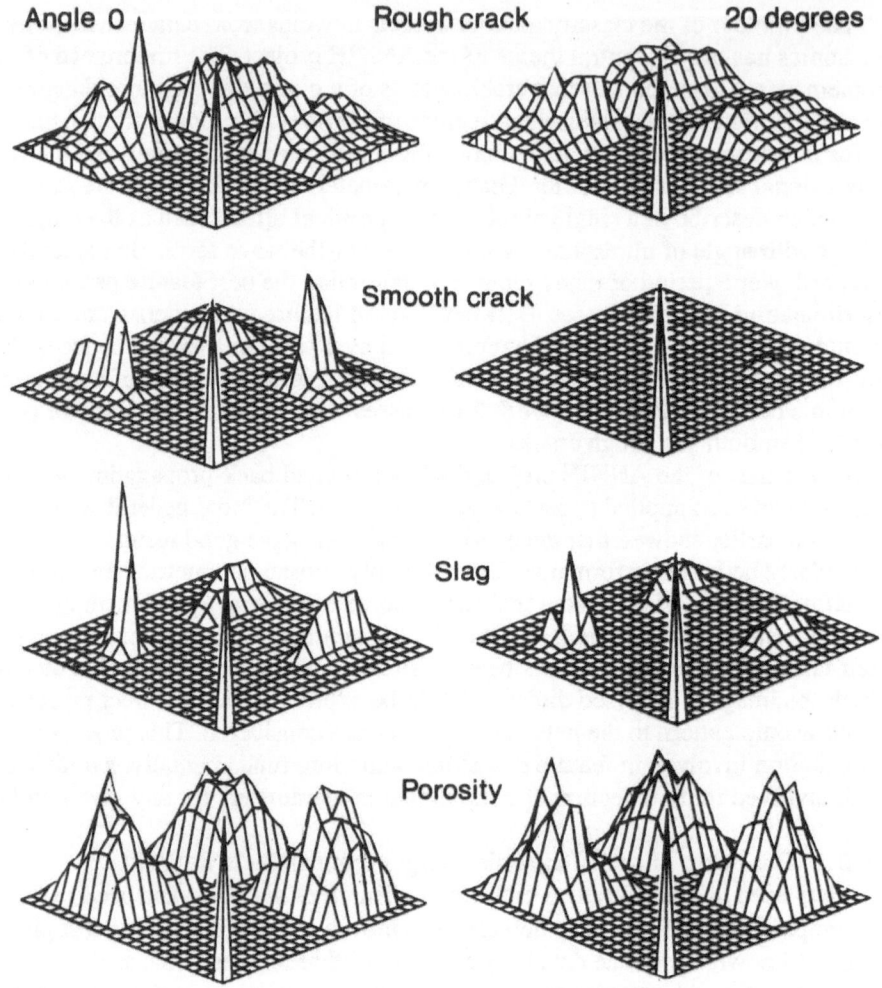

Fig 4.13 The ultrasonic images from four classes of defects in welds. The figure shows the response from two different angles of ultrasound, with the probe angle at 0° (left), and 20° (right) to the vertical. The elevations give the integral values over the full data in the x, y and z directions. The height gives the intensity of the ultrasound reflection

Examination of Figure 4.14 shows the difficulties to be encountered by any classification method. The rough cracks and the porosity have a very similar texture, or frequency of peaks across the image. Similarly the smooth cracks and the slag have a smoother texture. The differences between the two types can in both cases only be seen through the angular dependence. The smooth cracks act as specular mirrors and reflect the ultrasound preferentially into one direction. In most cases the cracks are aligned roughly parallel to the surface so the maximum signal is received in the receiver. The slag in contrast is rather more isotropic. Rough cracks are thought to correspond to a series of specular regions each with its own angle. The ultrasound is

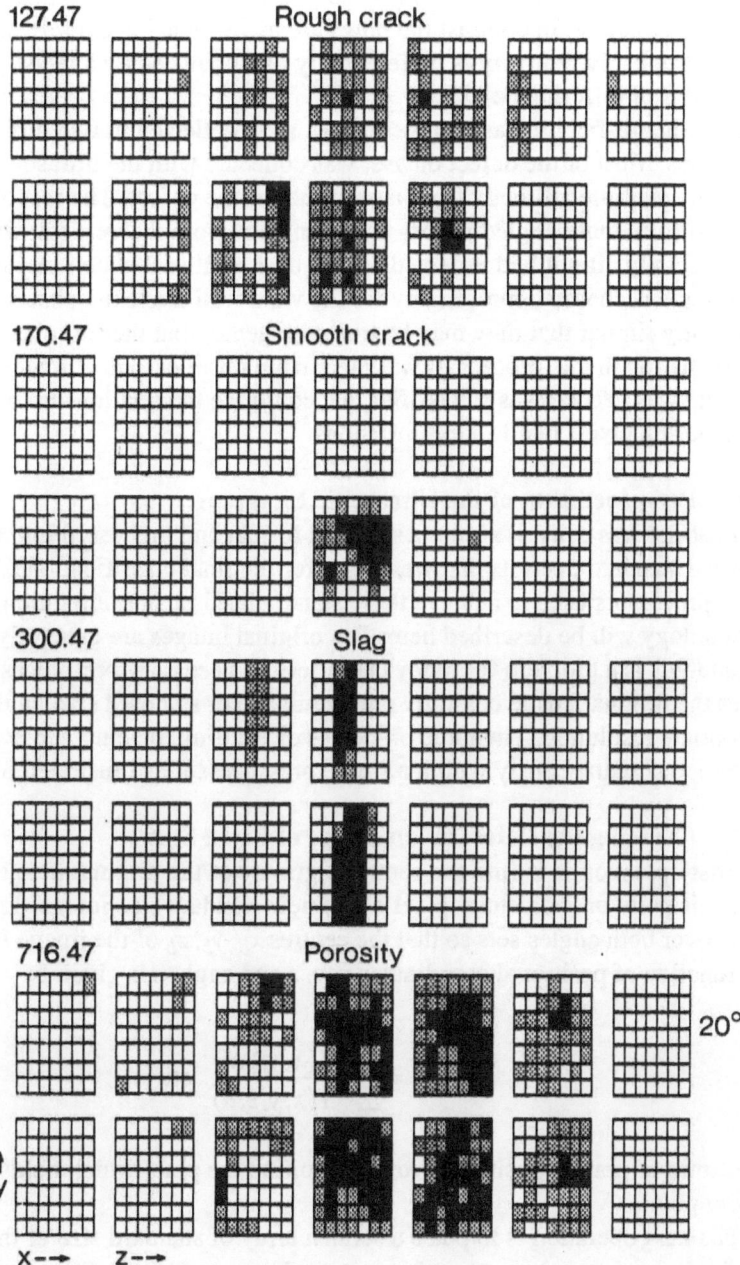

Fig 4.14 Another representation of the defects shown in Figure 4.13. Each section shows an (x,y) plot of the intensity of the ultrasound reflection. The images across the figure show these sections as a function of the depth z into the material. The two angles are placed above each other. This representation shows all the pixels of images from a series of 7x7x7x2 images

therefore preferentially reflected into one of the receivers, although all will usually record the structure. In contrast the pores are more rounded and reflect the ulstrasound more isotropically. Porosity is preferentially distinguished by a tendency to be split up into *islands* of differing depth.

A further difficulty was that the images were collected in two differing scattering geometries. Most of the defect dataset was collected with the ultrasonic reflected amplitude measured directly from a probe above the defect. The rest of the dataset was collected from more angled probes at 38° and 55° from the vertical, which measured the signal only after it had reflected off the back wall of a plate sample. When the probe angle is chosen appropriately for the weld wall angle the observations are sufficiently similar that they may be treated together, but there is clearly an approximation in this process. It was clear from observations of many such images that what was needed was a classifier that could see the relationship between neighbouring pixels in all four dimensions.

4.8.6   Pre-processing of the Ultrasonic Images

Each dataset was converted into a standard format and processed into a variety of modified standard datasets that were used for the final classifications. The details of the preprocessors used to achieve these standardised images are unimportant but the methodology will be described here. The original images are generally not of uniform size and this is a problem for many classifiers. A necessary pre-processor therefore places the original image centrally onto an image of standard size, with the most appropriate resolution. The three processes of centring, placing and averaging were performed simultaneously within a single preprocessor as illustrated in Figure 4.15.

4.8.7   Centring the Defect on an Image of Fixed Size

The first operation is to find the centre of gravity of the original image. The idea is to place this point on the central pixel of the new standard size image. The integration is made over both angles sets so that the centres x_b, y_b, z_b of the image $I(x,y,z,a)$ defined as a function of positional coordinates x, y, z and angle a is given by

$$x_b = \frac{\sum x.I(x,y,z,a)}{\sum \ I(x,y,z,a)} \ \ \text{etc} \tag{4.7}$$

The centre of gravity position is rounded to find the pixel which holds the centre of gravity point.

The next operation is to place a scratch array of standard size of the image so that the pixel containing the centre of gravity is the central pixel of the new image. Ideally the dimensions of the new array should be odd. Areas of the original image not covered by the new array are truncated and lost. Areas of the new image which are not covered by the original image are set to zero.

This placing process will in general change the centre of gravity of the image, since some of the original image may be lost. The new image centre of gravity may need to be recalculated by repeating the placing process until convergence.

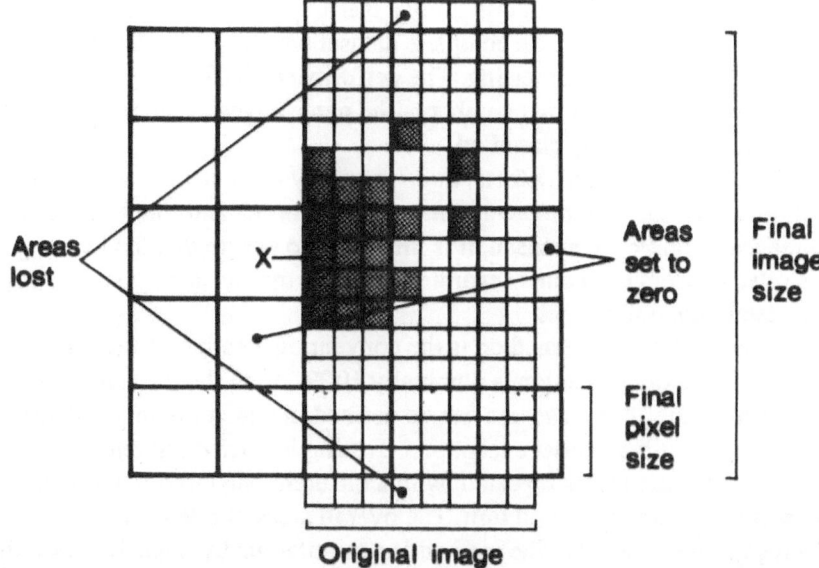

Original image

Fig 4.15 The processing of the raw ultrasonic image to give an image suitable for presentation
to a neural net classifier. First the centre of gravity of the original image (the small squares),
which can be of arbitrary size, is located. This position, marked X, is centred on a square
scratch array of defined size. Lastly pixel averaging is applied over the scratch array to give the
final array size (heavy lines)

The final process is the averaging of the pixels in the scratch array to give the final
image. The averaging may be different in all axes, but should ideally be an odd
number so that the central pixel will remain at the centre of gravity. The objective is
generally to find the largest degree of averaging which correctly describes the
essentials of each class, enabling a correct classification to be made.

4.8.8 Truncating and Normalising
The upper and lower limits of the intensity levels may be set to provide a lower
discrimination level of significant defects, and to truncate any overloading. In the
present work a lower level of around 1% of peak intensity was chosen as a lower limit
and there was no upper limit.

The absolute intensity level of ultrasound reflections cannot be relied upon. For
this reason it is sensible to normalise the image in some way. The method used in
these studies was to normalise it so that the peak intensity was a defined value; 999
was the value chosen. By this method the images used did not vary in intensity as the
image size was changed. This is not the case for the alternative method of
normalisation to a constant area of the image.

4.8.9 Choosing a Defined Learning Fraction of the Data
In analysing sets of data of this kind, it is essential to make a sharp distinction between
the data used for training the network and the data used for testing the classification

performance of the network. In this study a learning fraction was defined which gave the fraction of the images in the complete dataset which were to be used for training. The remainder were used for testing. The performance of the network can then be generally described by just two numbers - the percentages of the test and training images which are correctly classified.

In the case of ultrasonic defects, there are many more members of some classes than of others. In particular there are many more smooth and rough cracks than there are pores and slag. For this reason, it is important to ensure that the learning fraction of files is selected from each class separately. By this means each class is always fairly represented in the training set.

The *leave one out* (loo) method is the conventional name for the procedure to be used when the learning fraction is nominally 100%. If all the data are available for training, a meaningful test success can be defined in which training is carried out using all the examples in the dataset except one. Testing is carried out only on this one example, and the process is repeated, with a different one left out, until all examples in the dataset have been left out in turn. The overall success rate is then evaluated from the average of these results. The method is computationally intensive since the classification process is multiplied by the number of examples. However the method gives the success rate most appropriate to a commercial assessment, since it closely reflects the success rate of a new measurement made using all the existing data for training.

4.8.10   The Parameters of Some of the Datasets Considered in Detail

The main variables are the sizes of the original scratch array (s_x, s_y, s_z), which defines the measurement volume which is analysed in the classifier, and the averaging array size, (a_x, a_y, a_z) which determines the resolution volume of the data which is to be analysed. Clearly the number of pixels in the final array presented to the classifier is the simple function of these $(s_x/a_x, s_y/a_y, s_z/a_z)$.

The criterion for choosing the measurement volume was that it should be large enough to cover the larger defects, particularly the porosity defects. This meant that many of the smaller defects such as slag had to be *padded out* considerably.   The criteria for the resolution volume are less easy to evaluate. Information is lost by an averaging and some analysis was performed on unaveraged data. However, many algorithms give poor results with large image arrays, quite apart from computational considerations which make them impracticable. The final choices were made according to the test dataset success rates, see Table 4.1, with the choices guided by our human intuition on the information content remaining in the data.

4.8.11   Conventional Pattern Matching

As an example of conventional image analysis methods, a standard pattern matching algorithm was applied to the four-dimensional datasets. In this method the training dataset was used to define a single prototype or average image for each class. These four prototype images were stored and compared with each of the test images in turn. If the prototype image for class c is described by $I_{c,x,y}$, and the test image for example

e by the function $I_{e,x,y}$, then the pattern matching may be performed by evaluating the least squares difference

$$D_c = \sum_{x,y} \left[I_{c,x,y} - I_{e,x,y} \right]^2 \qquad (4.8)$$

The class with the minimum value of D_c then defines the class of the example. Figure 4.16 shows the four prototypes of the four-dimensional images. The sums of the squares of the differences between corresponding pixels in each prototype and the test image are evaluated and the class with the lowest sum selected. The method was run in an averaging mode where the images of each class from the whole of the training set were averaged together.

Table 4.2 shows some of the results obtained by pattern matching. It is seen that with these centred datasets the results are always quite good. The centre of gravity placing of the image has provided an element of spatial invariance not unlike that given by the receptive fields considered later. This is particularly notable for the smallest 5x5x5x2 arrays where 86% success was obtained.

Table 4.1   The datasets used in the direct image processing

Dataset	Field size	Average size	Final size	Best success in test %
.40	11x11x11	1x1x1	11x11x11	75.7
.41	33x33x33	3x3x3	11x11x11	90.9
.47	14x14x28	2x2x4	7x7x7	93.9
.49	10x10x20	2x2x4	5x5x5	92.4

Table 4.2   Pattern matching results

Dataset	Array/ Field size	Learning fraction %	Number of iterations	Success (training %)	Success (testing %)
.40	11x11x11x2	100	1	-	75.75
.41	11x11x11x2	100	1	-	80.3
.47	7x7x7x2	100			83.3
.49	5x5x5x2	100	1	-	86.4

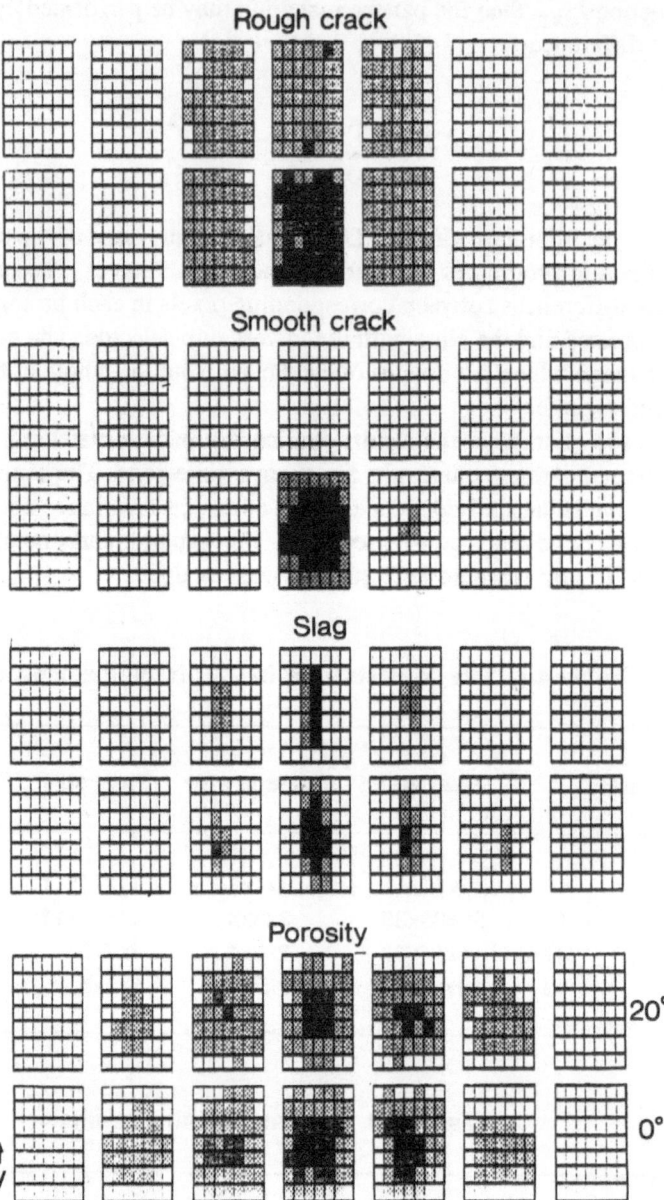

Figure 4.16 The pattern matching prototypes for the 4 classes of defects. The patterns show
the average over a typical training set. Each section shows an (x,y) plot of the mean intensity
of the ultrasound reflection. The images across the figure show these sections as a function of
the depth z into the material. The two angles are placed above each other. These patterns were
generated from a series of 7x7x7x2 images

4.8.12 The Adaptive Receptive Field Method

The reasons for the poor results using the conventional pattern method, relative to conventional feature parameter classification methods (which achieve nearly 95% success rates) are clear enough. Although the images are roughly centred through their centre of gravity position, there remains a good deal of spatial variance in the position of the peak and valleys in the image. The class significant relationships between adjacent pixels are smoothed out. Receptive field methods (see section 4.8.14 below) supply this spatial invariance by pattern matching on a smaller receptive field which is scanned across the image in a raster. In this adaptive method receptive fields are defined for each class with initial values defined from, for example, the positions in the image where there is a local peak. The receptive field is then refined iteratively by scanning over the training examples in each class and finding the position where the difference between the image and the field is a minimum. The portion of the image under the receptive field is then added to the field of that class. In testing, each of the class fields is scanned over the image in turn. The class whose field gives the best fit somewhere in the image is chosen.

The adaptive receptive field method has a number of adjustable variables equal to the number of pixels in the field, say only 54 for a 3x3x3x2 field. This is much less than many of the neural net methods considered later, and means that the problem of overfitting is much reduced. In fact the results are almost independent on the size of the training set and much of the optimisation of the method could be performed with the otherwise artificial training set of all the available examples.

Table 4.3 shows some of the results obtained using the adaptive receptive field method. Examples of the receptive fields generated are shown in Figure 4.17.

Table 4.3 Adaptive receptive field results

Dataset	Array Field size	Learning fraction %	Number of iterations	Success training %	Success testing %
.47	5x5x5x2	100	1	90.9	-
			2	90.9	-
			3	90.9	-
			4	92.4	-
			8	92.4	-
.47	3x3x3x2	100	4	-	89.39
	5x5x5x2	100	4	-	87.87
.47	3x3x3x2	100	4	-	93.90
.41	3x3x3x2	100	4	-	81.82
		50	4	89.69	79.08
	3x3x5x2	100	4	89.4	
	5x5x5x2	100	4	92.4	

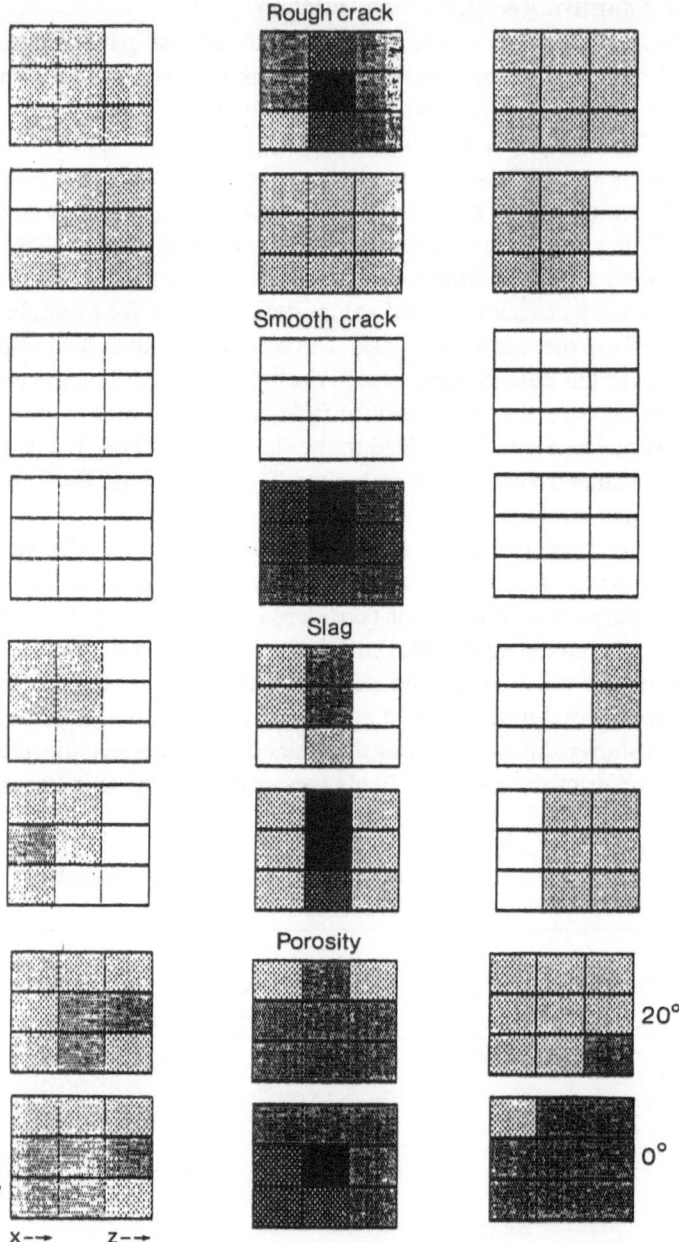

Fig 4.17 A typical adaptive receptive field generated from the ultrasonic images. There is a 3x3x3x2 field for each of the 4 classes of defects. The fields are shown after 4 iterations over the training set. Each section shows an (x,y) plot of the mean intensity of the ultrasound reflection. The images across the figure show these sections as a function of the depth z into the material. The two angles are placed above each other. The fields were generated from a series of 7x7x7x2 images

The method is seen to give generally good results with the best results in the 90% range. Appreciably better results are obtained if the dataset is split into different classes for the two geometries represented by the direct series and backwall reflected series. This method gave 93.9% success on *leave one out* testing. In practice there is never any doubt about the geometry of the measurement so that the use of different classes for the two geometries is no drawback, except for the fact that the effective size of each of the datasets is reduced.

4.8.13 The Standard Back-propagation Method
In the straightforward approach of the back-propagation method each pixel in the image is presented as a separate input to the network. It is well known from our own work on generic images and from many other studies that the problem with this approach is the disparity between the number of adjustable weights in the net and the number of distinguishable pixels in the dataset. With the 7x7x7x2 four-dimensional images, and with only two hidden units, and four output units, such a network has some 694 adjustable weights, and so is capable of learning the classes of, say, 33 training examples by rote.

Although back-propagation carries no direct information on the neighbourhood relationships between pixels in the image, these are implied indirectly through the training set. For example the smooth cracks correspond to a set of high values of adjacent pixels along a line in the image. The degree to which this information can be used by the method depends strongly on the number of training images. Some of the results obtained are shown in Table 4.4.

Table 4.4 Results using direct back propagation

Dataset	Array/ Field size	Learning fraction %	Number of hidden units	Success training %	Success testing %
.49(A)	5x5x5x2	50	2	100	75.76
			3	100	78.79
			4	100	81.82
.47(A)	7x7x7x2	50	4	100	78.79

The back-propagation results were obtained using two distinct codes. Most of the results were obtained using a commercial package, ANSIM, which coded the original minimum descent algorithm. A learning rate of 0.1 and a momentum term of 0.6 were used. The defect of this code was that each dataset had to be run sequentially so that batch work was impracticable. ANSIM runs will be denoted by 'A' in the results tables below.

Some of the results were obtained using a PROFAN compatible code (see chapter 3) which could run a *leave one out* series, or a set of learning fractions stored as defined files, in one batch. This code also had the advantage that the learning and testing success rate could be continually monitored during training. It also had a continually varied set of presentation probabilities, designed to optimise the learning rate. The current degree of fit of the ith example, e_i, is recorded and every few hundred iterations a set of presenting probabilities, p_i, are evaluated proportional to $p_0 + 1/e_i$ so that examples which have been nearly perfectly learnt already are not presented frequently. A minimum presentation frequency p of about 10% was always used to ensure that changes in the network brought about by learning the difficult examples did not upset the easy examples. This code was used in the so-called *banana* suite and will be denoted by 'B' in the tables.

These results show a consistent performance around 80% success in test. The reasons for this relatively indifferent performance are clear. The nets are all large with around 1000 and 2700 variable weights for the series denoted .49 and .47 respectively. With the present size of only 66 images, the number of training examples is insufficient to prevent overlearning.

4.8.14 Hybrid Methods Involving Receptive Fields and Back-propagation

Receptive fields provide a systematic way of extracting significant features from an image. The idea is to take a small section of the image, say a 5x5 section of a 25 x 25 pixel image, and to move it about the image in a raster fashion, at each point evaluating some measure, P, of the 'overlap' of the field with the image at that point. Receptive fields may be classed as either 'multiplicative' or 'least squares'. In a multiplicative field, the portion of the image $I_{x,y}$ is multiplied by the corresponding pixels in the receptive field $R_{x',y'}$, and is then summed, giving

$$P = \sum \left[I_{x,y} R_{x',y'} \right]$$

There is a strong link between such receptive fields and neural networks. Receptive fields detecting features such as orientated lines are known to exist in the visual cortex. The multiplicative field product is very similar to the sum of products of input vectors and weights of any neuron and the fact that the scanning over the image can generally be done in parallel makes fast processing possible. The method of receptive fields is used by Fukushima (1988) in his neocognitron neural net model, and applied to character recognition. However, in the ultrasonic application the spatial positioning of the features is of very little use. The neocognitron method has not therefore been pursued. The method used in this section was developed to combine the advantages of receptive fields with those of back-propagation. The shared weights method may be considered to belong to this class of methods, but has been included in a separate section due to its wide acceptance. The methods used in this section have all been developed for the present problem.

With the direct back propagation method of the previous section, the network is required to learn the various types of variance present in the images. Large numbers of examples are required for this, and the consequently large networks make the results vulnerable to overfitting. Using a moving window, rastered over the image, the network has a much smaller number of inputs equal to the number of pixels within the window. The number of examples could also be effectively increased in size since areas within several windows may be extracted from a single image. This is achieved by pausing the window at the position where the central pixel in the window was high, and also intense enough to be at least half the maximum intensity in the image. From the dataset of 11x11x11x2 images, this method could produce a second dataset of say,3x3x3x2 window images. These images can then be fed directly into a back propagation network as illustrated in Figure 4.18.

Table 4.5 shows some of the tests which were made to assess the performance of this method against the degree of learning as measured by the total RMS error between the targets and the outputs of the network. With this particular dataset a peak test performance of 96.97% was obtained at an RMS error of 0.08. The performance degraded very slightly as learning continued, (see also Figure 4.19). This result is typical of the learning curves generally seen. It implies that the network generalises best when the details of the training set are still not perfectly assimilated. In subsequent results quoted below the training was generally stopped when an RMS error of 0.08 was achieved.

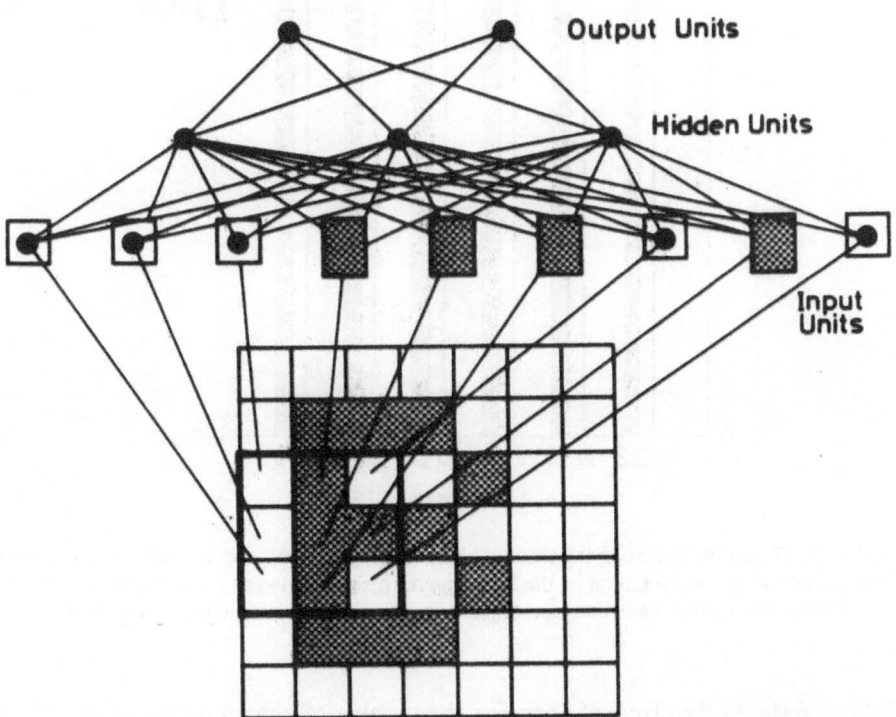

Fig 4.18 The moving receptive field method. The field, shown by the heavy square, is swept across the image in a raster mode. Pixels within the field are presented as inputs to a back-propagation net.

Table 4.5 Results using receptive field back propagation: RMS error

Dataset	Field size	Learning fraction %	RMS error	Success training %	Success testing %
.41(A)	3x3x3x2	50	0.32	84.85	81.82
			0.16	96.97	87.88
			0.08	100	96.97
			0.04	100	93.94
			0.02	100	93.94
			0.01	100	93.94

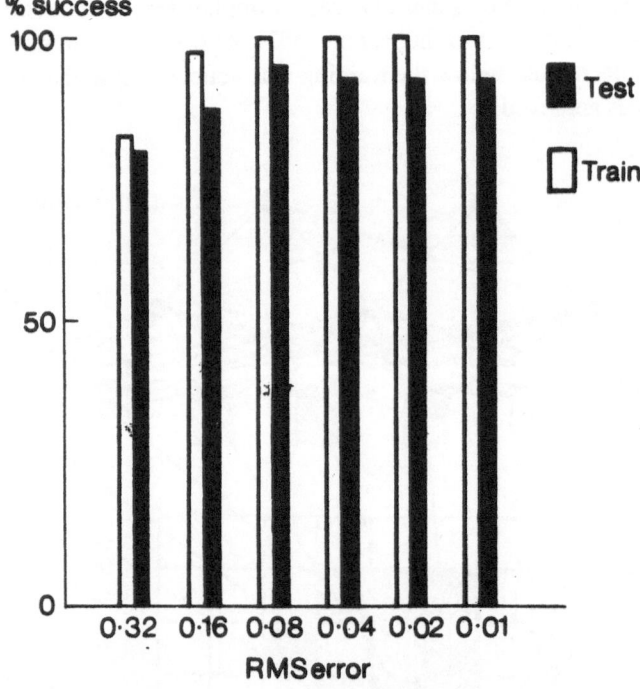

Fig 4.19 The % success in classifying the real ultrasonic images from test datasets as a function of
the degree of training as measured by the least squares residual between the network targets and
outputs. The results show that convergence is achieved with a residual of order 0.04

Table 4.6 shows the effect of changing the number of hidden units on the
performance. Only with one hidden unit was the net unable to train to 100%. The test
performance thereafter is not a strong function of hidden unit number. The best
performance was given for 4 hidden units, and this number was used in subsequent

studies. Performance as functions of receptive field size and learning fraction are given in Tables 4.7 and 4.8 respectively.

The method thus gave appreciably better results than direct back-propagation. With the network having only 100 or so adjustable weights, it is much less prone to overfitting. The best results, shown in Table 4.8, gave a *leave one out* testing performance as high as 93.9%.

Table 4.6 Receptive field back-propagation results: number of hidden units

Dataset	Field size	Learning fraction %	Hidden units	Success training %	Success testing %
.41(A)	3x3x3x2	50	1	84.85	78.79
			2	100	90.91
			3	100	90.91
			4	100	90.91
			5	100	90.91
			10	100	90.91
			20	100	90.91
.41(B)	3x3x3x2	100	4	100	90.91
			6	100	93.90
.47(B)	3x3x3x2	100	2	100	88.23
			4	100	91.84
			6	100	89.79

Table 4.7 Receptive field back propagation results: receptive field size

RMS error cut off = 0.01
Number of hidden units = 2
Average of 10 50% learning fraction datasets

Dataset	Field size	Learning fraction %	Success train %	Success test %
.41(A)	3x3x3x2	50	100	84.24
	3x3x9x2		100	75.15
	5x5x5x2		100	86.36
.47(B)	3x3x3x2		100	81.63

Table 4.8 Receptive field back propagation results: learning fraction

| RMS error cut off = 0.08 | | Number of hidden units = 4 | | |
Dataset	Field size	Learning fraction %	Success train %	Success test %
.41(A)	3x3x3x2	100	100	92.47
.41(B)	3x3x3x2	100	100	93.91
		50	100	84.24
		33	100	73.18
		25	100	69.39
	5x5x5x2	100	100	87.88
.47(B)	3x3x3x2	100	100	90.91

4.8.15 The Shared Weights Back-propagation Method

The shared weights development of the back propagation algorithm builds spatial invariance directly into the architecture of the network. The hidden layer weights are arranged in a two or three-dimensional grid, as in the image, and may be considered as a receptive field. The connections from the image to the hidden layer are duplicated for every possible position of the hidden layer within the image. However the weights to the layer are not independent, but are constrained to a single set of values. The code used was that from the ANSIM suite of neural networks. The method gave the results shown in Figure 4.20.

4.8.16 Comparison with the Feature Parameter Method Results

The best of the results obtained from the direct image classification methods (see Table 4.9) are compared with the best results obtained from the classification of feature parameter data in Table 4.10 and Figure 4.21.

Table 4.9 The best results for each method in direct image classification

Method	Dataset	Field size	Hidden units	Success testing %
Pattern match	0.49	-	-	86.4
Adaptive field	0.47	3x3x3x2		93.9
Direct back-prop	0.49	5x5x5x2	4	81.8
Shared weights	0.47	5x5x5x2	4	90.0
Receptive field BP	0.41	3x3x3x2	4	93.9

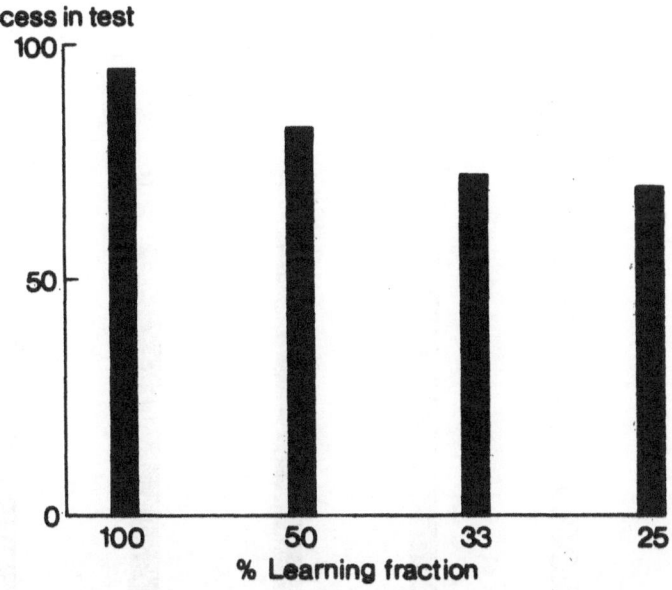

Fig 4.20 The % success in classifying the real ultrasonic images from test datasets using the shared weights back-propagation method, after training from a variable fraction of the total dataset available. The 100% figure is derived from the 'leave one out' training procedure

Table 4.10   A comparison of classification results from image and from feature parameter data

Method	Dataset size	Conditions test %	Success
Direct back-propagation	Features	4 hidden units (hu)	94.0
Receptive field BP	Image (.41)	6 hu, 3x3x3x2 fields	93.9
k-nearest neighbour	Features	*k*=3	94.4
Adaptive field	Image (.47)	3x3x3x2 field	93.9
Weighted min distance	Features		91.6

Table 4.10 shows a comparison of the two sets of best results for the success on testing using the *leave one out* method. No method gives perfect results. The reason is that the scatter in the measurements places some defects outside the general trend of their class. The general conclusion is that the results from the image and the feature data are comparable. There are several statistical classifiers available which can extract the information available in the raw data; whether the image is processed into features

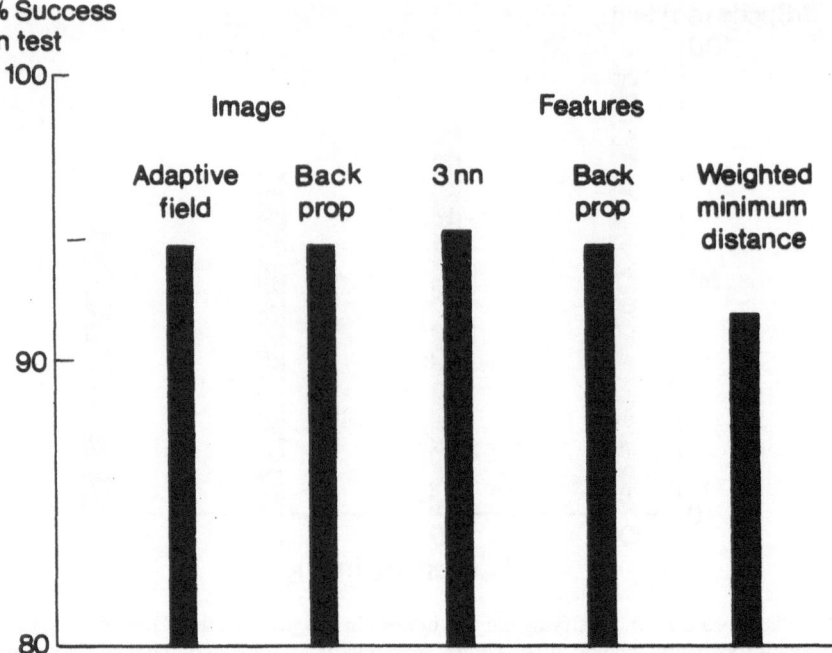

Fig 4.21 The % success in classifying defects directly from real ultrasonic images and from extracted features. The results are for test defects after training by all the other defects according to the 'leave one out' training procedure. The results show that the direct image methods are as effective at classifying the defects as the feature extraction methods.

or not is less important than the degree of class discrimination obtained in the measuring process.

The direct image classification is left with several significant advantages over feature extraction in practical implementation. In particular, the receptive field back-propagation method is very rapid in testing. The whole process of classification from the raw data takes only a fraction of a second. The results presented have been obtained with little expenditure of human judgement. This contrasts with the case of feature parameter extraction where expert decisions were used, and are still being used, to define the optimum features.

Finally, it should be noted that these methods have been implemented in a hardware demonstrator. This can characterise a defective weld in real time using a robust industry-standard ultrasonic scanning data acquisition system, but running the adaptive receptive field software developed in ANNIE to classify the defects.

4.9   ALOC Defect Detection

4.9.1   ALOC Data Acquisition

The ALOC amplitude-transit-time-location-curve-technique is used in recurrent non-destructive ultrasonic testing on thick-walled components in nuclear power plants. A schematic diagram of such test equipment is shown in Figure 4.22.

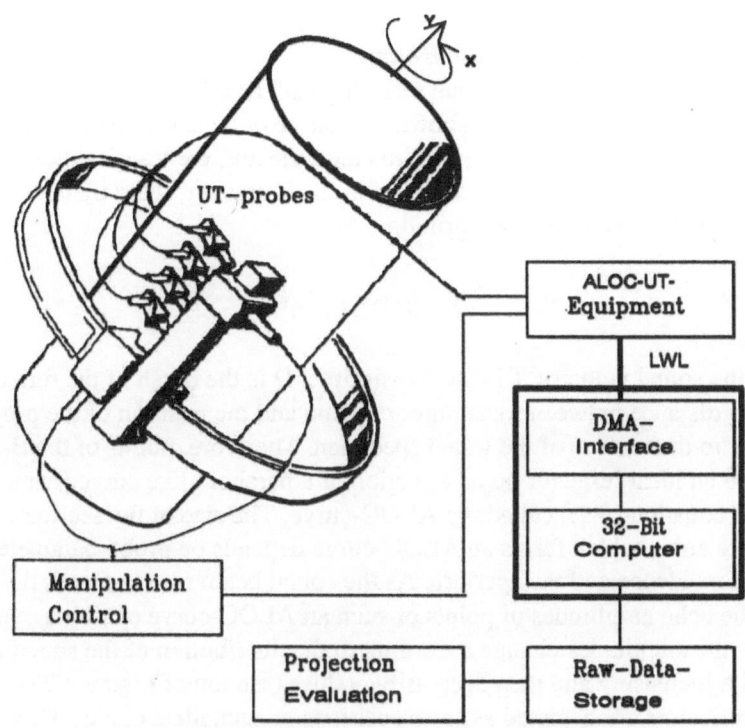

Figure 4.22   Diagram of test equipment

The ALOC data under consideration are sampled during ultrasonic testing (UT) inspections based on the pulse/echo method. The inspections make use of the single probe method where a single UT-probe transmits an ultrasonic sound pulse and subsequently receives the echo reflected from some point in the material. During the inspection the probe is moved along straight-line scanning paths across the test specimen. The ultrasonic pulses are generated at equally spaced distances, and the transit times and amplitudes of the received echos are measured by the ALOC device. The ALOC device allows transmitter pulse frequencies of up to 5 kHz. The received signals are filtered (narrow or broadband) and logarithmically preamplified with 80 db. These signals are analogue/digitally converted with a sample rate of 20MHz (resolution of transit time 16 bit, resolution of amplitude 8 bit). A halfwave peak detector first searches for the maximum of each of the digitised halfwaves of the rf-signal. These data are then passed to a compression unit, which finds the relative maxima of the A-scan using a special algorithm implemented in very fast ECL circuits. By means of this special algorithm, the ALOC equipment is able to detect all relevant information from the A-scan and to record these data with a high resolution, avoiding the need to record the full A-scan. This reduces considerably the amount of data to be processed and stored by the computing device. The compressed data, amplitudes and transit times together with the probe position are transferred to the computer system which stores them on a mass storage device. During tests of large vessels the computer has to cover data rates of about 80 kb/sec over long periods.

4.9.2   Flaw Representation by the B-scan
Displaying the data sampled along one scanning path as a B-scan (see Figure 4.23) an ideal point-shaped reflector will be shown as a curve of a characteristic shape: for example, while the probe is moving towards the reflector, the transit times of the echoes generated by this reflector become shorter. Moreover, these transit times change according to the following formula

$$cT = 2(D^2 + x^2)^{\frac{1}{2}} \tag{4.9}$$

where c is the sound velocity, T is the transit time, D is the depth of the reflector and x is the lateral distance between transmitter position and the position of the projection of the reflector to the surface of the tested specimen. Therefore, points of the B-scan belonging to an ideal reflector lie on a section of a parabola-like curve (namely the solutions of equation (4.9)) called the ALOC-curve. The size of the section of the parabola-like curve which forms an ALOC-curve depends on probe-parameters like the angle of incidence and the aperture. As the sound beam moves across the *ideal* reflector, the echo amplitudes of points of such an ALOC-curve exhibit a dynamic behaviour: the amplitudes change according to the distribution of the sound pressure initially with increasing and then decreasing values (see top of Figure 4.23). Real, non-ideal reflectors are depicted as superpositions of such ideal curves. However, parts of these more complex curves (eg the echos from the tip of a crack) are of the simple form described by equation (4.9), except for noise influence which leads to missing and shifted points.

4.9.3   Image Enhancement by Hough Transformation
The knowledge of the equation for an ideal ALOC-curve can be used to provide contrast-enhancement of the picture through suitable Hough transformation of the B-scan. This operation transforms the B-scan into the parameter space of equation (4.9), ie the two-dimensional space with coordinates (x, D), where each point of the B-scan is mapped to all possible ALOC-curves (x, D), on which it may lie. This also means, that the Hough transformation maps into a space with real-world coordinates. In our case, the mapping was adjusted so that the Hough transformation of the B-scan has a resolution of approximately 1mm x 1 mm. The pixels of the transformation are *weighted* with the sum of the absolute amplitude levels of the corresponding B-scan pixels. Long-lined ALOC-curves with high point densities and with clear amplitude

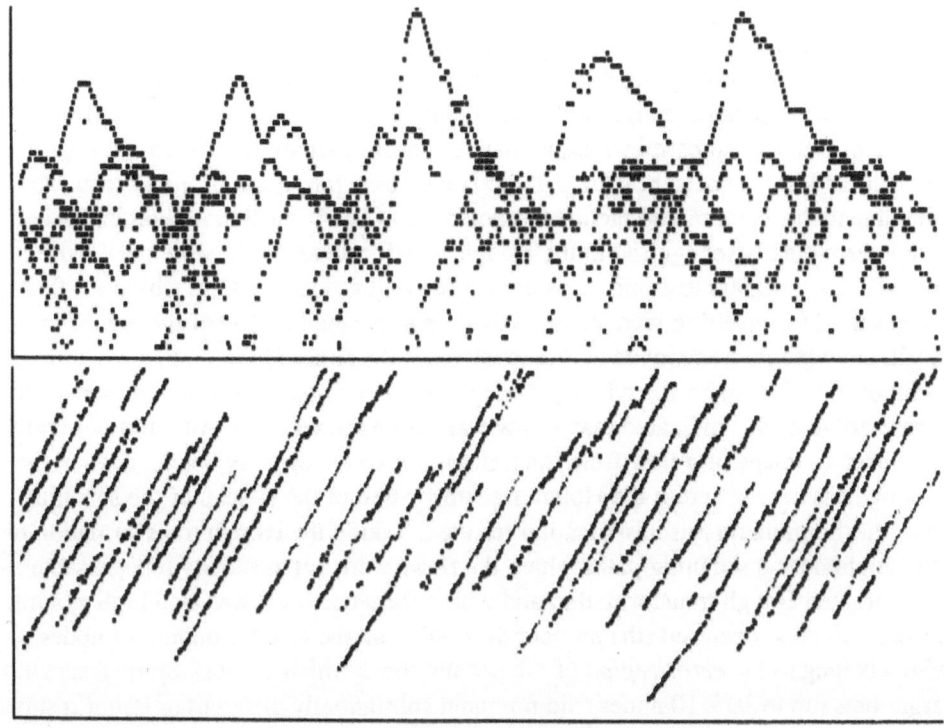

Fig 4.23   B-scan of a scanning track

dynamics are mapped by the Hough transformation to nugget-shaped regions of high intensity.   Detection of these nuggets is the task which remains to be performed by a suitable neural net.

4.9.4   Flaw Detection by a Hybrid Neural Net

A possible way to find these nuggets is to move a window across the transformed image. A neural net should then be capable of deciding whether or not a nugget is within that window at a particular moment. This window must be at least large enough to completely contain a nugget. However, it would be better to make the window a little bit larger to enclose some *context information*. This reasoning leads to a window size of 60 x 66 (=3960) pixels. If one would directly train a net with these *window* images, this would mean that for example a suitable three layered back-propagation net with 5 hidden and 2 output nodes would comprise approximately 20,000 weights which must be trained. This has two major disadvantages. First it requires a lot of computing time. And second, compared to the number of weights the amount of training examples is relatively small. Experience within the ANNIE project and elsewhere have shown that these kind of over-estimated nets have only minimal capability for generalisation, which is a key issue in pattern recognition. Therefore the

window data were compressed by reduction of the pixel resolution to a picture of
20x22 pixels, giving a resolution of about 3 mm x 3 mm.

Encouraged by promising results when applying Kohonen's self-organising feature
map (SOFM) to pattern recognition in artificial ALOC-datasets, a 5x5 SOFM was
trained with the compressed window images. During training the window images were
taken at randomly chosen positions in the Hough transformation of the B-scan. To
label the nodes, compressed pictures - named A,B,C,D,E - of five individual flaws,
which were clearly recognised in the Hough-transformed B-scan, were used. After
approximately 40,000 learning steps the net reached a stable state in which the five
images used for labelling were mapped to a separate small region of the net.   For
recall, the window was moved continuously over the entire Hough transformation. If
an image residing under the window were mapped to a node labelled with one of the
characters A-E, this indicated that a flaw was inside the window. All windows which
contained an image of a flaw from the training set were correctly detected when the
window was moved across the Hough transformation of the B-scan of the scanning
path. Furthermore no misclassification resulted. Taking the Hough transformation of
two neighbouring scanning paths where the flaws were not as clearly recognisable as
in the trained Hough transformation and where the noise level was also higher, some
misclassification occurred (the number depending on the exact grouping of nodes
which belong to the *error region* of the net and those which do not). Some tests with
larger nets (up to 10 x 10 nodes) did not yield substantially different or better results.

This exploitation of the SOFM uses only a very small and incomplete part of the
information inherent to the Kohonen net. The only information used for further
decision making is which node of the net showed the highest activation upon
presentation of a certain image. However, a Kohonen net, viewed as a way of data
compression, provides much more powerful information when using the complete
activation pattern built from values of the activation levels of all of its nodes. In our
case this information consists of a real valued, 25-dimensional (5x5 nodes) vector.
This vector of net activation can be viewed as the result of a second data compression,
reducing the size of a compressed window (440 pixels) by another factor of
approximately 17. These activation vectors were used to train a back propagation net
to make the final decision between *pictures with flaw* and *picture without flaw*. The
series of data processing steps is shown in Figure 4.24.

In order to build a dataset for training and recall of the back propagation net, we
selected 45 window images from the Hough transformation of three neighbouring B-
scans which we had already used to train and test the SOFM. Fifteen of the 45
windows contained, as we knew, the complete projection of flaws, while the other 30
windows contained either only small sections of flaw data or no flaw data at all. These
45 pictures were presented to a (trained) Kohonen net and the resulting 45 activation
vectors were first normalised and scaled and then used for training and test of a back-
propagation net. We used three-layered back-propagation nets with five, four or three
hidden units without getting significantly differing results.

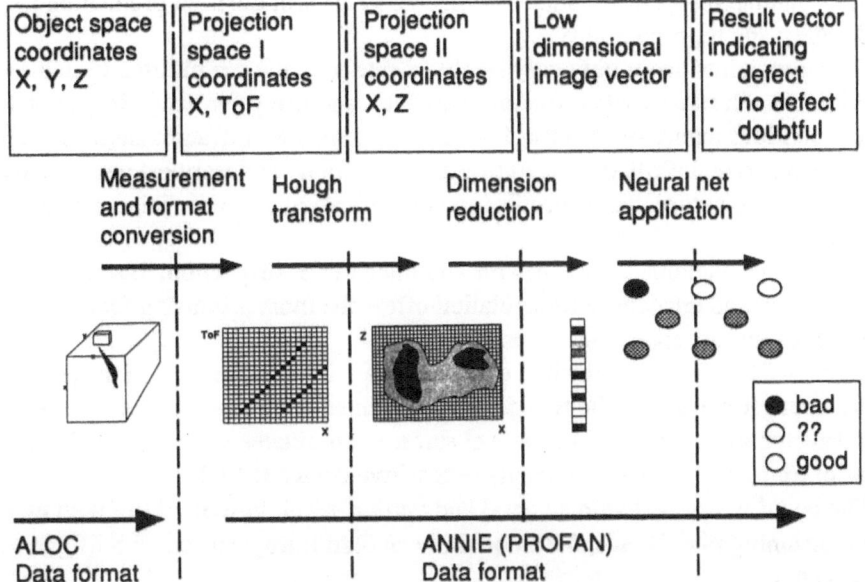

Fig 4.24 The data processing using a Hough transform, Kohonen mapping and back-propagation to detect a flaw from ALOC ultrasonic testing

If we used all of the 45 activation vectors to train the back-propagation net, no misclassifications occurred upon recall. If we used a randomly chosen part of the data collection (33% of the total data set, 5 pictures with flaw data and 10 images without) some (between 2 and 5) of the pictures not containing the complete data of a known flaw were misclassified.

4.9.5 Realisation with ANNIE-PROFAN Tools

To realise the described neural detector we implemented the specialised Hough transformation and a suitable SOFM and integrated them in the PROFAN demonstrator. To normalise and scale the activation vectors, preprocessors implemented and integrated for other applications were used. The classification task was done by back-propagation with shared weights which also found application elsewhere.

4.10 Solder Joints Inspection with Neural Networks from 3D Laser Scanning

4.10.1 Introduction - Problem Definition

Advanced technologies in the manufacture of printed circuit boards (PCB) require new methods and strategies for inspection and quality control. Due to the increasing trend towards surface mounted technology (SMT), miniaturisation and higher packing densities, human inspection cannot be considered to meet the high production targets of reliability and reproducibility.

Defective joints occur only very rarely - a rate of a few per million. Reliable detection of such rare events is therefore difficult and tiring for visual inspectors, due to the repetitious nature of the task. For these reasons, flexible automated inspection systems have to be installed in factories. In this section we describe a neural network approach for the automated inspection and classification of solder joints on printed circuit boards.

Among the available techniques for automated PCB inspection, 3D-sensing using laser scanning and telecentric triangulation offers the most promising features as regard to resolution, speed and costs.

Figure 4.25 shows the principle of the 3D scanning system used in these studies. The scanner provides both the two-dimensional intensity image measured by the photodetector and the three-dimensional surface coordinates which are evaluated by means of triangulation from the position-sensitive device (PSD).

The board is scanned with a lateral and vertical resolution of 70 µm with up to 1 MHz scanning rate. Hence, for a board of standard European size, 6.5 Mbyte of data are provided.

On populated boards in conventional through-hole technology, the following defects have to be detected:

- solder bridges
- pin not visible
- excess solder (solder sphere)
- insufficient solder
- blow out.

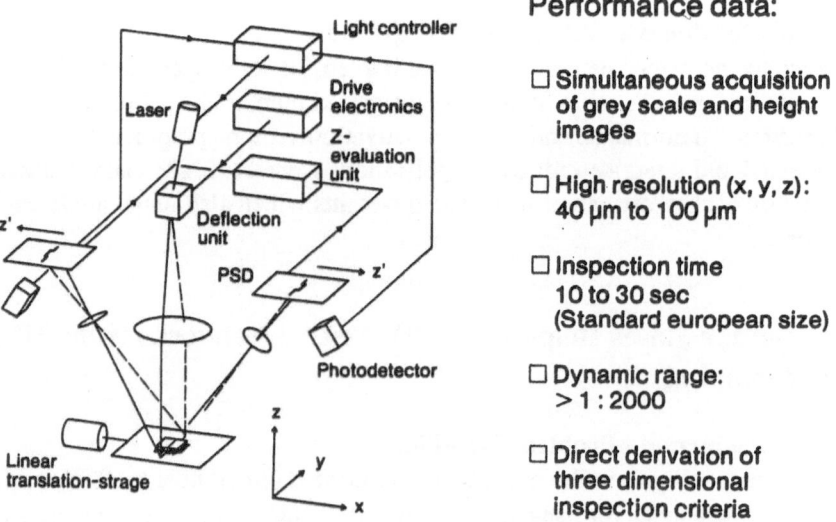

Performance data:

☐ Simultaneous acquisition of grey scale and height images

☐ High resolution (x, y, z): 40 µm to 100 µm

☐ Inspection time 10 to 30 sec (Standard european size)

☐ Dynamic range: > 1 : 2000

☐ Direct derivation of three dimensional inspection criteria

Fig 4.25 Principle of 3D scanner

Figures 4.26 and 4.27 show some examples of 3D data for the different joints. A typical correct solder joint is given in Figure 4.26a. The problem of using conventional pattern recognition techniques and extracting typical feature parameters is that there is a large variety of correct soldered joints which in detail depend on the components spectrum and soldering process used in the factories. Figure 4.26b and c represent joints of an integrated circuit and a component, where the pin is clinched strongly, but which are still good. Typical defects are presented in Figures 4.27a to c.

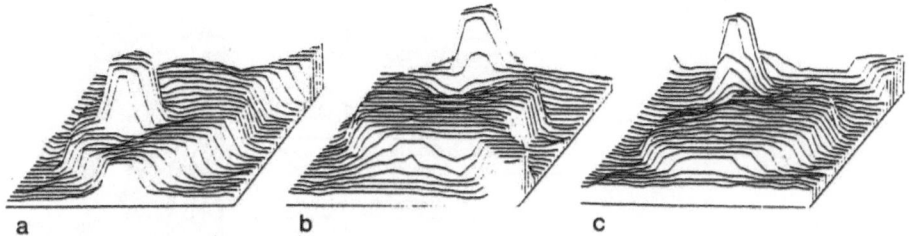

Fig 4.26   3D data of correct soldered joint

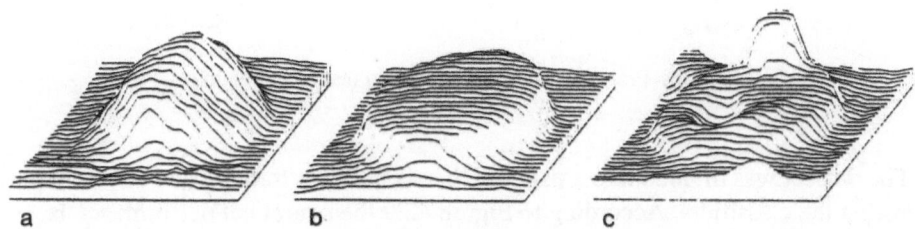

Fig 4.27   3D data of defect soldered joint

4.10.2   Classification Based on Feature Extraction

In a first approach to applying neural networks we compared their performance with conventional classifiers in feature space as illustrated schematically in Figure 4.28. From 3D scans suitable feature parameters were extracted to enable the classification of joints. Six feature parameters were used to distinguish between good and defective joints and these were processed by conventional classifiers and neural networks. For a dataset of more than 300 joints of all former classes the following methods were examined:

- minimum distance classifier
- *k*-nearest neighbour classifier with various *k*
- back-propagation
- learning vector quantisation.

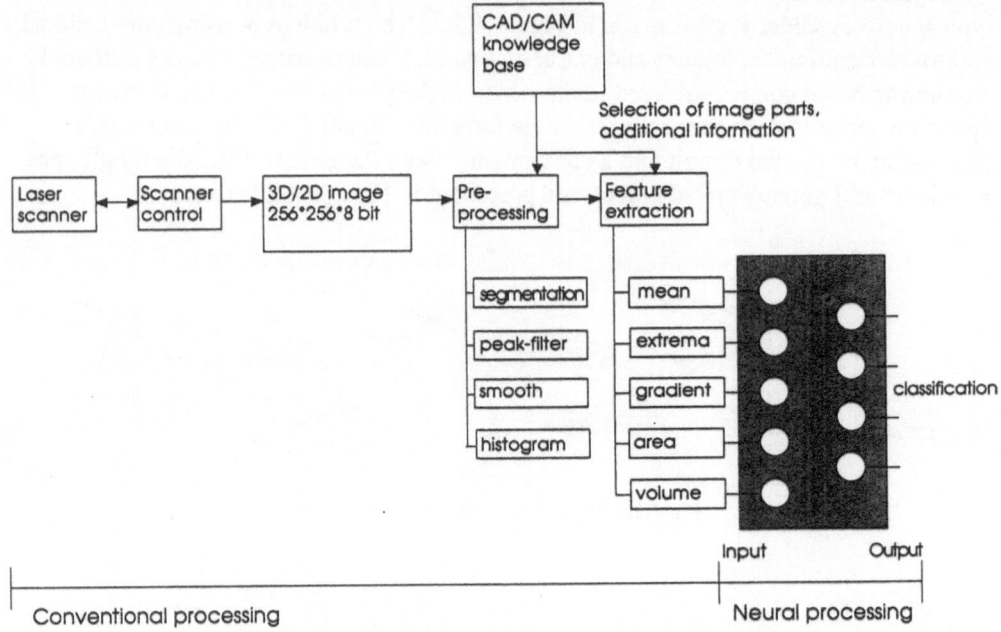

Fig 4.28 Processing principle based on feature extraction

The dataset was divided into a part which was used for training and a second part for testing the classifiers. According to Figure 4.29 the neural net performance is similar to the best conventional approach but did not exceed it. On the other hand the feature extraction task dominates the processing time and limits the success rate, so that neural networks with their fast data processing give no special advantages. These results were in line with the generic studies described earlier.

4.10.3 Classification Based on Direct Data Processing
The approach of more direct data processing from the images avoids the computation involved in feature extraction, and more fundamentally, avoids the expert work involved in the selection of optimal feature extraction for any given problem. The idea is to use the neural network to find suitable features directly from the training set. The processing principle is explained in Figure 4.30. The scanner data are segmented via a CAD interface which provides the on-line positional information of solder joints within the image data for the board under inspection. Each pixel of the joint is presented as a separate input to the neural network.

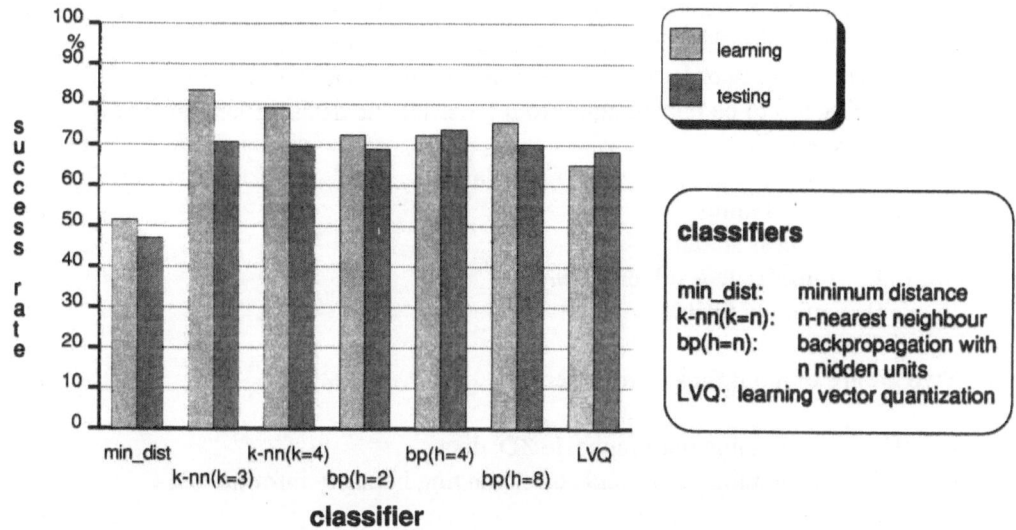

Fig 4.29 Performance of feature based classification

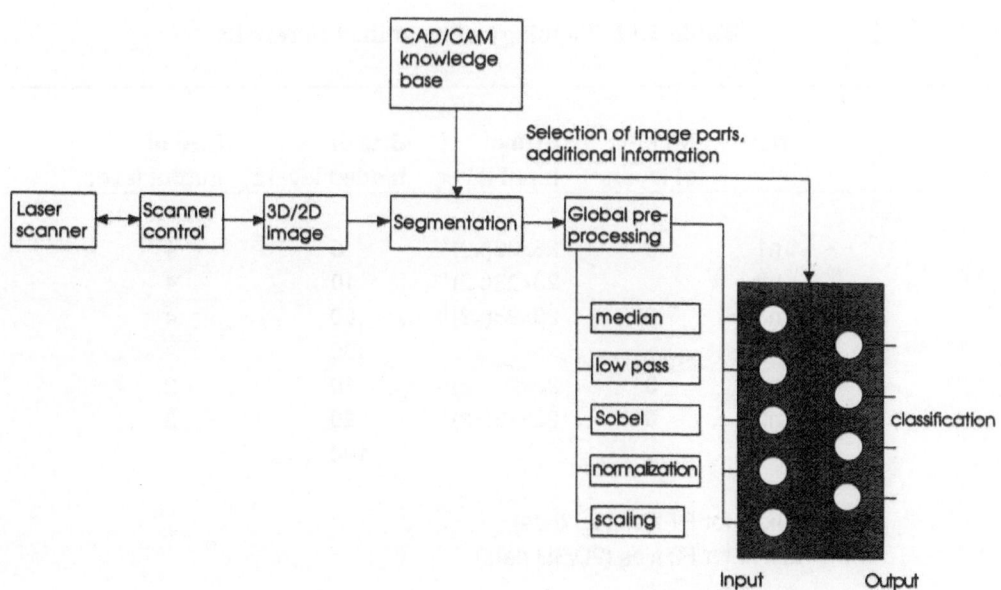

Fig 4.30 Principle of direct data processing with neural networks

Examination of appropriate networks

A fully connected feedforward network using an error back-propagation learning algorithm is suitable for this kind of problem. However, the number of adjustable weights is quite large. In order not to overfit the net the number of examples for network training has to be big enough. From the available dataset four classes of solder joints were built:

- (c1)   good joints
- (c2)   excess solder
- (c3)   insufficient solder/blow out
- (c4)   pin not visible.

Two groups of input files were generated:

- (F1)   containing only range, ie 3D, data
- (F2)   containing range and corresponding intensity information.

Each input file contained 100 solder joints of equal class numbers, and images were partly moved and rotated to force the network to learn simple geometric variances. Prior to presenting the data to the network a scaling was performed to the range of the neural activation level.

After some experimentation we examined the networks listed in Table 4.11 in detail.

Table 4.11 Topology of examined networks

Net	Number of layers	Size of input layer	Size of hidden layers	Size of output layer
n1	3	23x23(x2)[1]	5	4
n2		23x23(x2)[1]	10	4
n3	4	23x23(x2)[1]	20	4
			100	
n4	3	23x23(x2)[1]	10	2
n5	3	23x23(x2)[1]	20	2
			100	

[1] 23x23 for F1 files (3D data)
23x46 for F2 files (2D/3D data)

All four joint classes were presented to networks n1 to n3 (4 output nodes). For n4 and n5 only classes c1 (good) and c3 (blow-out/insufficient solder) were used (only 2 output nodes). Four different files were used for the two sets of learning and testing.

File1 and file2 contain different solder joints of all four classes, where joints are also moved and rotated. File3 only contains the classes *good* and *blow-out*, and joints in file4 are not moved.

The percentage success for the individual classes are shown in Tables 4.12 and 4.13. Numbers in bold show training results.

Table 4.12 Percentage success for range data input

Network	Class	File1	File2	File3	File4
	c1:	**100**	100	94	96
	c2:	**100**	100		100
n1	c3:	**100**	64	50	80
	c4:	**100**	100		44
	all:	**100**	91	72	80
	c1:	**100**	100	96	88
	c2:	**100**	100		92
n2	c3:	**100**	76	48	76
	c4:	**100**	92		24
	all:	**100**	92	72	70
	c1:	**100**	92	90	100
	c2:	**100**	100		100
n3	c3:	**100**	60	60	72
	c4:	**100**	88		20
	all:	**100**	85	75	73
	c1:	92	92	**100**	96
	c2:				
n4	c3:	76	92	**100**	72
	c4:				
	all:	84	92	**100**	84
	c1:	88	88	**100**	92
	c2:				
n5	c3:	60	88	**100**	76
	c4:				
	all:	74	88	**100**	84

Table 4.13 Percentage success for range and intensity data input

Network	Class	File1	File2	File3	File4
	c1:	**100**	76	92	68
	c2:	**100**	88		100
n1	c3:	**100**	48	38	40
	c4:	**100**	100		100
	all:	**100**	78	65	77
	c1:	**100**	84	96	92
	c2:	**100**	96		100
n2	c3:	**100**	40	48	56
	c4:	**100**	100		100
	all:	**100**	80	72	87
	c1:	**100**	84	94	92
	c2:	**100**	92		80
n3	c3:	**100**	52	52	56
	c4:	**100**	100		100
	all:	**100**	82	73	82
	c1:	96	88	**100**	100
	c2:				
n4	c3:	76	84	**100**	80
	c4:				
	all:	86	86	**100**	90
	c1:	92	100	**100**	100
	c2:				
n5	c3:	52	80	**100**	88
	c4:				
	all:	72	90	**100**	94

The main conclusions of this work are:

(i)   training always reaches 100% success
(ii) there is perfect classification of *pin not visible* for 2D/3D images
(iii) excellent success is achieved for *excess solder* in 3D images but invariances
      have to be learnt by translation of training data
(iv) poor results for *blow-out* are obtained with networks which have to separate
      all classes (4-class nets)

(v) the performance rises, when the classification is reduced to 2 classes and
     images contain 3D and 2D information

(vi) the defects *excess solder* and *pin not visible* can be classified very well, but
     they require different network types. Success rates for *blow-out/ insufficient
     solder* are not sufficient.

Examination of pre-processing

Since the classes *excess solder* and *pin not visible* can be detected very well, the next
stage is to concentrate on the detection of *blow-out* defects and to examine the effect
of pre-processing. Pre-processing aims to enhance defect specific information in the
image and to reduce superfluous information or noise. An analysis of the available
defect joints of the classes blow-out and insufficient solder shows that a simple
mathematical description of the defect criteria is impossible. Humans often specify the
joints by a fuzzy linguistic description like 'the image looks rather inhomogeneous or
'the image contains some dark/light areas'. Some defects can be seen in the 3D range
data, but some can be observed only when looking at both range and intensity images.
Therefore, both images were always presented to the neural networks.

For data pre-processing the following basic algorithms were implemented:

- median filter
- mean filter
- data reduction
- Laplace filtering
- Sobel filtering
- normalisation
- thresholding
- scaling.

The network dimensions were chosen according to those of network net4 (3D/2D
images, 2 classes). Data were pre-processed and then fed to the network. For all tests a
training set of 80% of available joints was used, the rest was taken for recall. Training
always reached perfect (100%) classification with below 100 cycles. Figure 4.31
shows the percentage success for testing.

We concluded from these experiments that:

- scaling by vector yields a good success for classifying defects, but poor results
  for correct solder joints. The results get even worse when applying
  normalisation because the more dynamic intensity data eliminates information
  from the range data
- averaging the data in order to reduce the number of weights in the networks
  also seems to eliminate important local information which is necessary for
  classification
- the Sobel operator applied as a preprocessor gives the best success rates of
  95% for *blow-outs*, which is enough for factory demands.

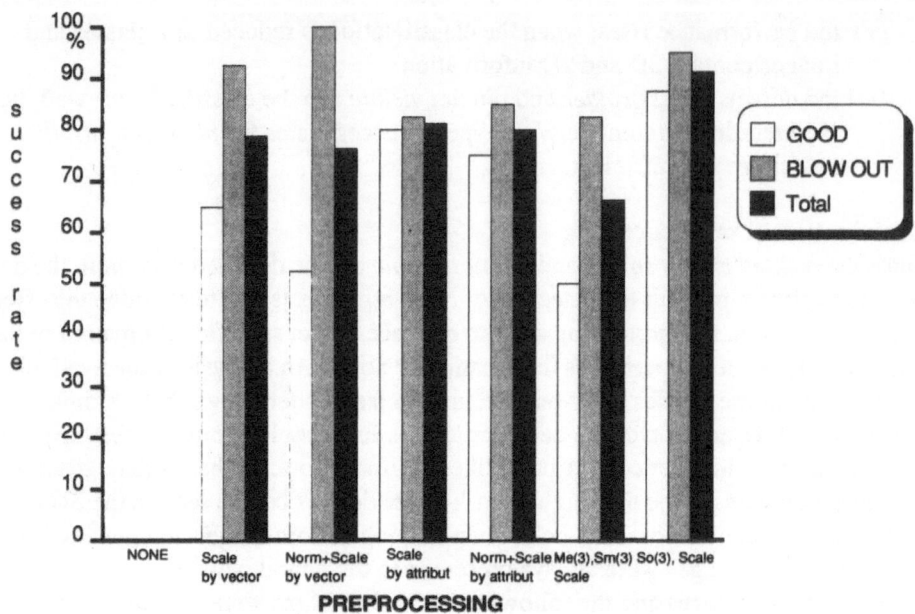

Fig 4.31 Performance of different pre-processors

Figures 4.32 shows two examples of images processed by the Sobel operator. The pin of the component and the border of the joint can be seen clearly. In the *blow-out* image, additional lines and patterns appear which are characteristic of the defect.

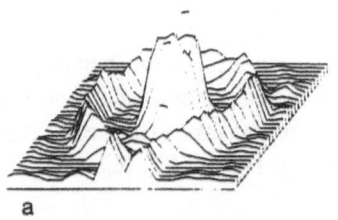
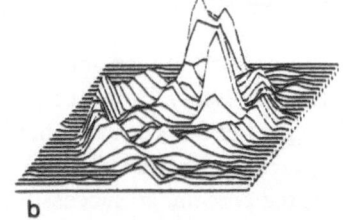

Fig 4.32a Sobel processed data
of correct joint

Fig 4.32b Sobel processed
data of blow-out

Shared weights networks

One problem of implementing standard back-propagation networks for these direct image classification tasks is the large number of nodes in the networks and the limited number of examples, which makes it very difficult for the networks to generalise. Shared weights networks reduce the number of nodes by defining series of weights between layers which are constrained to be the same (ie 'shared') and effectively form receptive fields. We applied an advanced shared weight algorithm on Sobel preprocessed solder joint data. Best results were achieved using 2 receptive fields of the 5x5, which are moved independently across both images. The activation of each field is stored in one node of the receptive layer, which is fully connected with the output layer. Figure 4.33 shows the percentage success rate achieved with different fractions of the data used for training.

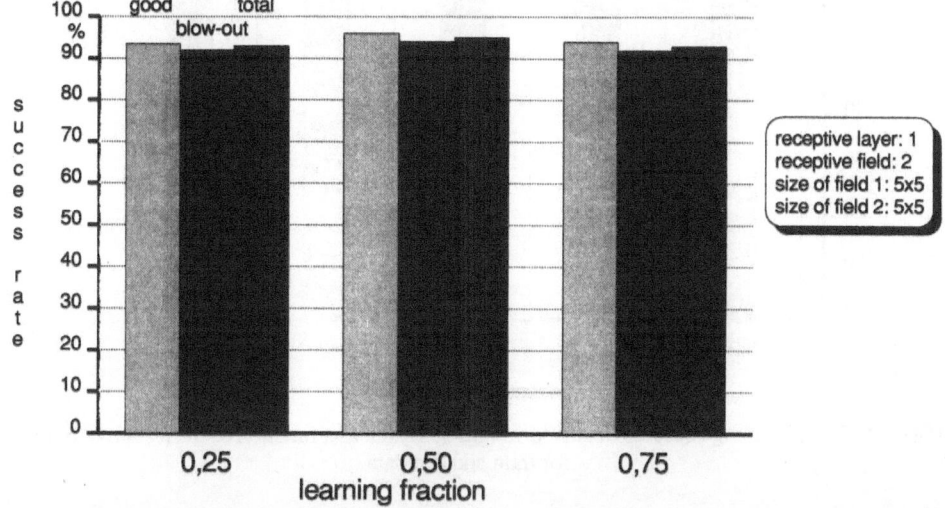

Fig 4.33 Performance of shared weight networks as function of the learning fraction

In another approach the images were reduced to half of the original dimensions by averaging each label in a 2x2 spatial box and then feeding them to a network with one receptive layer, so further reducing the number of adjustable weights in the network. Table 4.14 and Figure 4.34 show the results using different receptive field sizes.

Table 4.14 Shared weights results on direct solder joint data

One receptive layer	Two receptive fields	RMS error cut off: 0.0001
Line search start: 0.1	Tolerance: 0.1	Learning fraction: 50%

Run file	File size	Number of weights	Success train %	Success test %
.aac	1x1x2	52	69	76
.aab	5x5x2	106	98	88
.aag	7x7x2	202	100	98
.aad	9x9x2	330	100	96
.aac	11x11x2	490	98	94

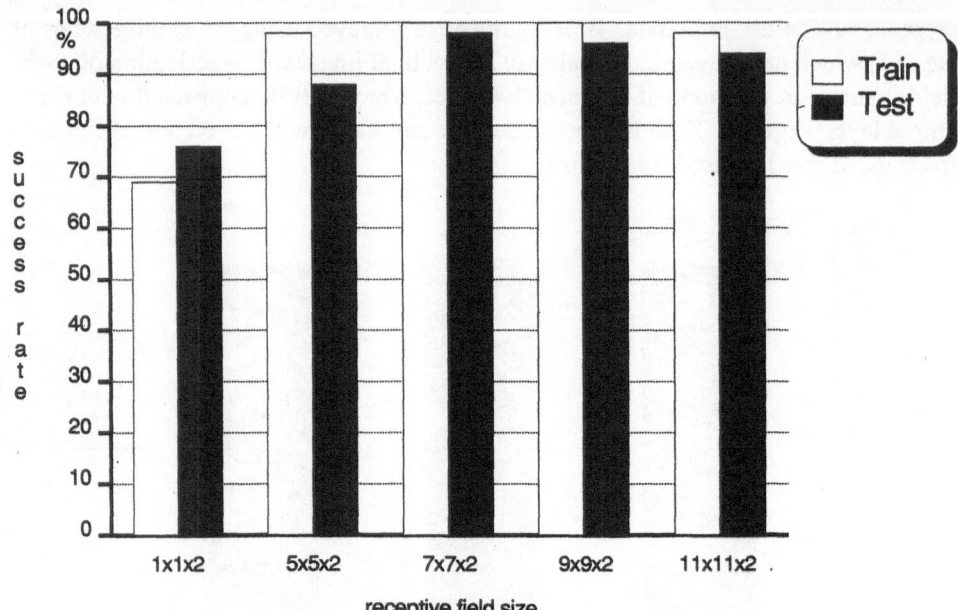

Fig 4.34 The percentage success of shared weight networks as a function of the receptive field size for train and test data

4.10.4 Conclusions for the solder joint classification problem

Figure 4.35 shows the best results of all the networks investigated for defect detection. Excellent success rates were achieved for the defects *excess of solder* and *pin not visible*. The defect *blow-out/insufficient solder* can be detected with 95% accuracy. However different networks and different pre-processing of data are necessary. For industrial exploitation in the factory the processing time on a PC is still too large, so either parallel neural hardware (a neurochip) or some high speed coprocessors are necessary. The most promising neural approaches now are going to be implemented on a multiprocessing PC environment with several RISC i860 coprocessors, which will enable the classification of a solder joint in less than 20 ms.

4.11   Conclusions

4.11.1 Objectives

The objective of the pattern recognition part of ANNIE was to develop prototype systems using neural network techniques for the characterisation of defects in the field of non-destructive testing. This has been met for some real industrial problems.

By specifying a common file structure, and developing a set of pre-processors and classifiers using this data format, it was possible to build a system (PROFAN, see chapter 3) which is not specialised to the particular images but can be applied quickly to new problems. This enabled rapid progress to be made in parallel on several applications, with a maximum reuse of software.

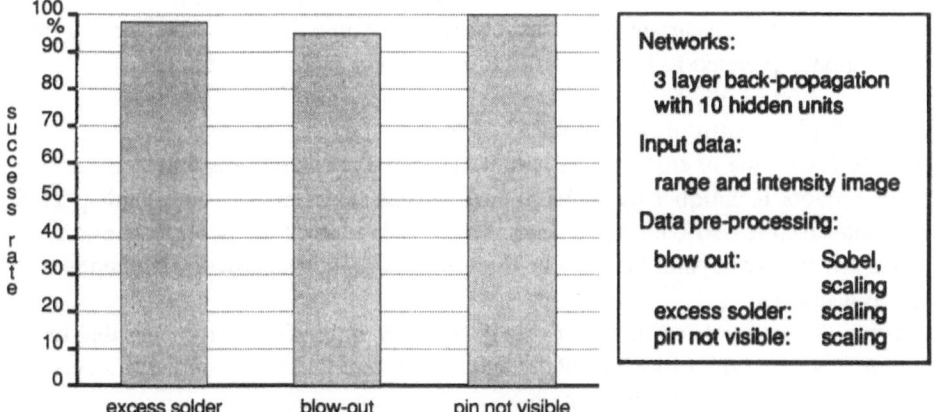

Fig 4.35   Overall defect detection performance

Following initial studies of classification methods, in the chosen applications of the present study the major target was to use the full data in its original *image* form rather than to process the data to extract *feature parameters*.

4.11.2   Comparison of Techniques

A common result of the study was that several classifiers which were found to be appropriate for this type of data when using generic images gave similar results when tested on the real data. Examples misclassified using one method tended to be misclassified also by others. Generally the most successful classifiers for image data were the shared weights and receptive field versions of the back-propagation method, and the adaptive receptive field method. All these methods perform as well or even better than conventional back-propagation by having many fewer adjustable weights and so avoiding the overfitting problem.

For the ALOC data a considerable image enhancement was achieved using the Hough transformation, and further processing with Kohonen and back-propagation networks yielded very good detection of defects.

In the application to classify defects in welds from direct ultrasonic diffraction data, all three back-propagation classifiers gave around 94% success in classifying test data on a *leave one out* basis. This is comparable with the best results with feature parameter data. This degree of success was only possible by describing the ultrasonic reflectivity as a function of three positional dimensions and one angular dimension. The classification is fast enough that it has already been implemented in an on-line demonstrator.

In the classification of solder joints on printed circuit boards direct data processing reached even better results than extracting feature parameters especially for *blowout* defects which could not be classified with sufficient accuracy with conventional techniques. Again the best solutions were found using standard, receptive field and shared weights back-propagation that would characterise these defects with the required accuracy of above 90%. In order to achieve this number it was again necessary to use multidimensional data combining range images with intensity images. The method is planned to be installed into the factory in the near future.

4.11.3 Implementation: Key Points

The PROFAN suite of software has enabled general investigations and comparisons of neural network techniques by problem size, system performance, costs and speed for many industrial classification problems. The capabilities of the developed and tested system áre clearly beyond the state of the art of existing classification methods in several instances.

From the experience obtained in studying these applications, it is possible to draw some useful general conclusions about how and where to implement neural networks successfully in pattern recognition.

The most important observation is that in all cases it requires considerable understanding of the problem if the success is to be maximised. This understanding has to exist in all aspects of the problem: the physics of the process giving rise to the pattern recognition problem; statistical information about the dataset; the resolution of the detection device; the noise inherent in the measurements, etc. Any methods which might help this understanding should be used. Both formal analysis methods, and visualisation tools can be of great value. Only with such understanding can the best approach to the problem be determined.

By understanding the problem it is possible to optimise the choice of preprocessing and of data representation which will enhance the signal being sought. This choice frequently turns out to be more important than the subsequent choice of network or classifier.

In addition, understanding will allow the problem to be broken down into manageable parts. Useful combinations of techniques can then be employed, such as hybrid networks (Kohonen plus back-propagation as in the ALOC case) or successive filtering with different networks (as in the acoustic emission problem). It might also indicate the best classification structure for the problem under study.

If the understanding of the problem is sufficient that key features can be deduced either analytically, or conceivably by intuition, then a feature extraction followed by classification process may well be the preferred route. On the other hand, if the complexity of the problem is such that simple rules cannot be derived, neural network methods have the potential advantage that they can deal with data at a less processed level. In most of the cases considered here, there was some element of the problem when broken down for which that was true; in some cases this might be the only route available. Consequently, neural networks have become a useful tool to use in providing part of an overall solution to pattern recognition problems.

References

Burch S F and Bealing N K (1986) NDT International 19, 145-153

Chatfield C and Collins A J (1980) Introduction to multivariate analysis. Chapman and Hall

Duda R and Hart P (1973) Pattern classification and scene analysis. John Wiley & Sons, New York

Fukunaga K (1972) Introduction to statistical pattern recognition. Academic Press, New York

Fukushima K (1988) Neocognitron: A hierarchical neural network capable of visual pattern recognition. Neural Networks 1(2), 119-130

Huang W Y and Lippmann R P (1987) IEEE First International Conference on Neural Networks, San Diego CA, USA, 1987, IV 485-494

Kohonen T (1988) Self-organisation and associative memory. Springer-Verlag, Berlin, second edition

Kohonen T, Barna G and Chrisley R (1989) Neural networks from models to applcations. (Ed) Personnaz and Dreyfus, IDSET, Paris, 160-167

Rumelhart D E, Hinton G and Williams G (1986) Learning internal representations by error propagation. In Parallel Distributed Processing: Explorations in the Microstructure of Cognition, Bradford Books/MIT Press, Cambridge, Massachusetts

Tou J T and Gonzalez R C (1974) Pattern recognition principles. Addison-Wesley

Chapter 5

Control Applications

5.1   Introduction

The control task started with an overview of the application of neural networks in
control to aid the choice of suitable demonstrator projects. The generalised application
was chosen to provide insights into a wide range of technical problems based on:

- motor sensor interaction
- signal classification
- sensor fusion
- control.

   The main application which incorporates all these features was an automatic
guided vehicle (Lernfahrzeug). This is described in sections 5.4-5.5 below. A further
application concentrated on sensor fusion in a position finding system, and is
described in section 5.7.
   The concentration on control tasks was influenced by awareness of problems of
using conventional processing techniques in the control area such as response time,
creating mathematical models for complex processes, adaptability to system and
environment changes, learning/teaching efforts.
   The variety of problems in the control area addressed by neural networks is
impressive:

- trajectory planning
- collision avoidance
- security control
- autonomous mobile robots
- robot arm control.

   The next generation of control systems will produce architectures and procedures
with self-orienting automatically adjusted free parameters in dynamic environments
incorporating learning procedures. The implementation of neural networks in these
architectures can start at the level of sensor signal interpretation and preprocessing,
sensor fusion, feature extraction, classification, model behaviour representation, sensor

motor association; and last but not least in the area of learning, from supervised learning through to reinforcement learning.

The multiple integration possibilities of neural networks in control provide the following main advantages:

- low costs for programming and learning due to training on examples or by learning sets
- tremendously short reaction time with high robustness during real time use.

Thus the following sections cover aspects such as control fundamentals, neural network integration, simulations, comparison of networks, learning procedures.

5.2   Overview on Control Technology

5.2.1   Introduction

Control technology has grown tremendously in almost every field of technical endeavour. The normal advances in theory and practice have been greatly accelerated by high performance requirements. As the need for speed and accuracy grew, so too did the complexity of the control system. Empirical and trial-and-error design methods soon became unacceptable. New design methods were developed to meet these requirements. Some mathematical and graphical methods offered by the developers of control theory were soon accepted by practitioners and became standards by wide acceptance in the industry. The microprocessor technology soon found its acceptance and digital control became commonplace even in consumer products. The expected impact of *modern control theory* based on *state space* methods over the *classical control theory* based on transfer functions (which covered linear, non-linear, sampled-data and digital systems working with either deterministic or stochastic (random) input) never became apparent even after so many years except in a few cases such as the aerospace industry. New design methods are therefore expected to prove themselves in actual practice before they can displace well-accepted techniques. Except for a small number of quite specialised systems, modern control theory has really not been widely used in modern control applications. By now history has verified that classical methods may be augmented by *modern control theory* but will not be replaced by it. This may again be the case in practice today when the VLSI microprocessor technology has grown enormously and the control system designer is often confronted with situations such as whether to choose a 32-bit micro controller, a 32-bit digital signal processor (DSP), an application specific (motion control) IC (ASIC) or a 32-neuron neural processor to implement the controller!

In control parlance, various names such as *process, plant, controlled system*, etc. are used for the devices to be controlled. A process or plant input/output configuration, such as that of Figure 5.1, is often used to highlight the process inputs (subject to our control) and disturbance inputs (undesirable or unavoidable inputs which are beyond our control) that cause the process or plant to react or respond.

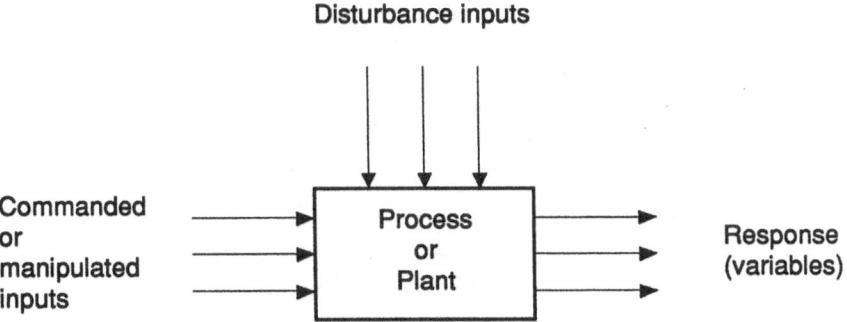

Fig 5.1 Representation of process or plant configuration

Another classification involves two broad areas of application: *servomechanisms* and *process control*. One of the definitions of a servomechanism is a feedback control system in which the controlled variables are mechanical motions, forces or torques. In the so-called process industries (petroleum, chemical, paper, cement, steam power, food, etc) one often encounters the need to control temperature, flow rate, liquid level in vessels, pressure, humidity, chemical composition and the like. Such applications are generally considered process control. When the desired value is more or less fixed, the main problem is to reject disturbance effects. In this context, the controller is normally called a regulator.

5.2.2   Control Theory and Principles
A control system is normally needed because a process or plant is often subjected to disturbance inputs. The inputs can be cleverly manipulated or generated so as to counteract and eliminate or reduce the effects of the disturbances. This implies that the response variables associated with the plant can be made to behave in some fashion as desired and required. Therefore a control system might be required even if disturbances were not present. Thus in general the need for a control system arises either for command following (feedback) or disturbance rejection (feedforward or sometimes both feedforward and feedback).

5.2.3   Classical Control
The principles of classical theory can be easily understood by analysing a typical industrial feedback control system as shown in Figure 5.2.
   The dynamic stability of a closed loop control system critically depends on the delays (transient and transport delays) and output-input gains (mechanical, electrical and others) distributed in the elements forming the closed loop. Thus, determining the permissible gains in the controller of a control system irrespective of whether it is digital or analogue is the main emphasis of control system analysis and design using classical control theory.

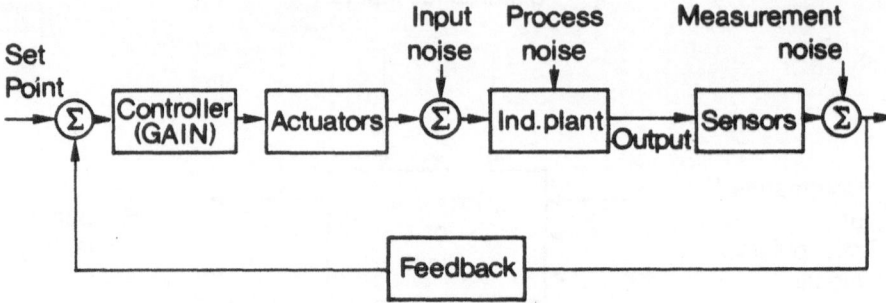

Fig 5.2 Analysis of typical feedback control system

Time response methods are useful for analysis and design of low order systems.
Second order systems enjoyed a specialist treatment because the majority of the
servomechanisms exhibited dominant second order behaviour (higher order effects
being weak). Settling time, time constant, steady state errors, overshoot, speed of
response, are typical specification parameters in the time domain.

Frequency response methods on the other hand (root locus, Bode plots, Nyquist
plots, Nichols chart, phase plane trajectories, etc) provide an easy graphical method of
determining the stability and performance of single input-single output (SISO) plants.

Systems are often described by cascaded first or second order transfer functions
but these models do not account for time-varying plant parameters or for the modelling
of complicated disturbance patterns.

Simple *lead* controllers, *lag-lead* controllers, *proportional-integral-derivative*
(P-I-D) or *three-term* controllers, feedforward control etc, are common in control
system design and analysis for SISO systems. They can be implemented in analogue,
digital or in both as needed. Laplace transformations and

Y transformations are the mathematical tools operators use to analyse stability
issues in continuous/analogue systems and discrete/digital systems respectively. The
P-I-D control shown in Figure 5.3 is a good example of a system which we will look at
more deeply in order to obtain a feel for a classical control systems design.

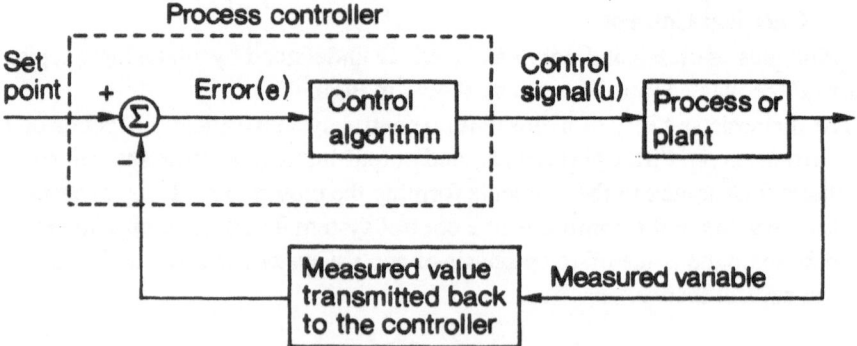

Fig 5.3 Basic elements of P-I-D control

The error (e) between the set-point value and measured value is fed into the control algorithm. An output control signal (u) is then fed to the process.

A P-I-D controller uses a three-term algorithm to calculate its output control signal (Figure 5.4).

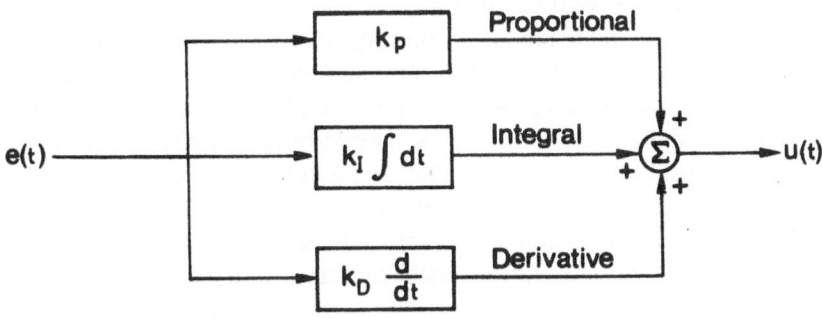

Fig 5.4 P-I-D control algorithm

The algorithm is based on the error (e) between the controller set point and the measured value of the process variable. The algorithm consists of a proportional term which affects the speed of response, a derivative term which governs the oscillatory behaviour and an integral term which corrects any steady state errors. The three gains K_P, K_I and K_D are the three parameters which are adjusted when tuning a P-I-D controller.

The discrete version of the P-I-D controller (with any feedforward) can be described in the block diagram form shown in Figure 5.5.

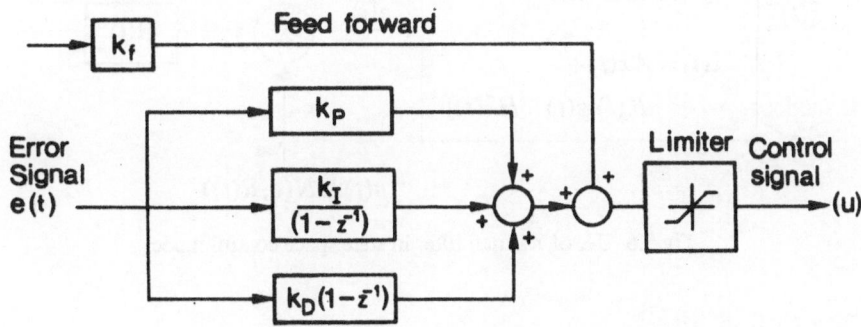

Fig 5.5 Block diagram for discrete P-I-D control

5.2.4 Modern Control

Modern control theory can present an inhibiting new kind of terminology to the *non-control* community: buzz words would include Kalman filters, Weiner filter, ARMA (auto regressive moving average) predictor, optimal control, self-tuning, robustness, disturbance rejection, parameter sensitivity, state space, multivariable control, inverse Nyquist array etc. The mathematics can easily turn back even some of the control engineers to try their trusted three-term controllers rather than a multivariable controller!

Multi input-multi output (MIMO) problems (eg simultaneous speed and torque control of an automatic drive transmission) are conveniently handled in the *state space* framework. In a state variable feedback controller or an optimum controller like a linear quadratic regulator (LGR), all state variables are required to be measurable. Accessibility, measurement devices and cost are therefore the limitations. When these states are not available or when the outputs are corrupted by noise, a Kalman filter provides estimates of the unknown variables and noise corrupted measurement as illustrated in Figure 5.6. These can then be used within an LQG control law (where G stands for Gaussian representation of the noise). In practice, however, engineering constraints often intrude and a state estimate will usually be utilised by a sub-optimal P-I-D controller.

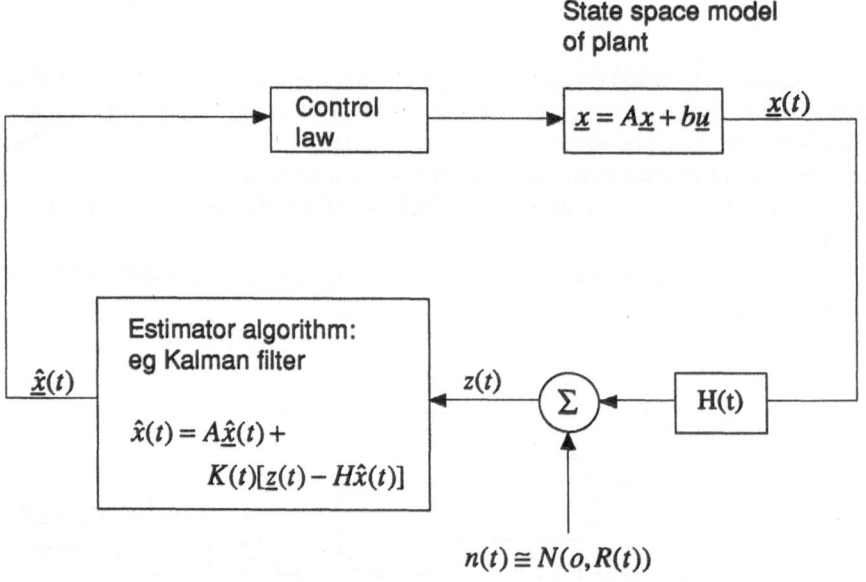

Fig 5.6 Use of Kalman filter in state space control model

Multivariable processes have multiple inputs and outputs that exhibit cross-coupling such that changes in an input (manipulated) variable causes responses in both the intended output and also in the unintended output.

If the cross-coupling is not too strong and/or performance need not be optimum, the controls using individual feedback loops may be acceptable. However, the individual controllers may sometimes 'fight' each other excessively. Then it may be necessary to use *multivariable control theory* that deals with the process cross-coupling by designing a cross-coupled controller to obtain improved performance.

5.2.5   Programmable Logic Control

A relatively small number of basic logic functions (AND, OR, NOT etc) together with counting and timing elements can be used to build up a controller for a plant or machines. The *programmable logic controller* (PLC), a microprocessor-based general purpose control device provided a *menu* of such basic functions that can be configured by programming (using ladder diagrams) to create a logic control system for many applications. This software or programmable approach to logic control has many advantages, and although originally applied only to large systems has recently become economic even for small ones. A car washing machine is a good example which uses a mini PLC as a cost effective controller. Some PLCs can also handle analogue signals and closed loops controls including motion controls.

5.2.6   Adaptive Control

Adaptive control systems can take a number of forms but, as the name implies, they are intended to provide improved performance by adapting themselves to changing conditions. An ordinary feedback control system can accommodate a considerable tolerance for changes, thus adaptive control systems are only needed for extreme varying conditions.

Flight control systems for aircraft that operate over wide ranges of altitude and flight speed are a good example since the vehicle's natural frequency may vary over a 10:1 range while damping ratios change by 20:1. So, a controller designed for one flight condition would be grossly inadequate at others. An industrial robot is another example where natural frequencies vary over a wide range due to wide changes in inertia caused by varying arm configurations and different payloads. Damping also varies depending on the end effector interaction with objects and the environment.

The most common successful approach to adaptive control is a preprogrammed or scheduled scheme, which measures the environmental factors that are causing the changes in the controlled system's parameters, and continuously adjusts the controller parameters to accommodate the current situation. This is sometimes called gain scheduling. In the aircraft, the vehicle's natural frequency, damping ratio, etc, depend mainly on the altitude and Mach number which can be calculated with reasonable accuracy from measurements. It may then be possible to design a controller to change its parameters to give a good performance over the entire flight.

The so-called *self-adaptive* systems which deal with the same problem but require no environmental measurements are also possible and can take a variety of forms.

The *model-reference* approach, see Figure 5.7, uses a model of the system to be controlled to generate the required plant response. The model is placed in parallel with the process and the error between the actual and desired response is fed through an adaptive block.

The controller parameters are changed to ensure satisfaction of a predetermined control strategy. One disadvantage is that the basic design assumes noise-free measurements.

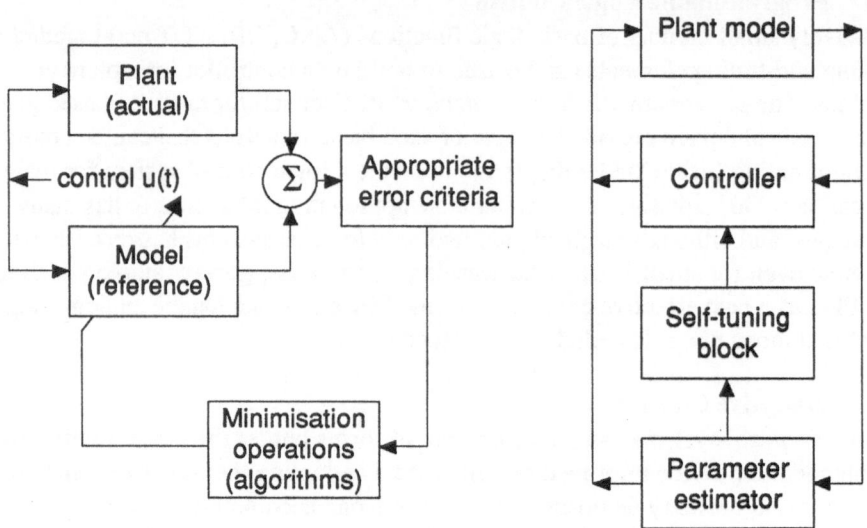

Fig 5.7 The model-reference (left) and self-tuning methods (right)

A *self-tuning* regulator is another kind of adaptive control which employs a parameter identification scheme based on (eg least squares techniques) to identify a process model. The control law is dependent on the adaptive system model parameters. The self-tuning procedure is performed online and updates the model parameters and control law continuously. (This method, illustrated in Figure 5.7) is different from auto-tuning, where the constant P-I-D control parameters are determined automatically by the controller from the system response to some specified inputs).

The development of any modern controller involves a procedure which ranges from modeling and defining control objectives through to simulation and implementation. Computer simulations of systems are used extensively to provide pre-design insight of complex system interactions, specifications and recommendations. It can be used to test, verify and assess a proposed design hypothesis.

5.3   Use of Neural Networks for Control Purposes

Generally, neural networks could be used in control for many purposes. The main fields are:

- neural networks as a direct part in the control loop
- neural networks for optimisation of the control loop behaviour or the result of the control process
- neural networks for improving the adaptability of the control electronics and for putting the control system into operation (start-up problem)
- neural networks for safety reasons.

5.3.1   Neural Networks as a Direct Part in the Control Loop
The following parts of a control loop may consist of neural networks or may be supported by neural networks.

Control unit. It is possible to use a neural network based control unit. In this case completely new control algorithms and/or control strategies are very likely. Very complex and heuristic control strategies especially could be performed by neural networks.

Observer/predictor. Neural networks are able to predict the behaviour of a process signal and they are also suitable for observing a process. It might be possible to generate a neural network model of a process by teaching a net without any knowledge of the process itself.

Inverse kinematics. A big problem in robot control for continuous movement is the fast computation of the inverse kinematics. Neural networks are able to learn the relationship between the path in world coordinates and the movement of the robot axis. For the future, hardware neural networks are also fast enough for real time computation of the inverse kinematics.

5.3.2   Neural Networks to Optimise the Control Loop Behaviour or the Result of the Control Process
The optimisation of control loop behaviour is either done while putting the system into operation or during operation. The second means an adaptation of the system parameters to changing environmental circumstances or to changing control loop parameters.

Neural Networks for identifying the control loop parameters. Neural networks can be used to identify the plant parameters. This should be possible with less information about the plant structure than is needed for conventional methods.

Neural networks for adaptation of control loop parameters. Instead of only identifying the plant parameters it should also be possible to influence the control loop parameters directly. This could be trained in a self-learning process. It should also be possible to adapt a control loop to environmental changes (for example temperature change).

Complex control strategies with neural networks. Neural networks are intended for the realisation of complex control strategies and are able to perform sophisticated optimisation strategies.

Optimisation of the guidance signal. In many cases it is necessary to optimise the guidance signal, for example in cutting metal sheets with a robot system to minimise rubbish. Neural networks could be very useful for such optimisation problems.

5.3.3   Neural Networks Supporting the Putting of Control Systems into Operation
One of the most difficult tasks for engineers is to put control systems into operation. In this stage there should be some optimisation of the control unit parameters, but in many cases they stop their optimisation efforts when the system works just well enough. Neural networks could support this task enormously.

Adaptation of control systems to different tasks. For example with robot systems there could be a lot of different models of mechanical systems with differences in kinematic and mechanical parameters. Neural networks could support the adaptation of the control electronics to the mechanical systems.

Support of the user by providing additional information. The operator of a control system suffers in many cases from having too little information. Self-organising networks especially could support him here, as they are able to provide additional information about systematic connections within the process by mapping the process parameter field to one with less dimensionality which is more intelligible for the operator.

5.3.4   Neural Networks for Safety Reasons

In many applications special care must be taken to avoid accidents caused by faults or by collision with obstacles. Neural networks are able to contribute to a solution of this problem.

Choosing the best sensor signals. In some cases there are redundant signals available in a control process. The problem is to choose the best to use, and the best choice may change depending on process parameters. A neural network can be used to make this choice.

Sensor fusion. It is also possible to use not just the best suitable signal, but all the sensor information available and match it together in a special sensor fusion process. This is a typical task for neural networks, where the loss of a part of the signals should not affect the result!

Collision avoidance. Neural networks are able to learn information. For example the position of obstacles can be taught to a network and the network can be used to avoid the collision of robot systems (or parts of robot systems) with these obstacles. This is a very important problem in controlling robot movement in reprocessing plants.

5.4   Lernfahrzeug System (NeVIS)

The ANNIE Lernfahrzeug-project was inspired by previous work on remote controlled vehicles. These vehicles have been equipped with ultrasonic range sensors used as part of an emergency stop system. Conceptual studies for an autonomous mobile vehicle based on the remote controlled prototype had been performed before the ANNIE project started. We concentrated in our first investigative steps on the problem of collision avoidance which represents a sensor-motor association problem in the control area. We started developing a closed-loop software simulation system able to demonstrate the performance of different kind of neural network models in the given task of collision avoidance. The combination of short system response time and trained collision avoidance strategy is what we call collision avoidance with reflexive behaviour.

As with other ANNIE tasks, the development was centred on the production of a simulation environment (NeVIS, see chapter 3). This provided a framework in which to try out many different neural architectures and data processing techniques in a comparative way.

In our simulation the vehicle has 9 ultrasonic sensors on board and can perform 8 discrete translatory and rotatory movements per step. System state space therefore spans nine dimensions and action space is represented by a 4 bit vector. This model could be extended in a way where in addition to the sensor values of the overall position of the vehicle, the target vector from global path planning, and other characteristics, could serve as input elements to a neural network. The neural network output could be expanded correspondingly to allow a wider range of movements. Figure 5.8 gives an overview of the simulation concept.

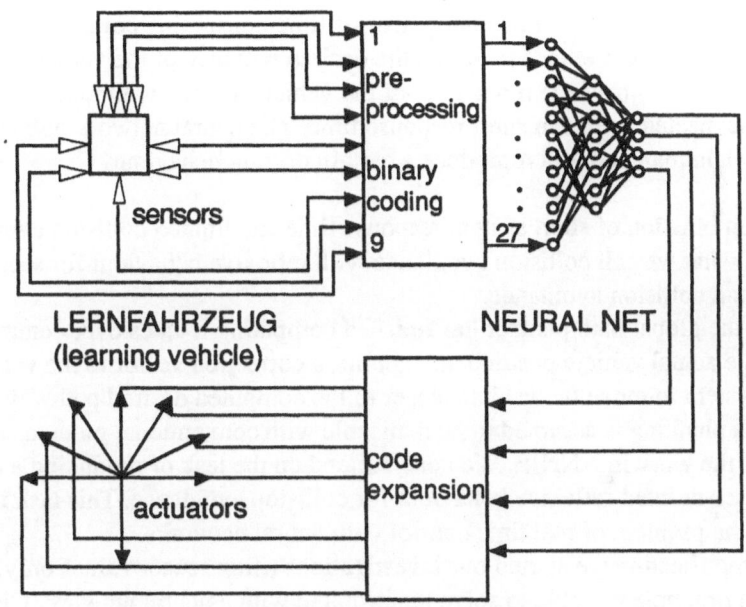

Fig 5.8 Simulation concept

For future technical real time tasks of automatic guided vehicles (AGVs) operating in environment space, system space and reaction space, information processing techniques with high tolerance and the ability to handle complex unexpected situations are demanded. Numerical processing techniques with their relatively slow response times are mostly inadequate.

The background of the *Lernfahrzeug* (autonomous mobile robot) was to create a simulation system for neural network integration and testing in an application field with a high degree of relevance to realistic situations. Using a sensor system for object contour lines and a locomotion set, different neural network types could be applied and investigated. In our case we started with *feedforward networks* (back-propagation) followed by an associative memory system converging ultimately on more sophisticated networks with self-organising behaviour like *reinforcement learning models.*

Conventional approaches apply global path planning strategies to the problem of collision avoidance and use artificial intelligence methods (expert systems) for problem solving. This is very time-consuming and implies high knowledge of the environment which the vehicle is exploring.

Our approach divides the problem into two asynchronously manageable tasks according to the following assumptions:

(i)   a global path planner computes the desired pathway given the location of the vehicle in global coordinates and a goal position based on a stored global map of the environment (not necessarily complete)

(ii) while the global path planner performs time-consuming computing with conventional methods to define the desired pathway of the vehicle, a neural network controls the movement of the vehicle to cope with random unexpected events in short response time. The neural network will be trained by a human and will reproduce a certain driving behaviour.

The combination of short system response time and trained collision avoidance strategy is what we call collision avoidance with reflexive behaviour for security purposes and collision avoidance.

When the global path planner has finished computing it takes over control, acquires the actual vehicle position and applies a correction vector to the vehicle actuator system to move the vehicle closer to the computed desired pathway. The global path planning is assumed to be realisable with conventional methods and was not part of the work in ANNIE. We concentrated on the task of producing a neural network user-defined reflexive behaviour for collision avoidance. This task is highly related to the problem of real time control with neural networks.

For simplification we started our investigations using sensor values only, to show that it is in principle possible to solve the problem with a neural network. Given a certain point in system state space $(S_1,...,S_9)$ the neural network has to learn the mapping to the action space $(M_k; k = 1...16)$ in such a way that the consequent feedback does not place the vehicle at a point in system space which is marked as failure (one where any of the sensor values S_n are set to the emergency stop distance).

For our investigations we implemented a software simulation tool (NeVIS) to generate training data and to make learning and recall of the applied neural network transparent for detailed evaluation. While we used logical sensor and actuator representations for simulation purposes it is planned to connect in the near future a trained neural network directly to physical sensor units and to the effector electronics of a real vehicle. Initially, ideas to produce a demonstration integrated hardware system were rejected because production of a real vehicle moving in a hall would have been too costly. Besides this, there would still have been the need for simulated training data generation since creation of the necessary amount of training data with real obstacle generation and positioning would have been too time-consuming.

5.4.1   Objectives

The overall aim and purpose of the simulation (NeVIS) was to implement and investigate the following:

- learning reflexes and strategies for avoiding collision with simple and complex obstacles
- control of robot vehicle movements in a structured environment, subject to local perturbations; operation in unknown environment with reflex behaviour
- performing cooperative behaviour of multiple mobile robot vehicles
- training by supervised learning with input/output vector association without algorithms for programming or mathematical models
- clear separation of the Lernphase (training of the network) and the Kannphase (recall of the network) for investigation of learning and recall performances using a bottom-up approach with physical sensors and neural network software simulators.

5.4.2   Simulation Concept

The simulation concept has been worked out in many discussions and after information exchange with experts in the area of autonomous guided vehicles (AGVs). The concept survey, illustrated in Figure 5.9, shows five main working fields with different realisation possibilities. It took intensive pre-investigations to work out a whole concept which fitted together and had a high chance of success.

5.4.3   Simulation System Implementation

A number of simulation systems were developed in the course of the project. The earliest versions enabled the exploration of a variety of data representations and network types. The final version used back-propagation learning on more powerful hardware.

5.4.4   NeVIS I

NeVIS I is the first example of a collision avoidance simulation system trained on 10 predefined primitives. The applied neural network paradigm follows Rumelhart's back-propagation model. This model is well known, so the evaluation of our approach can be performed on a common knowledge base. The NeVIS I simulation software uses the features of the Neuralware Professional II simulator for neural networks.

The aims of this implementation were to investigate:

- neural network parameters and learning algorithms and to acquire experience in choosing suitable dimensions and parameters for neural networks
- preprocessing concepts for sensor data to obtain logical sensor representations
- concepts for interfacing neural networks and application simulation
- analysis of possible applications in the field of autonomous robot systems.

The following steps were undertaken:

- implementation of neural network topologies with the neural network simulator
- design of a realistic system concept
- software development of the graphic interface
- training of single primitives
- simultaneous training of ten representative primitives
- implementation of simulated sensor noise.

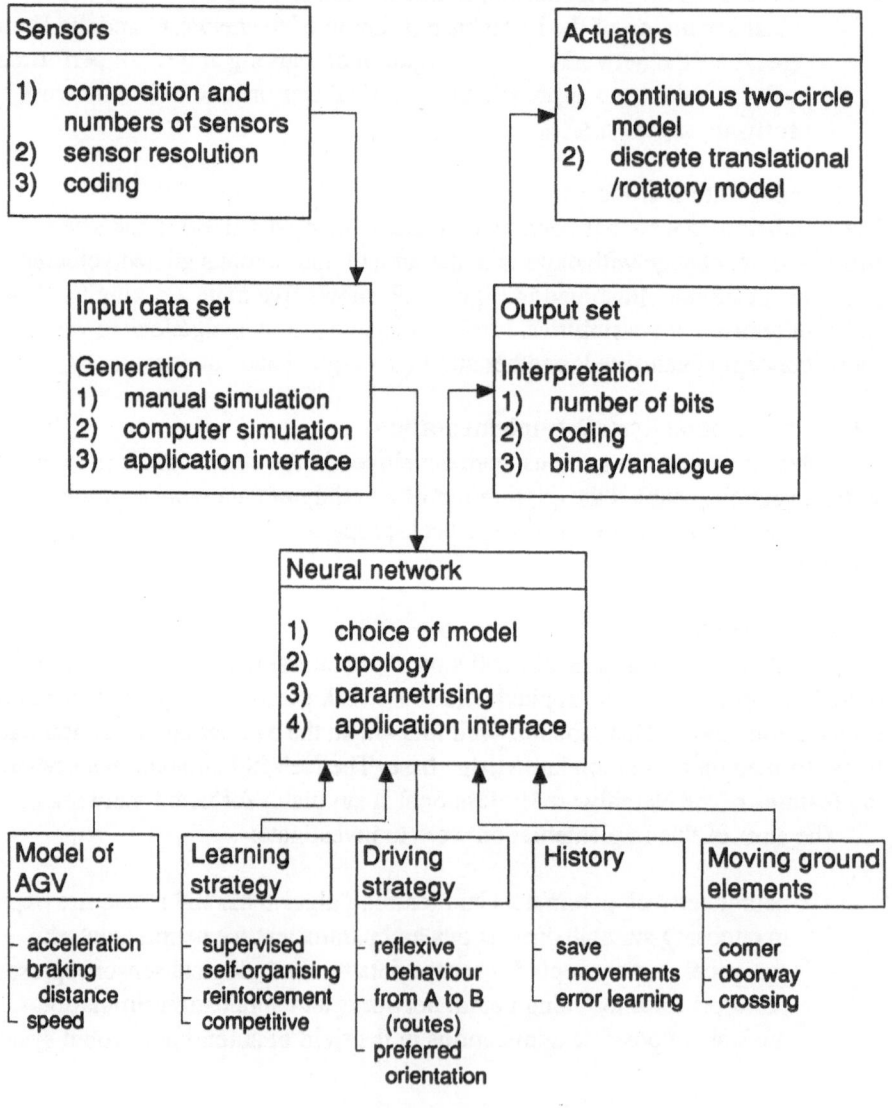

Fig 5.9 Simulation concept survey

Simulation principles

NeVIS I was developed to investigate the possibilities of using back-propagation networks in order to control an automobile robot system able to recognise its environment by ultrasonic sensor signals.

The number and configuration of the ultrasonic sensors plays an important role in the movement control task. On the one hand the number of sensors cannot be arbitrarily large, on the other there have to be sufficient to acquire significant information about the vehicle's environment.

For distinction of a limited set of obstacle primitives 9 sensors appeared to be adequate. Since the vehicle should move forward only with low velocity, one sensor in the rear seemed appropriate for preliminary investigations. The overall configuration of the sensor system is shown in Figure 5.10.

To eliminate minor noise on the physical sensor inputs and to lower the frequency of incoming sensor data the sensor cones have been partitioned into 15 areas.

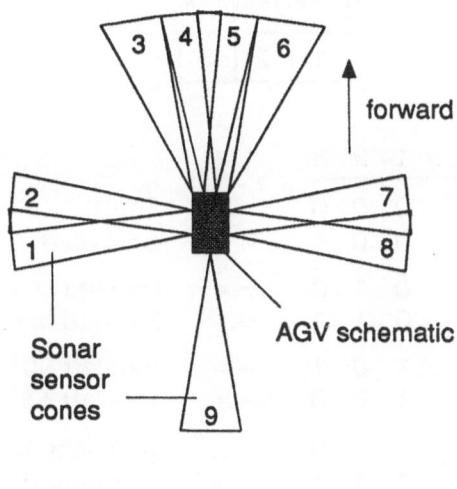

Logarithmic resolution
$$y = 12 - \log(16 - x) * 9.96$$

Area	Distance (m)	Coding
15	12	0001
14	9	0010
13	7.245	0100
12	6	1000
11	5.034	0101
10	4.245	1001
9	3.578	1010
8	3	1100
7	2.490	0110
6	2.034	0011
5	1.622	0111
4	1.245	1011
3	0.899	1101
2	0.578	1110
1	0.279	1111

Sensor solution and coding of
sensitivity areas

Fig 5.10 Automatic guided vehicle sensor system configuration

The resolution of the sensors has been discretised in a logarithmic way shown in the table in Figure 5.10. Grading the resolution of the sensors in a logarithmic method was selected to give a specific kind of sensitivity along with the area coding. Since the sensors range up to 12 metres and the vehicle was assumed to move indoors, information about the environment close to the vehicle has more influence on driving decisions. This is taken into account by logarithmic partitioning where there are more areas in the near range of the vehicle than there are far away.

We took into account that the learning process of a network is more efficient if input patterns that are similar correspond to output patterns that are similar. So we designed a coding of the sensor blocks where, most of the time, the code of block n differs from that of block $n + 1$ in only 2 bits. We could not use the well-known Gray code because we needed an overall strategy declaring that more bits set to 1 in an input vector corresponded to an obstacle at nearer range.

The locomotion vector comprises 8 different moving actions: stop, fast forward, slow forward, turn right 45°, turn right 90°, turn left 45°, turn left 90°, slow backward. In Figure 5.11 the actuator moving directions are shown as well as the data coding of the 3 bit output vector.

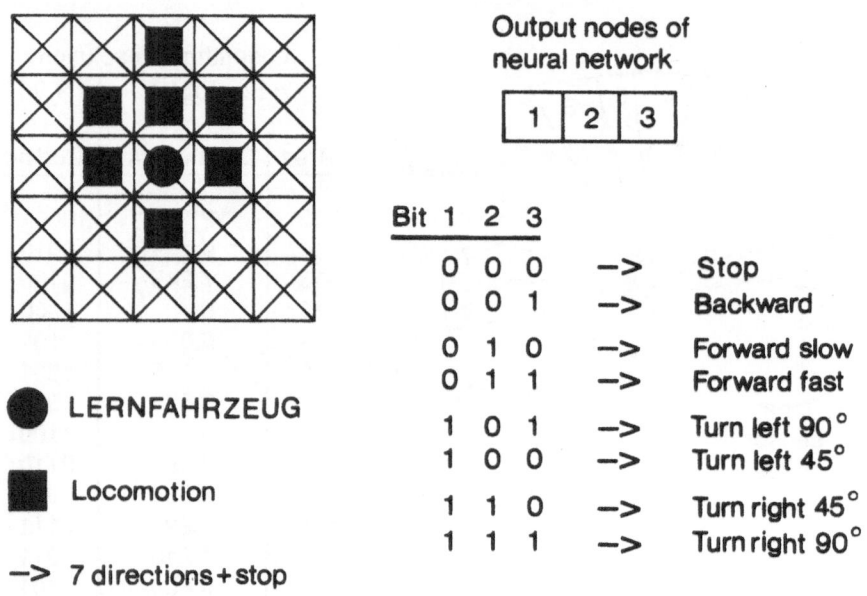

Fig 5.11 Locomotion field and locomotion coding

Initially, the NeVIS simulation system was trained to recognise and to avoid collision with 10 predefined primitives. The primitives were chosen with possible enlargements of the simulation for all forms of obstacles in mind. The neural network consisted of 36 input neurons (4 bits for each sensor, 9 sensors), a variable number of hidden nodes and 4 output nodes, representing the possible driving directions. It was trained by the error-backpropagation algorithm. The training datasets were generated manually for each predefined primitive and appended together for training of the network.

A trained network is able to avoid collision with an obstacle by giving the correct 3 bit output value after presentation of the actually measured sensor data. The simulation consisted of two parts. The first part was the dimensioning and parameterising of the neural network (back-propagation) including the training with the manually created learning datasets. The other part was the software for displaying the network's output in recall.

Some of the 10 predefined obstacle primitives are shown in Figure 5.12. The vehicle was trained to avoid the primitive and to reach the upper right corner of the simulation screen. Therefore the edge of the screen was considered to be a wall for the vehicle sensors.

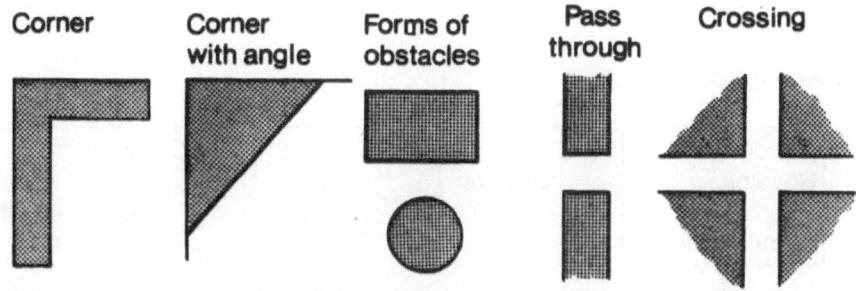

Fig 5.12 Selected obstacle primitives

The complete simulation run is shown in Figure 5.13. First the input dataset has to be generated by manual simulation then the binary coded learning data file has to be created. A suitable network with the correct amount of input and output nodes has to be built. After the network training is completed, an obstacle primitive can be selected and the recall data files can be processed to control the performance of the trained network.

The input vectors for the Lernphase and Kannphase were presented without any noise factor. We wanted to test the claimed fault tolerance on our trained back-propagation net. Therefore we changed parts of the simulation concerning the binary presentation of input vectors. We tested each primitive with 0%, 10%, 20% and 50% noise 10 times. The noise factor represents the amount of bit inversions in relation to the total number of elements in the sensor vector.

Learning performance
Training has involved many different parameter values concerning learning and momentum terms as well as different dimensions. We chose different numbers of layers and hidden nodes. We observed very strong *dependency* of learning convergence and choice of parameters. Therefore we tried out different combinations of parameter values. There is also a strong dependency of learning speed on the choice of initial weight.

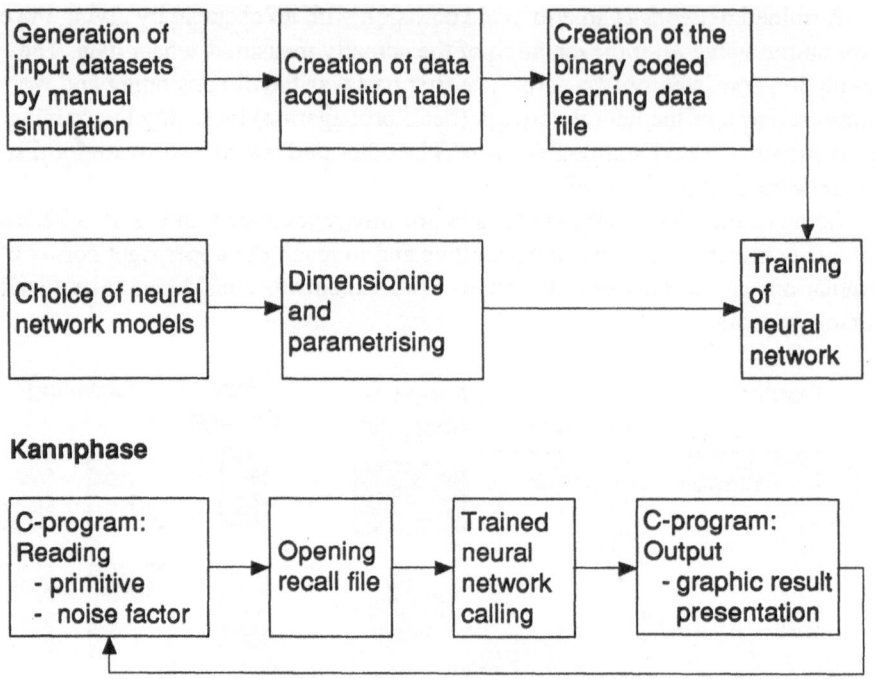

Fig 5.13 Complete simulation run

Training with analogue data was investigated as well and, in further developments, we created a complete software package allowing investigation of performance with trained as well as untrained situations.

We were able to demonstrate that back-propagation networks were able to control autonomous vehicle movement, reproducing characteristics of the networks teaching input. These networks build up a nucleus of correctly learned vectors (*Lernfeld*) and another area of unlearned but correctly associated vectors (*Fangfeld*).

Results
The evaluation of the performance for the 10 primitives was carried out by defining different obtsacle courses consisting of 0 to 3 corners, which are the most difficult patterns. We awarded one point for each correctly driven corner, one point for good driven lines and lastly, one point if no slow/fast errors occurred. The results are listed in Table 5.1.

Table 5.1 Results with noisy input patterns

Noise Obstacle	0%	10%	20%	50%
1	100	67.5	47.5	37.5
2	100	54	40	30
3	100	65	32.5	22.5
4	100	48	34	24
5	100	44	30	26
6	100	75	55	40
7	100	100	95	75
8	100	45	42.5	40
9	100	100	100	75
10	100	70	55	47.5

From these investigations we concluded that:

- a significant range of fault-tolerance was demonstrable, because 10% noise in the Lernfahrzeug-simulation is much more noise than one would expect in reality
- it is possible to see which primitives are learned in a stable way and those which have a few unstable vectors
- our trained net is able to tolerate up to 10% noise. That is because very often with a 10% noise factor only occurred slow/fast errors occurred, meaning that the car is mostly driving in the correct direction, although turning at different distances to the obstacle.

5.4.5 NeVIS II (NeVIS reinforcement)

The intention of the NeVIS II simulation was to show the abilities of neural networks to be trained by reinforcement learning. The algorithm used was developed by Williams (1988). The algorithm has been investigated in detail by other partners in the ANNIE project. The software is similar to the NeVIS III software described in the next section and differs only by use of a different algorithm.

Simulation principles

The creation of a proving ground was done by moving various objects like circles, rectangles and blocks on a test area. Out of these objects the user was able to define an environment for training of the vehicle. The proving ground is saved in binary format and can be loaded for further modifications.

For recall with the NeVIS application both PCs were needed. One shows the neural network while the other shows the vehicle driving around in an unknown environment. The vehicle is able to drive in eight directions at any point. After each driving step of the vehicle the new sensor information was detected and sent to the network. Included in this sensor information there is an evaluation of the last step. These data are coded and fed into the network, which changes its weights according to the evaluation of reward or penalty. Some milliseconds later the actual driving reference is available at the output nodes. They are decoded and transmitted back to the vehicle which drives in the corresponding direction.

The aim is to increase the number of rewards and to minimise the number of penalties. This is ensured by the use of the Reinforcement Learning algorithm. Depending on the complexity of the proving ground and the performance of the computer a good driving behaviour is reached after some hours.

Results

The results have been very encouraging. In our simulation we counted the number of steps without a crash to the wall. As can be seen in Figure 5.14, a good driving performance is reached after 3000 cycles. (A cycle stands for a restart after hitting an obstacle.) At this state an average of 300 moving steps were made without any mistake. Two restart principles were implemented. The first was to put the vehicle to its user defined initialisation point and the other was to put the vehicle two steps back, before it made a wrong association.

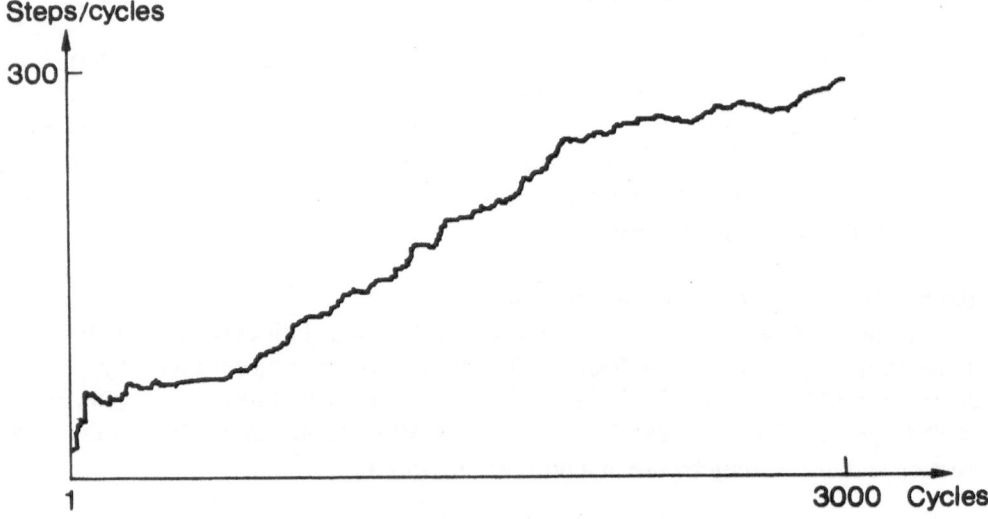

Fig 5.14 Learning curve of reinforcement learning

A number of investigations have been made to improve the driving behaviour. The most important thing is to find a good evaluation set which describes the condition for reward and penalty. Another result we recognised was that the best driving performance was reached without any hidden layer.

An important role concerning convergence speed lies in the choice of learning rate and penalty rate. Good values are smaller than 0.3 for the penalty rate and a learning rate of 1.0 or 2.0. In a later implementation we could choose a factor which changed the parameter set and the evaluation set according to the learning state and behaviour. This increased the convergence speed considerably.

The performance of a trained network to control the vehicle in an unknown environment was not as good as that achieved with back-propagation.

5.4.6   NeVIS III

For NeVIS I, training data had been collected manually to obtain a limited number of training sets for first tries in a short time. Since the results achieved were encouraging, we implemented a simulation system for computer-based interactive training data generation. The system allowed fast generation of voluminous training sets.

For performance investigation purposes the simulation system displays learning and recall features for neural network processing. Embedded in the simulation system is the possibility of generating various user-defined moving grounds (driving courses) for training and recall. This feature was of high importance because we could now test the recall performance of a neural network in an environment different to the one used for training data generation.

For demonstration purposes the system was designed to run on two interacting computers, one showing the neural network recall in operation, the second displaying the movement of the vehicle in a selected moving ground. The complete simulation system was called NeVIS III (Neural Vehicle Information System), and aimed to demonstrate the following:

- investigation of global driving strategies and desired targets under realistic simulated conditions
- automated generation of learning data
- interfacing and integration of various network paradigms
- creation of moving grounds by use of a specially developed software program
- coupling of two computers: PC1 with the graphic surface, and PC2 with the neural network
- training of reflexive behaviour.

The steps involved in the demonstration were:

- creation of a moving ground
- saving and loading of moving grounds
- driving control of the vehicle by a windrose
- automated sensor data acquisition
- menu controlled user interface
- training of neural networks.

Simulation principles

The conceptual layout of the system is the same as that shown previously in Figure 5.8. Sensor data from the vehicle are preprocessed and fed into a chosen neural network. The output of the neural net controls the vehicle actuator system that moves the vehicle to a different location by overlay of translatory and rotatory components.

A complete simulation run consists of various phases. First a moving ground has to be configured, which will be used in the learning and recall phase. Subsequently a learning file has to be created by leading the vehicle around obstacles a few times. With every movement the sensor data and the appropriate moving data are written into a file. In the learning phase this file is presented to the neural network for training. When the neural network was converged down to a small global error (<1) the net is trained sufficiently and can be used in recall mode. The vehicle can then be positioned arbitrarily on the moving ground. The vehicle will drive around obstacles controlled by its derived reflexes incorporated in the trained neural network structure.

Results

Figure 5.15 shows the results of training some networks with data generated on a moving ground. The training has been stopped after 800,000 presentations and the learning file consisted of 1,000 patterns. The networks were initialised with small random values between 0.1 and -0.1.

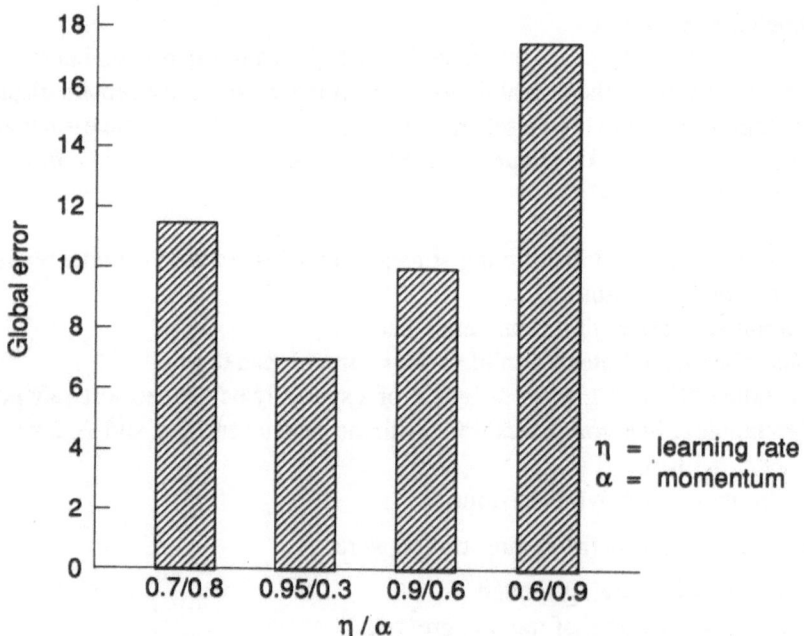

Fig 5.15 Some results experimenting with NeVIS III

Because of the large amount of training data acquired, there was a high chance of contradictory input/output representations being included. We therefore stopped training after a sufficiently high number of presentations. The remaining network error at that time is the quantity plotted for a number of different network parameters.

The speed and quality of convergence depend on different factors. Good convergence rates can be reached if the training patterns contain no contradictory input/output pairs. Ensuring this is the case is not easy because the training data generation has to be done by driving in a predefined moving ground a few times in a a supposedly similar way. It can happen that at the same location a turn of the vehicle is done once to the left and at another time to the right. Changing directions without any obstacle in the sensor cones has the same effect.

These restrictions could be eliminated by adding strategic inputs for training like actual and goal positions of the vehicle or externally defined direction vectors for moving, etc.

Another kind of data generation gives good results as well. This is to train on single, standard obstacles. Twenty different obstacles were created and appended together with the training dataset (see examples in Figure 5.16). The advantage of such an approach lies in the explicitly trainable near-range collision avoidance. It is possible to train the collision avoidance of obstacles by driving towards them from different sides and turning with different angles.

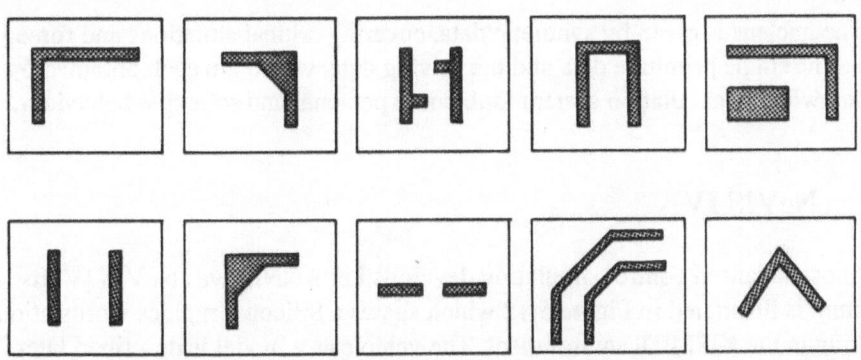

Fig 5.16 Examples of single primitives used for training

Convergence tests have been performed with training sets created by the procedure described above. The training was stopped after 500,000 presentations and the training dataset consisted of 350 patterns.

Summarising the results, the following conclusions and advantages were demonstrated:

- nucleus area (Kernfeld) with over 1000 accurately learned representations
- a wide associative space of identical behaviour (Fangfeld) for less accurate but similar situations

- fast convergence by use of suitable learning parameters
- short response time in the recall mode (Kannphase).

A third way of data generation has been implemented in software. This, so-called synthetic data generation, has been performed to obtain learning data without any contradictory I/O-representations. Another advantage of this approach is that it is possible to create as many data as desired. In Figure 5.17 a pictorial representation of the different training datasets is given.

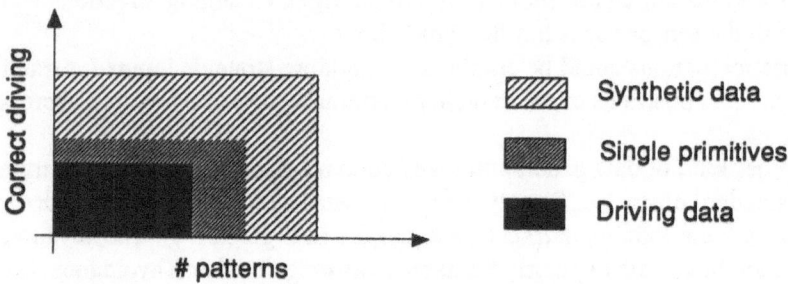

Fig 5.17 Survey of training datasets

The nucleus is given by synthetic data, covering critical situations and forced turns. The single primitive data and the driving data, which are each obtained by driving with the simulation system tool, cover personal and reflexive behaviour.

5.5   NeVIS IV

The most advanced control simulation developed in ANNIE was NeVIS IV. Its structure is illustrated in Figure 5.18 which shows a Silicon Graphics workstation working in the KISMET environment. The vehicle as a model is described later. The control of the vehicle movements is achieved by a back-propagation network running on either the MicroVAX II, the PC or as a second process on the Silicon Graphics.

Control is achieved by two moving values. The first value defines the direction to move. Values higher than 0.5 correspond to turning right, while values lower than 0.5 correspond to turning left. The other value represents the change of speed. Values higher than 0.5 correspond to acceleration and values lower than 0.5 correspond to braking.

These two values are represented by two output nodes of the neural network. The 13 input nodes of the network consist of 12 nodes for actual, normalised sensoric values and 1 node for the actual speed of the vehicle. Therefore the network works with analogue values instead of the binary input/output samples used in NeVIS I - NeVIS III. The number of hidden layers and units could be varied.

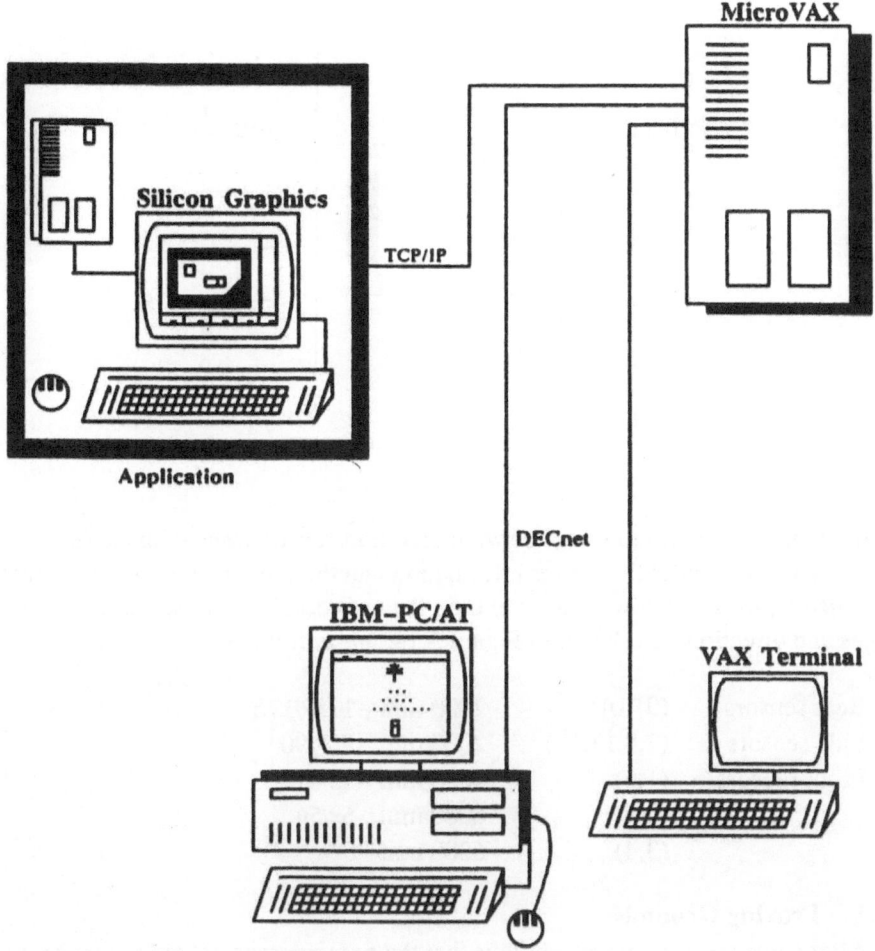

Fig 5.18 System configuration (NeVIS IV)

The number of sensors and their location on the vehicle are easy to change. The working area (the maximum ranging distance of the sensors), and the sensor detection angle can be changed by parameters. The vehicle is shown in Figure 5.19. The proving grounds are generated by an integrated CAD graphic package.

5.5.1   Teaching

For generation of learning data, the vehicle has to be moved by the user, ie with pointing by mouse clicks, inside an existing proving ground. Single points of this trace get saved in so called KISMET - *Teachfiles* in IRDATA-Code. These describe the driving route of the vehicle in the particular proving ground.

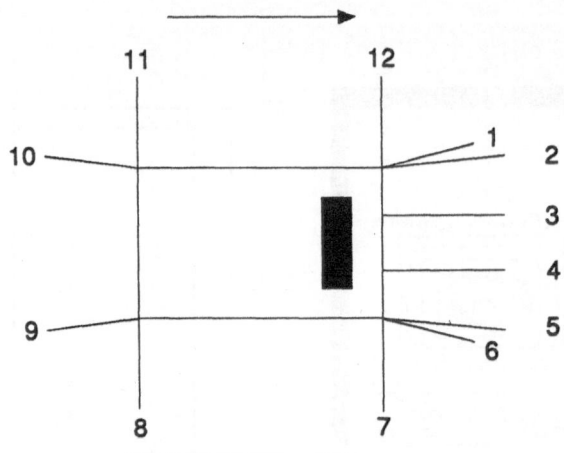

Fig 5.19 Vehicle with sensors

In the next step these *Teachfiles* were driven in *repeat mode* with the sensors switched on. In parallel the sensor information and the movements were then written into *learning files* which were used to train the network. Typical sensor configuration, ranges and directions, as shown in Figure 5.19, are as follows:

Rear sensors	(9,10)	2000 mm	$-175°/175°$
Side sensors	(7,8,11,12)	2000 mm	$-90°/90°$
Front sensors	(1,6)	4000 mm	$-15°/15°$
	(2,5)	6000 mm	$-5°/5o$
	(3,4)	6000 mm	$0°$

5.5.2 Proving Grounds

After some test runs it was obvious that the network was not able to generalise sufficiently by training over a single proving ground. Therefore it was necessary to train a reduced set of single primitives with different starting positions (as for NeVIS III). By appending and mixing all learning files an overall learning file for the network was built. Figure 5.20 shows the training primitives used and some starting positions. Each of the obstacle avoidance vectors shown are *Teachfiles* resulting in different learning files.

5.5.3 Training of the Neural Network

The network consists of 2 output nodes and 13 input nodes (12 input nodes which represent the actual sensor information; the 13th representing the actual speed). All input values are normalised to a value between 0 and 1. The scheme in Figure 5.21 shows the learning procedure and parameters. During learning, convergence files and an error file were created. To increase the learning speed, the graphics were minimised to ASCII output of necessary parameters.

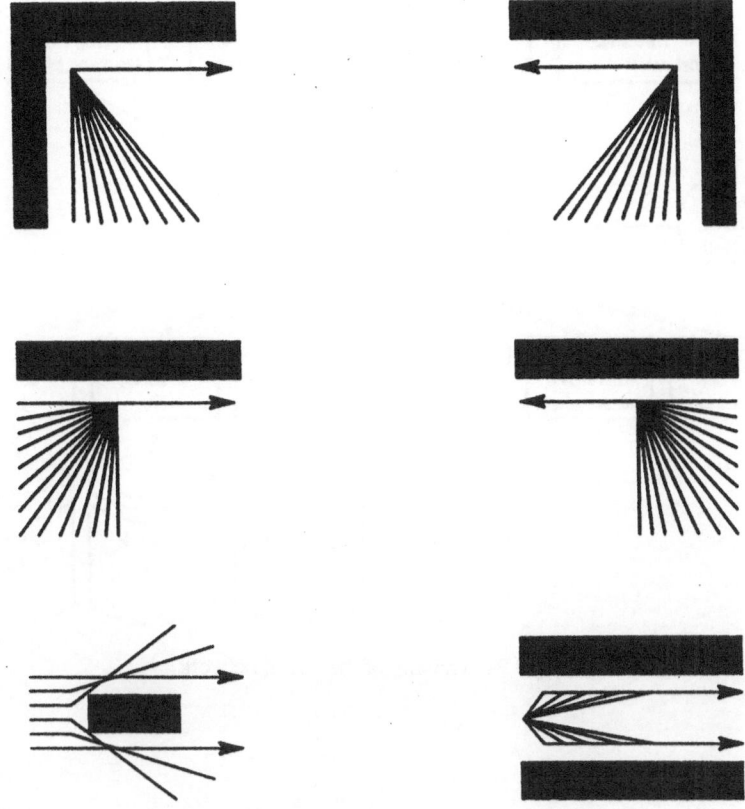

Fig 5.20 Single proving grounds

5.5.4   Results

The training of the network has been done with different numbers of hidden units, layers and parameters. However, after some tests it became obvious that it would be impossible to examine all kinds of configurations. Therefore we concentrated on a so-called standard configuration with a learning rate of 0.9, a momentum rate of 0.6 and a variable number of nodes in the first hidden layer. No hidden nodes for the second and third layer were used. In the beginning we also tried using nodes in all layers, but the performance of the vehicle in recall worsened and the learning time increased rapidly.

Table 5.2 shows an extract of our test configurations and the following figures (Figures 5.22-5.25) show the global error curves and a split into acceleration errors and direction errors.

Figure 5.22 shows some global error curves of the networks described in the table. The networks N2, N4 and N6 had the best error values after 900 training cycles. Therefore only these were trained further.

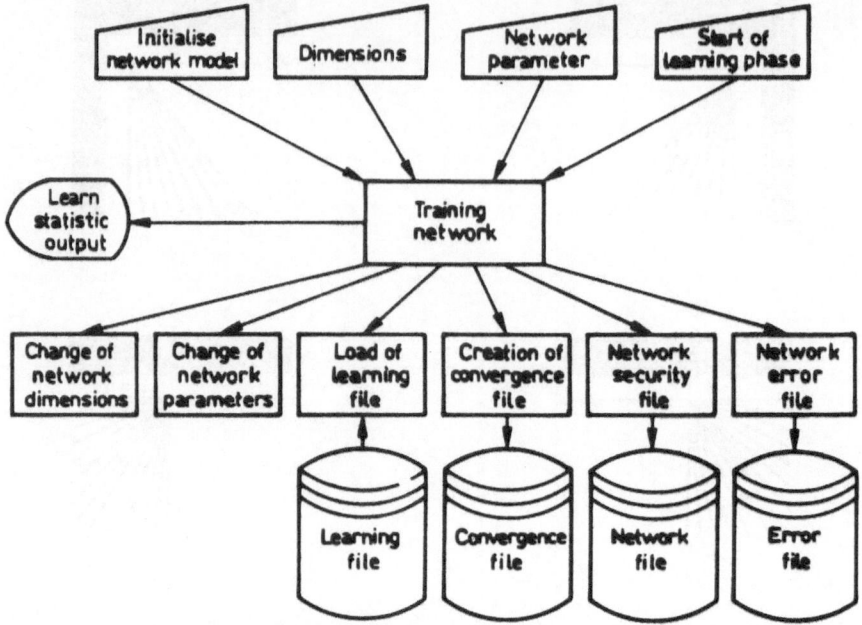

Fig 5.21 Training of the neural network

Table 5.2 Extract from test configurations

Input nodes N1	N2	N3	N4	N5	N6	N7	
1 Hidden layer	19	21	23	25	27	24	26
2 Hidden layer	0	0	0	0	0	0	0
3 Hidden layer	0	0	0	0	0	0	0
Output nodes	2	2	2	2	2	2	2
Learning rate	0.9	0.9	0.9	0.9	0.9	0.9	0.9
Momentum 0.6	0.6	0.6	0.6	0.6	0.6	0.6	

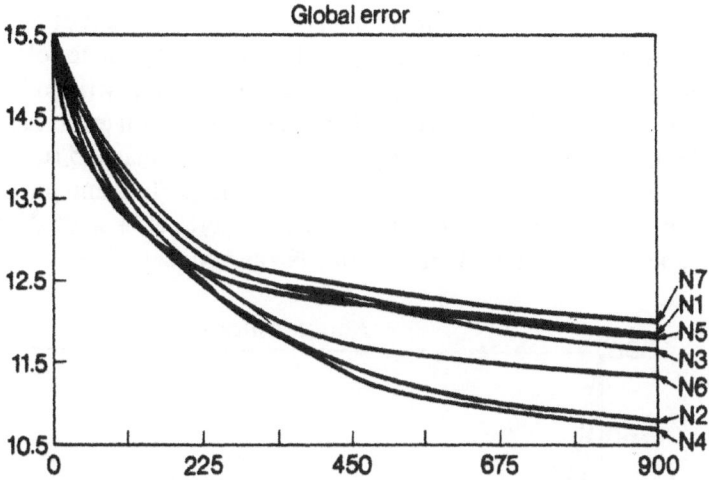

Fig 5.22 Some global error curves of networks as a function of the number of training cycles

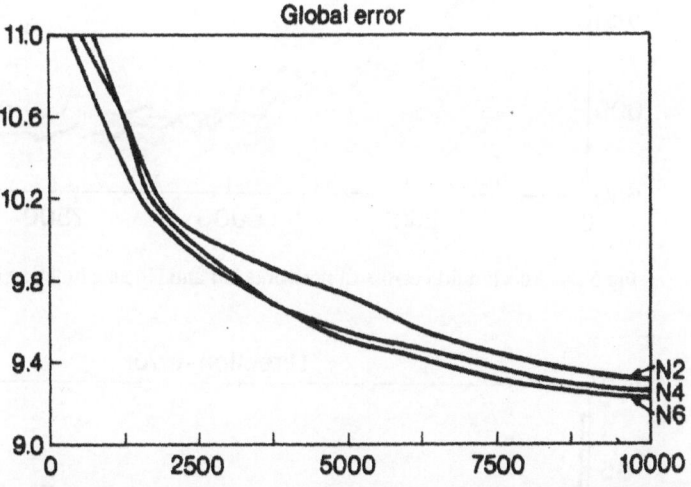

Fig 5.23 Error curves of networks N2, N4 and N6 as a function of the number of training cycles

We also trained some networks with more hidden layers and other parameter sets. The error curves of these networks have been much worse than those with only one hidden layer. Therefore we stopped the investigation of networks with more than one hidden layer.

The network N6 shows better convergence after 10,000 training cycles although N2 and N4 were better after 900 cycles. The convergence speed decreased intensively after 2500 training cycles. In the next step we investigated which part of the output,

the *acceleration* or the *direction* node, caused the main part of the error. Therefore we
implemented a new software version, which had an option to select an *error file*.
During learning, for every learning vector a remark was written into this error file as to
whether the network made a direction or an acceleration error. An error for
acceleration was noted if the output node was larger than 0.5 (the network breaks)
instead of smaller (the network should accelerate). The same principle was
implemented for the *direction* output node. Figures 5.24 and 5.25 show the
acceleration and direction errors of the N4 and N6 networks.

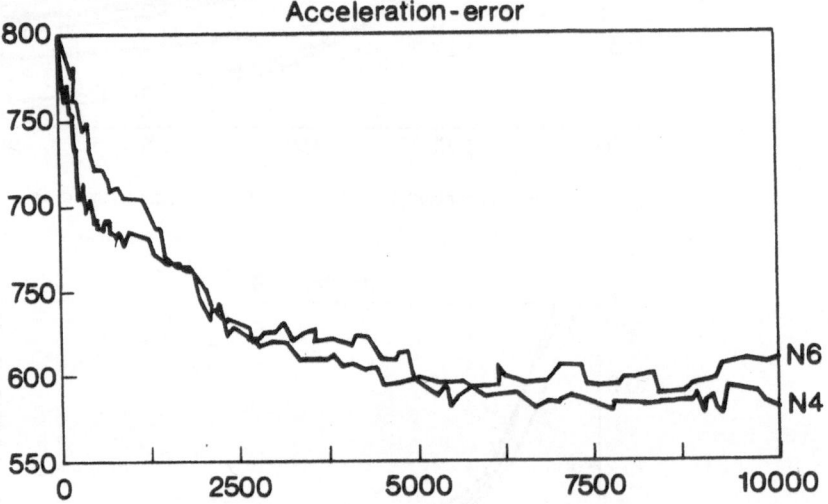

Fig 5.24 Acceleration errors of networks N4 and N6 as a function of cycle number

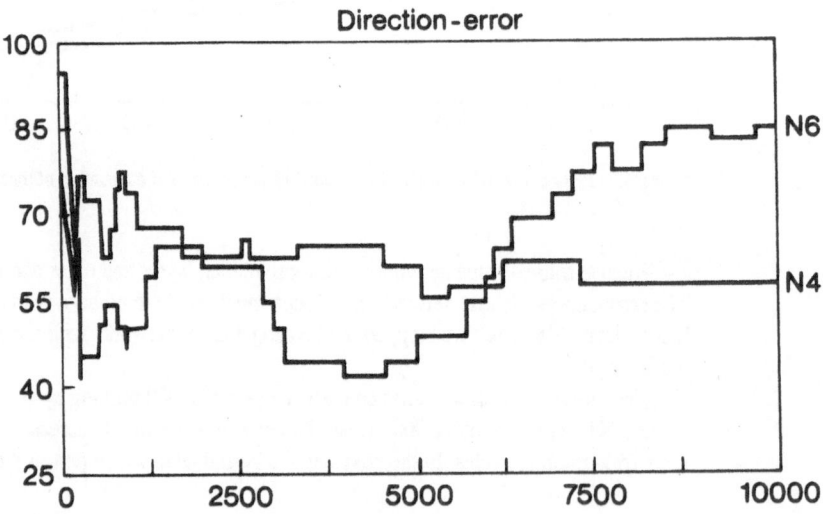

Fig 5.25 Direction errors of networks N4 and N6 as a function of cycle number

The acceleration error shows a different behaviour. Its function is not monotonic like the global error function. It is obvious that longer learning time does not lead to better values. The direction error in Figure 5.25 shows that there are fewer errors caused by directions than by accelerations, and that they appear more stable over a period of learning.

A detailed analysis of the direction-error curves shows the effect of overtraining of networks. During the first 500 learning file cycles there are peaks in both directions. It can be seen that direction learning is achieved first and then the acceleration. The network then loses some of its direction knowledge. It is important not to stop at the point when a first local minimum is reached as the driving behaviour in recall is not good at this moment. Here the influence of the correct acceleration becomes more obvious and the best driving behaviour can be seen after 4040 learning cycles. Further learning does not improve the direction behaviour. Because of some problems with the KISMET modelling environment we had a number of difficulties in obtaining good acceleration training examples. This would be the main target if improvements of the system were to be made. Unfortunately, there is no test like a benchmark which says definitely how good a network really is. We only have different proving grounds and starting positions to choose. The best measure we can use is therefore just the time of driving without crashing into any obstacle.

5.6   Methodology

The flow chart in Figure 5.26 summarises the methodology experience gained in developing software simulations for control of automobile robot systems. The flow chart, which gives an overview of principles and steps in developing neural network applications, should be useful for non-experts to recognise possible applications and to solve special tasks. No detailed explanations of each step can be given but some specific features to each point are mentioned.

The flow chart shows the importance of problem analysis and conceptual work necessary for application of neural networks to real problems. Many recent publication treat this means of problem-solving too enthusiastically, promising that there is no programming work necessary for users of neural computers. Although this statement is true, potential users must be aware of intensive efforts in developing concepts for preprocessing, isomorphic representations and in composition of a sufficient number of *representative* training data.

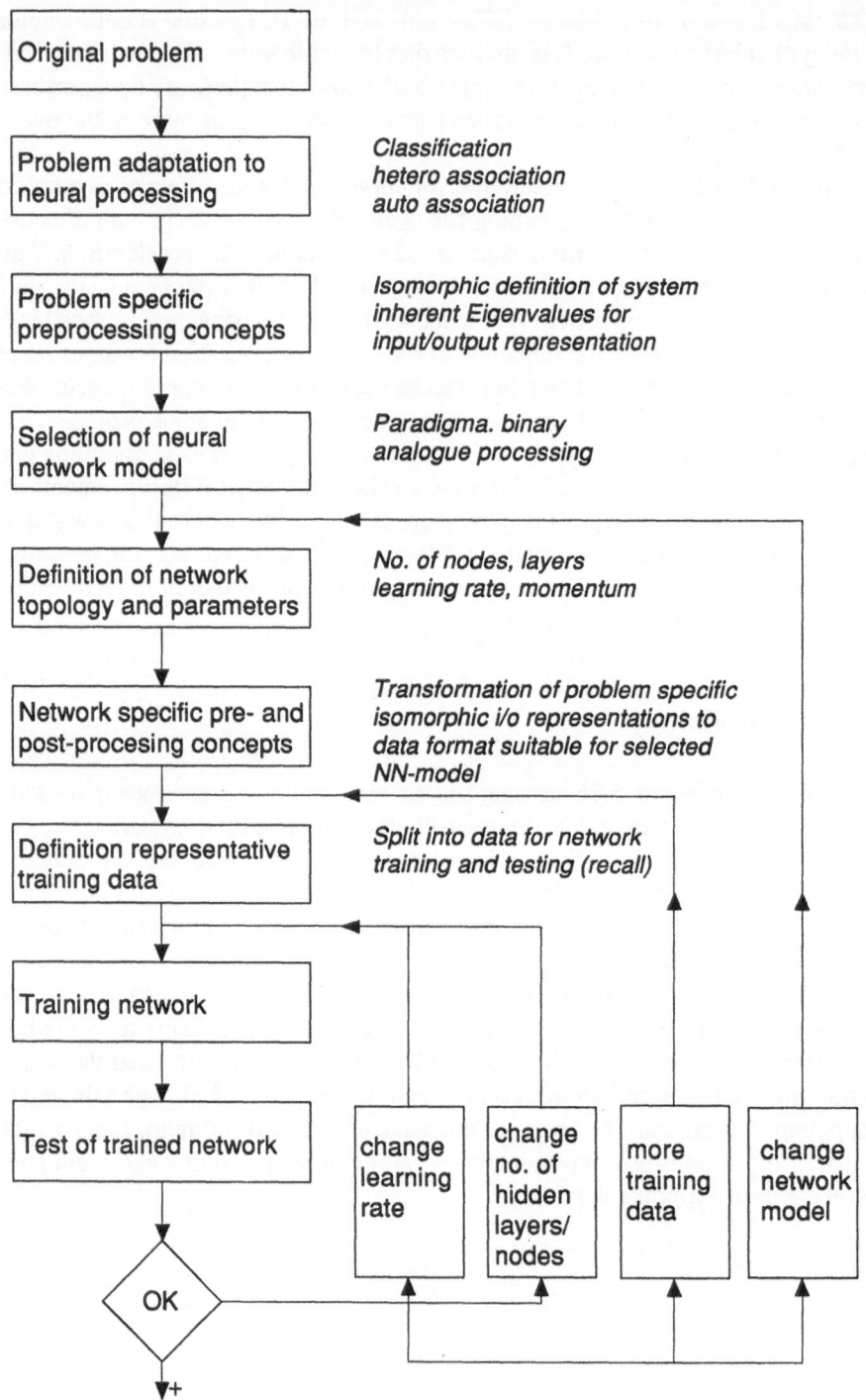

Fig 5.26 Neural network application guideline

5.7   Identification of a Moving Robot

Besides autonomous vehicle control, the other main control activity in ANNIE has been the investigation of the use of neural networks to identify a localised moving object, in particular a mobile robot in a dynamic environment, by combining information from both on-board and distributed off-board sensors. In this case the on-board sensors measure distance travelled (odometry) and the off-board sensors comprise a multiple-camera tracking system.

The relative position of a mobile robot may be obtained from its odometry. In the case of a wheeled mobile robot this information may be obtained from optical encoder discs mounted on the main drive wheels. Although the odometry provided in this way can have a very high resolution, the measurement of the position of the robot using this data is prone to systematic error. Therefore, to determine its absolute (x,y) location via its odometry is not always reliable. These errors are primarily due to wheel slippage, motor dynamics, and turning inaccuracies which although individually are small can accumulate over relatively moderate distances and become significant. This localisation problem can be overcome by obtaining an absolute measurement of the position of the robot by using an off-board sensing system to monitor the environment in which the robot operates. However, off-board sensors can also be subject to inaccuracies and incomplete knowledge and it can be difficult for an off-board sensor to identify events sensed in its environment. The aim of this project is to combine both off-board and on-board sensor information in order to overcome the limitations of each separate system. A neural network solution is sought since although statistics can be gathered about the errors generated by each system, no *a priori* error model of either system exists.

For the purposes of this project the off-board sensing system is supplied by another ESPRIT project SKIDS (Signal and Knowledge Integration with Decisional Control for Multisensory Systems). The SKIDS machine is capable of tracking multiple moving objects (events), and giving the absolute position of each event in the environment, including the mobile robot used for this project, the Robuter produced by Robosoft. For the sensing system to be able to localise the robot, the system must be able to identify which of the events tracked is that of the Robuter. This section describes the construction and analysis of a neural network system designed to allow interaction between the SKIDS machine and the Robuter to take place, so that the SKIDS track of the Robuter can be identified and its position determined.

5.7.1   Environment Description
The operating environment (see schematic diagram Figure 5.27) and equipment used for the tracking and localisation consisted of:

- 10m x 10m open space in which moving events are tracked
- SKIDS tracking system
- Robuter + RS232 radio frequency line + transputer array Sun4 workstation.

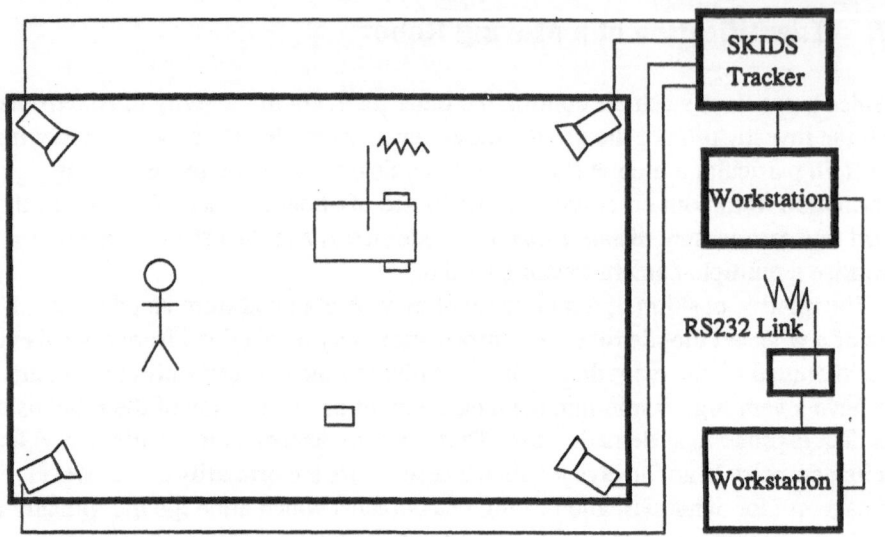

Fig 5.27 Schematic view of environment, showing the room with four cameras, the Robuter, which is
linked by radio telemetry to the workstation, and other objects in the field of vision

Moving objects are referred to as events. Track data is generated from these events
and preprocessed before presentation to the neural networks. Typically an events could
be either a human, a radio-controlled model car or the Robuter.

5.7.2 Nature of Data

As with any sensor systems, the SKIDS system and the Robuter odometry have
intrinsic differences in the resolution and accuracy of the measurements they generate
(see summary in Table 5.3). They also have dissimilar coordinate frames.

**Table 5.3 Resolution and accuracy of the Robuter odometry
and the SKIDS tracking system**

System	Resolution	Accuracy	Coordinate frame
Robuter	2.5×10^{-4}m	1%	Relative
SKIDS	0.1-0.2m	0.1-0.2m	Absolute

Coordinate frame

Although both systems sample from the same environment, the coordinate frames are
not coincident. The SKIDS cameras are statically fixed to the ceiling in the four
corners of the room, pointing down towards the centre of the room. The Robuter

coordinate frame is centred at the robot's drive wheels. When the robot is initialised it is assumed to be at an (x,y,θ) position of $(0,0,0)$ unless instructed otherwise. The coordinate system may be reset or arbitrarily adjusted at any point in time by appropriate commands to the on-board operating system. The Robuter may therefore be considered to have a relative coordinate frame and to be arbitrarily displaced from the SKIDS fixed frame of reference.

Resolution

Resolution is the smallest change measurable by the measuring system. The SKIDS CCD cameras have a 128 x 64 pixel resolution and 55º field of view. This gives a resolution of about 0.1 m. The odometry on the robuter is able to resolve movements of 2.5×10^{-4}m.

Accuracy

Accuracy refers to the actual measurement performance obtained in practice. The resolution of a system represents an upper bound on the accuracy of instantaneous estimates of position; ie the practical accuracy is limited to the resolution of the system. The Robuter has a relative measurement system and was estimated to have a 1% accuracy.

The SKIDS system has four statically fixed cameras, so in principle the accuracy is dictated only by the pixel resolution of the measuring system, and is estimated as 10-20 cm.

In summary, the local coordinate system of the Robuter has good resolution but is subject to global drift and thus its accuracy degrades. The SKIDS system has a far coarser resolution but is fixed and thus has an accuracy equal to its resolution.

5.7.3 Preprocessing of Data

Direct comparison of the raw data obtained from the SKIDS system and from the mobile robot odometry as described in the previous section is unlikely to yield good results for three reasons:

(i)   the two coordinate frames of the two systems are arbitrarily displaced from one another both in translation and rotation. Thus any attempt to directly compare coordinates, without preprocessing, from the two coordinate systems is difficult

(ii)   no attempt has been made to extract structure from the SKIDS data which may yield discriminant information (ie try to classify on the basis of behavioural characteristics)

(iii)   although the events for both sensor systems are time-stamped, they are not synchronised; thus, there is no explicit correlation in time between the data.

The role of preprocessing was thus to generate measures or features which are coordinate frame independent and are likely to yield structural information which would discriminate between a Robuter event and a non-Robuter event.

To allow the two sets of data to be handled in a uniform manner, the raw (x,y,θ,t) data obtained from the SKIDS tracker and robot odometry were initially stored in a common 4-column format file. The θ component of the track was given explicitly by the operating system on board the Robuter. In the case of the SKIDS system no explicit measurement of θ is available, so estimates of θ were obtained from the spatial (x,y) derivatives.

The preprocessing of this raw data takes the form of transforming the cartesian coordinates from the SKIDS and Robuter odometry data to polar coordinates and then taking the first and second temporal derivatives of both sets of coordinate data. Third derivative estimates were also considered as a feature, since the constant Robuter acceleration would yield a zero value from the odometry and humans would tend to have non-zero values due to the discontinuities in walking. However, the timing resolution of the SKIDS system effectively precludes this since at a sample rate of 3Hz the resolvability of such a spatial event is insufficient. The resulting data is stored in a 13 column format composed of 12 features and the original time stamp (synchronisation having also been performed off-line). The data required by the neural net is obtained by stripping data from the appropriate columns.

5.7.4   Network Design

The neural networks used for the project experiments were multilayered perceptrons using the generalised delta-rule for backpropagating errors to update weights. Two network structures have been implemented. An homogeneous, multiple input architecture and a single input network which is more appropriate for implementation in an array.

The first structure implemented (see Figure 5.28) was an homogeneous network with data from all SKIDS events being presented at the input along with the odometry data from the Robuter. The input structure of the network was initially formatted with an 8 unit input layer. This was constructed from a pair of odometry velocities acting as a reference signal and three pairs of units representing velocities of three SKIDS tracks. The output layer had 3 units formatted as a 1 from n coding to indicate which SKIDS track the network most closely associated with the Robuter odometry.

This network structure essentially performs a *relative* decision task: Which SKIDS data corresponds *best* with the odometry data?

The second network structure implemented is configured so that the input layer receives a reference measurement from the odometry and a *single* SKIDS track.

The presence of a Robuter event at the input is labelled by a 1 at the single output node and a non-Robuter event by a 0. In this case the network attempts to perform an *absolute* task, which is to answer the question: Does the SKIDS data correspond with the odometry data?

To obtain such information from a number of events, several can be combined to give the architecture shown in Figure 5.29. The structures described have different qualities which lead us to believe that different types of behaviour may emerge. It should be noted that there is no direct equivalence between the two structures in terms of the coding of the output.

Robuter		SKIDS tracking data					
Odometry		Skids 1		Skids 2		Skids 3	
F	G	F1	G1	F2	G2	F3	G3

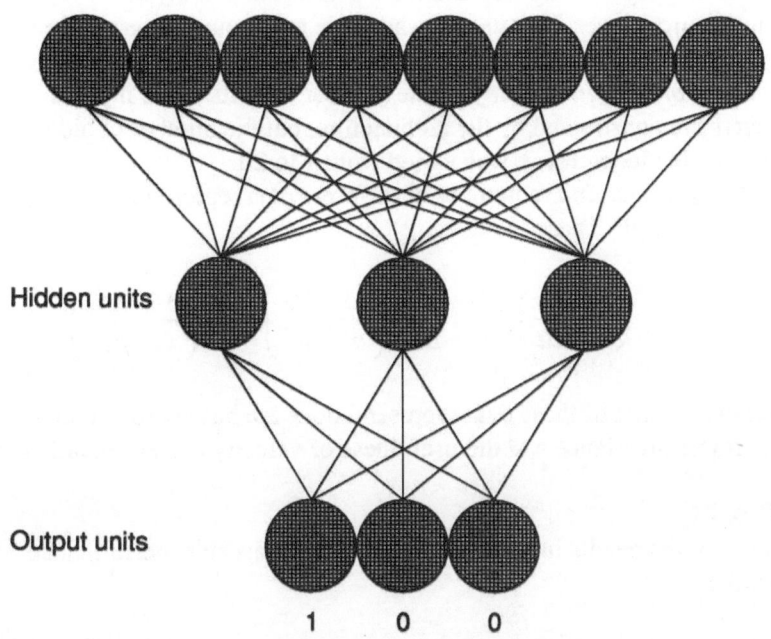

Hidden units

Output units

1   0   0

Fig 5.28 Multiple track input

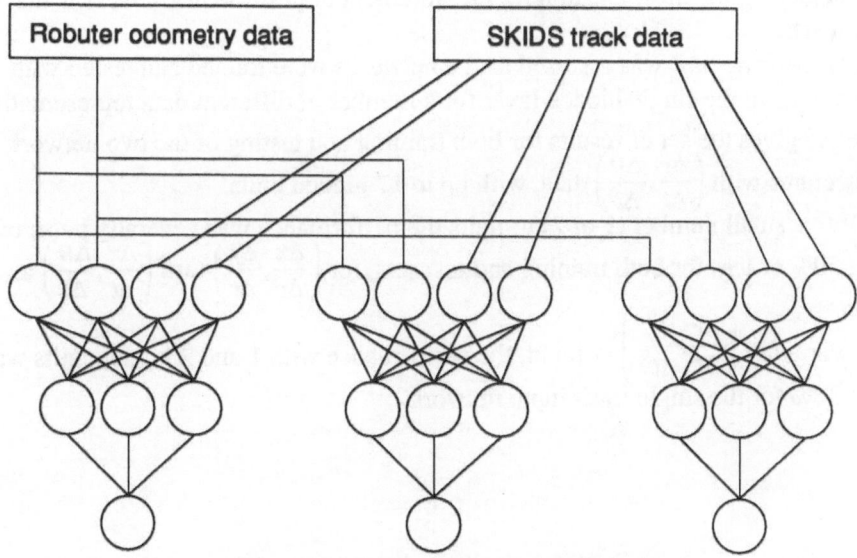

Fig 5.29 Array of single track input networks

5.7.5   Experimental Design

An experimental study was carried out to compare the different neural network architectures against a number of different fitness criteria. Factors and parameters which were considered included:

- *number of layers:* this was maintained constant at 3 layers - 1 input, 1 hidden and 1 output layer. No attempt was made to try more layers since the results obtained with 3 layers compared well with an optimal performance analysis
- *number of units in each layer:* the number of units in the input and output layers are constrained by the architecture, but the number of hidden units, in the single hidden layer, was varied from 1 to 20
- *data representation:* a number of different data representations were presented to the network. These included

$$\left(\frac{\Delta x}{\Delta t}, \frac{\Delta y}{\Delta t}\right), \left(\frac{\Delta r}{\Delta t}, \frac{\Delta \theta}{\Delta t}\right), \left(\frac{\Delta r}{\Delta t}, \frac{\Delta_r^2}{\Delta t^2}\right), \text{ and } \left(\frac{\Delta \theta}{\Delta t}, \frac{\Delta^2 \theta}{\Delta t^2}\right)$$

The relative merits of these data representations are based around the issues of coordinate frame invariance and the usefulness of velocity and acceleration measures.

5.7.6   Results

We will present the results in terms of three different aspects: performance, stability, and scalability.

Performance

Here we define the performance as the percentage classification accuracy on a set of data. In general the most meaningful measurement of performance fitness is that on the test sets.

This performance was assessed as the networks were trained and tested with 1-20 hidden units in the single hidden layer for a number of different data representations. Table 5.4 gives the set of results for both training and testing of the two network architectures with $\left(\frac{\Delta r}{\Delta t}, \frac{\Delta \theta}{\Delta t}\right)$ data, with up to 17 hidden units.

With a small number (1 or 2) of units the performance was generally poor, often being 50% or less for both training and test sets, for $\left(\frac{\Delta x}{\Delta r}, \frac{\Delta y}{\Delta t}\right)$ and $\left(\frac{\Delta r}{\Delta t}, \frac{\Delta \theta}{\Delta t}\right)$ as inputs.

However, with $\left(\frac{\Delta r}{\Delta t}, \frac{\Delta_r^2}{\Delta t^2}\right)$ as input, the performance with 1 and 2 hidden units was about 90% for the single track input network.

Table 5.4 Mean training and test performance (%) for single- and multitrack

$$\text{input networks using } \left(\frac{\Delta r}{\Delta t}, \frac{\Delta \theta}{\Delta t}\right) \text{data}$$

Number of Hidden Units	Single		Multi	
	Train	Test	Train	Test
1	37.9	34.6	28.5	25.8
2	59.0	57.0	55.8	56.6
3	96.6	91.9	95.1	94.2
4	96.4	90.1	91.3	90.1
5	90.9	86.0	88.5	83.9
6	86.4	82.4	94.5	91.6
7	95.6	91.6	94.6	91.9
8	96.7	88.8	95.2	94.5
9	96.9	89.4	94.0	93.5
9	96.9	89.4	94.0	93.5
10	95.6	88.7	93.3	94.1
11	96.8	91.4	95.1	94.4
12	98.1	91.7	97.6	92.4
13	87.5	78.7	98.2	89.1
14	98.9	88.1	98.7	87.9
15	98.4	91.1	98.6	87.2
16	98.9	87.3	98.7	84.2
17	99.4	88.5	98.9	83.9

With this data representation, therefore, there appears to be sufficient structure to form a simple discriminant with a single unit. This observation appears consistent with the hypothesis that coordinate frame independent measures provide a simpler learning task for the networks. This raises the question of whether a similar performance can be obtained with a single layer perceptron and thus whether having discovered this particular combination of features as a discriminant, a neural network can offer any significant benefits over conventional techniques using the same combination of features.

For most data representations the peak performance was observed over the range 3-10 hidden units for testing for both the single- and multi-input network.

For the training sets of data the performance over the range 10-20 hidden units improved typically to 90% for both multiple and single input networks. However, the test results from some runs over this range showed some degradation in performance, indicating an overfitting of the training data for the test data.

The best performances obtained from the multi-input networks over a range of hidden units were higher than the corresponding single-input network by 2-4% for a range of the data representations. The higher performance observed in the multi-input networks can be attributed primarily to the extra information at the input: with an input containing two track points which individually could be classified as a Robuter event,

the multi-input network effectively has the chance of taking the best estimate between
the two; conversely each individual single-input network must make the decision in
the absence of the other track information, and is likely to classify both as a Robuter
event.

Stability

The stability is taken to be the repeatability in terms of performance for a network
trained from a number of different sets of initial weight conditions. In the case of the
networks used this corresponds to different initial weights connecting the layers of
neurons.

   The stability of the single-input networks was better both in terms of the variation
in performance with respect to the number of hidden units and the average standard
error of the performance at a given number of hidden units.

   The $\left(\dfrac{\Delta x}{\Delta t}, \dfrac{\Delta y}{\Delta t} \right)$ was particularly unstable with respect to the performance from
different starting weights for all architectures.

Scalability

The scalability of a network architecture is considered to be the manner in which the
computational cost of processing, for a trained network, increases with the number of
events occurring in the environment. The time taken to train a network was not a
particular concern since this is off-line and is not time critical.

   It can easily be shown that the computational effort is smaller for a set of n single
input networks than for the multi-input method. The single input structure scales
linearly $(O(n))$ whereas the multiple input structure scales as $O(n^2)$.

   Thus for sequential computation the single networks are faster. With parallel
processing the time taken for the processing can remain independent of the number of
events for the single-track input networks, since the event data can be distributed on a
one event per network basis. In contrast the multiple input network is inherently
constrained to operate on all the tracks at once. The greater flexibility of the smaller
single-track input networks may lead to a more effective parallelisation of the network.

5.7.7   Discussion and Conclusions

The above results allow us to compare the relative merits of the two neural network
approaches which aim to solve through sensor fusion the problem of identifying
which, out of several moving objects, is a robot.

   It is important to note that we do not consider the absolute *performance* of a
network as the sole determinant for the choice of a network in a practical system, just
as one does not choose a motor for an engineering application purely on the basis of its
peak torque performance. We have considered also the stability and the scalability.
(There are additional aspects such as robustness which could also be considered).
Below we consider the effects of architecture and data representation as factors of the
choice to be made.

Architecture

The multiple-track input network has a slight advantage in mean performance over the single-track input networks which could be attributable to the extra information available at the input. However, the stability of the single-track input networks was found to be much better than that of the multiple-track input networks because, presumably, of the smaller number of interconnections in the system compared to an equivalent multiple-track input network.

The computation cost of the homogeneous network increases as the square of the number of events being tracked, whereas that of the single networks scaled linearly. Thus the single input networks are superior in terms of sequential and possibly parallel implementation.

Overall, therefore, the performance of the single networks was slightly poorer in terms of peak performance than the multi-input networks, but the simpler structure tended to give more stable results, presumably because generalisation is easier over the few number of variables present at the input.

Data representation

No data representation was found that performed well (> 70%) with less than 3 hidden units, with the exception of speed and acceleration which performed as well on 1 and 2 hidden units as with a greater number of hidden units. This result appears to indicate that speed and acceleration provide for a more simple characterisation of the events than the other combinations studied, and were found to be the most effective combination of measures tried in terms of both performance and stability. The (x,y) data appeared to be particularly prone to instability. It is considered likely that this stems from the coordinate frame variance of this data.

In general, it appears reasonable to conclude that careful consideration of both data representation and architecture can yield improvements not only in performance but also considerations such as stability, scalability and computational cost.

Chapter 6

Optimisation

6.1 Introduction

The published work on the application of neural networks to optimisation problems is dominated by the *travelling salesman problem*. Since Hopfield's paper (1985) there have been many attempts to produce solutions to this problem using neural and conventional methods.

Neural networks (as they are applied in optimisation) should be regarded as inherent approximation algorithms. As such, they should be compared with conventional heuristics. For heuristics used to tackle *hard* problems, a similar statement holds: in order to deliver reasonably *good* solutions they rely heavily on a knowledge of the problem's structure. That is why alternative heuristics for a problem are usually very diverse in nature. It is even more true for different problems. A lot of effort is needed to identify common methodologies in the area of heuristics. The task of comparing neural networks and heuristics in a set of representative hard problems would be a formidable task on its own, that could very well have lasted through the ANNIE lifetime, thus prohibiting the completion of the application task. Instead of performing the comparison work on the generic level, we decided to do it on the application level. That is, to compare conventional and neural network methods for the *crew scheduling problem*.

This chapter reviews briefly some conventional methods, followed by a discussion of the specific problem and its mapping onto a Boltzmann machine. The last sections discuss the potential for improvement of the methods using hybrid network techniques.

6.2 Conventional Methods in Combinatorial Optimisation

Many problems of great practical importance are concerned with the choice of a *best* configuration or a set of parameters to achieve some goal. For example, in the industrial or economic domains the goal could be the minimisation of costs (or the maximisation of profits) under restricted availability of some resources such as production capacity, requirements of quality, constraints in person-power, liquidity of the company, etc. Over the past few decades a hierarchy of such problems has emerged, together with a corresponding collection of techniques for their solution.

At one end of this hierarchy is the general nonlinear programming problem described by the following equations

$$
\begin{aligned}
\text{minimise} \quad & f(x) \\
\text{subject to} \quad & g_i(x) \geq 0, \quad i = 1, \ldots, m \\
& h_j(x) = 0, \quad j = 1, \ldots, p
\end{aligned}
$$

where f, g_i and h_j are general functions of the parameter $x \in R^n$. The techniques for solving such problems are almost always iterative in nature, and their convergence is studied using the mathematics of real analysis.

Nonlinear constraints are often found in technical processes, for example in the form of material balance equations, when the end product quantity depends on the quantities of the input products in a nonlinear manner.

When f and all the g_i and h_j are linear, we arrive at the *linear programming problem*. A striking change occurs when we restrict attention to this class of problems; the problem reduces to the selection of a solution from among a *finite* set of possible solutions.

The widely used *simplex algorithm* finds an optimal solution to a linear programming problem in a finite number of steps. Forty years of refinement has led to forms of the simplex algorithm that are generally regarded as very efficient. Additionally, a variety of other methods (some of them very recent) exist for solving the linear programming problem.

Usually, constraints depending on economic facts, such as restricted availability of raw materials, are linear. Problems from a wide range of application areas are modelled as linear programming systems. Typical examples include production planning, investment problems, installation of financial plans, mixture problems, transport problems, etc.

When we consider the linear programming problem and ask for the *best* solution, with the restriction that it has *integer-valued* coordinates, we arrive at the *integer linear programming problem*.

For integer programs, as for linear programs, a finite algorithm exists. But the resemblance stops there. The integer linear programming problem is in general very hard to solve (with the exception of some cases).

Problems from a variety of application areas may be modelled as integer linear programming problems. Typical examples include production scheduling, transport problems, and other problems with mixed cargo (eg cars) in which rational solutions are meaningless.

The complete state of affairs described so far is shown in Figure 6.1.

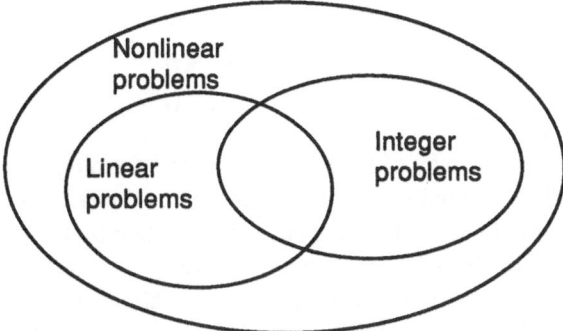

Fig 6.1 The landscape of optimisation problems

Optimisation problems seem to divide naturally into two categories; those with *continuous* variables, and those with *discrete* variables which we call combinatorial. The boundary between the two categories is the linear programming problem.

Linear programming is in one sense a continuous optimisation problem but, as we mentioned above, it can be considered as combinatorial in nature. We shall therefore give a definition of an optimisation problem general enough to include both types (and almost any other optimisation problem).

An optimisation problem is a pair (F,c), related by the mapping

$$c:F \to R$$

where F is any set representing the domain of feasible points and c is the cost function. The problem is to find an $f \in F$ for which

$$c(f) \le c(y) \forall y \in F \tag{6.1}$$

In the following sections we will focus on the linear and integer programming problems and the corresponding solution methods. We will not be concerned with the general nonlinear programming problem. In sections 6.3 and 6.4 we will examine linear programming and integer linear programming problems. Section 6.5 is devoted to the heuristic methods for solving integer linear programming problems. The emphasis there will be in the local search method, which will serve as a bridge for the neural network methods for solving optimisation problems, described later.

6.3   Linear Programming

In general, the *linear programming problem* can be formulated, in matrix form, as follows:

$$\text{minimise} \quad cx$$
$$Ax \ \leq b$$
$$x \ \geq 0 \qquad\qquad\qquad (6.2)$$

There is a useful *geometric* way of viewing linear programming. The constraints
$Ax \leq b$ define an n-dimensional polytope (assuming n is the rank of A). Feasible
solutions are represented by points in the subspace defined by the polytope.

Given an instance of linear programming a *basic feasible solution* (bfs) is defined
to be a corner in the corresponding polytope. The combinatorial nature of linear
programming is revealed by the following fundamental theorem: there is an optimal
bfs in every instance of linear programming.

In fact, the preceding theorem describes one of the best known algorithms for
solving linear programming, the simplex algorithm. The simplex algorithm starts from
any bfs and traces a path through the edges of the polytope, moving from a bfs to a
neighbouring bfs at no greater cost.

In practical applications, there are several modifications of the simplex algorithm.
They have their advantages in reducing the computing time for high linear
programming problems and in getting a very clear form of the computation steps.

Other methods like the decomposition method of Dantzig and Wolf (1960) can be
used for solving problems with many constraints, but with few of the variables present
in each constraint. This situation arises for example in the production problem, when
the products are fabricated in local separate plants, and each product is produced in
only one of these plants.

The simplex algorithm although efficient *in practice*, is *not* a polynomial time
algorithm. The reason is that the number of basic feasible solutions is in general very
big, and the simplex algorithm may have to trace through most of them before it finds
the optimal one. Khachian (1979) published a proof that a certain algorithm for linear
programming (the famous ellipsoid algorithm) is polynomial, thus resolving a long-
standing open question. Later on, even more efficient algorithms for linear
programming were constructed. Perhaps one of the most efficient is Karmarkar's
algorithm (1984).

6.4   Integer Linear Programming

The integer linear programming (ILP) problem can be formulated in similar fashion,
as follows:

$$\text{minimise} \quad cx$$
$$Ax \ \leq \ b$$
$$x \qquad \text{integer} \qquad\qquad (6.3)$$

The importance of the ILP problem stems from the fact that most of the well-
known combinatorial optimisation problems can be formulated in terms of ILP.

Typical examples include *job-shop scheduling* problems, the *knapsack* problem, *transportation* problems, the *travelling salesman* problem, etc.

Many interesting combinatorial optimisation problems are characterised by the appearance of only *yes* or *no* possibilities (ie a project has to be realised, a machine has to be purchased or not). Therefore, the permissible values of the variables are 0 or 1. The corresponding integer linear programming formulation of these problems forms a subclass known as 0-1 integer linear programming problems. The importance of this class comes from the well known fact that any general integer linear programming problem can be formulated as a 0-1 integer linear program.

As an example of the expressive power of the integer linear programming problem we give below an ILP formulation of the *travelling salesman problem* (TSP).

In the TSP we are given a set of $n + 1$ cities, nodes $0,1,...,n$, and intercity distances (d_{ij}). We are asked to find the minimum round tour that visits all cities, each one only once. If we associate the variable x_{ij} with each pair of cities (i,j), and let $x_{ij} = 1$ if the salesman travels from city i to city j in one step, and 0 if not, we can begin the formulation of the TSP by writing:

$$\min z = \sum_{\substack{i,j=0 \\ i \neq j}}^{n} d_{ij} x_{ij} \tag{6.4}$$

$$\sum_{\substack{i=0 \\ i \neq j}}^{n} x_{ij} = 1 \qquad\qquad j = 0,...,n \tag{6.5}$$

$$\sum_{j=0}^{n} x_{ij} = 1 \quad i = 0,...,n \tag{6.6}$$

$$0 \leq x_{ij} \leq 1 \quad i = 0,...,n, \quad j = 0,...,n$$

$$x_{ij} \text{ integer} \quad i = 0,...,n, \quad j = 0,...,n$$

The preceding equalities express the fact that the salesman must visit each city exactly once. It must be noted, however, that the above constraints do not capture all the complexity of the TSP.

A naïve method for solving integer linear programs is the following: neglect the integer constraints, solve the corresponding linear programming problem, and round the solutions to the closest integer. This is certainly a plausible strategy, but in many cases is not applicable. Often, in fact, rounding a fractional solution can be as hard as solving the combinatorial problem from scratch.

A general method for solving integer linear programming based on the idea of rounding is the *cutting plane* algorithm introduced by Gomory (1958). Again, the integer constraints are disregarded and the simplex method is used. If the solution at that level is not integer, we start adding constraints successively that do not exclude integer feasible points. The strategy is to continue adding constraints until the solution of the linear programming problem becomes integer. A linear constraint that does not exclude any integer feasible points is called a cutting plane.

As a matter of fact, Gomory's contribution to the cutting plane algorithm is an algebraic method for generating cuts obeying the aforementioned constraints. Gomory's is not the only cut generating procedure. Several varieties of cutting plain algorithms do exist in the literature described in detail in textbooks (see for example Hu, 1972).

The main disadvantage of the cutting plain method is its computational complexity. Besides the fact that generally it is not a polynomial time algorithm, practical realisations are not tolerant of rounding mistakes. Additionally, intermediate solutions cannot be used as permissible integer approximated solutions. If the method stops prematurely, there are no permissible solutions as for the example in the simplex method.

6.5   Heuristics

In order to solve a combinatorial optimisation problem for which an exact solution would require a prohibitive amount of computation time, approximation algorithms, or heuristics, are used. The term approximation has several meanings. Here, we distinguish two.

The first is based on intelligent search in the sense that we seek to obtain as much improvement over straightforward exhaustive search as possible. Among the most widely used approaches to reducing the search effort are those based on *branch-and-bound* or *implicit enumeration* techniques. Although these search techniques are of exponential worst-case complexity, they can be applied successfully to instances of reasonable size.

The branch-and-bound method is based on the idea of intelligently enumerating all the feasible points of a combinatorial optimisation problem, intelligently in the sense that it is hopeless simply to look at all feasible solutions. In general we try to construct a proof that a solution is optimal, based on successive partitioning of the solution space. The *branch* in branch-and-bound refers to this partitioning process; the *bound* refers to lower bounds that are used to construct a proof of optimality without exhaustive search.

The second notion of approximation is that of searching for a satisfactory (and not necessarily optimal) solution. The most widely applied technique, that of *local (or neighbourhood) search*, is described below and which is based on perhaps the oldest optimisation method, trial and error.

There are two major criteria for evaluating heuristics. First is the quality of the solution delivered. In this case, the quality is measured either by average criteria (average deviation from the optimum) or by worst case criteria. Of course, in the case that no solution exists, an approximation algorithm should fail. Second, the computational complexity of an approximation algorithm is a major criterion. Again, it is measured either in the average or in the worst case.

With regard to their applicability, approximation algorithms can be divided into two categories: tailored algorithms strongly depend on the specific problem under consideration thus having limited applicability, while general algorithms apply to a wide variety of optimisation problems.

The above mentioned local search (LS) (or iterative improvement) algorithms and their generalisation, simulated annealing (SA), can be viewed as general algorithms. In the rest of this section we will focus on these two general algorithms, which are related to the Hopfield network and Boltzmann machine respectively.

The application of local search algorithms, for a given set of solutions and a cost function for an optimisation problem, is possible after the definition of a *generation mechanism*. The generation mechanism defines a neighbourhood R_i for each solution i, consisting of all solutions that can be reached from i in one transition (ie a small perturbation).

The algorithm can be formulated as follows:

(i) Starting off at a given solution, a sequence of iterations is generated, each iteration consisting of a possible transition from the current solution to a solution selected from its neighbourhood.

(ii) If this neighbouring solution has a lower cost, it replaces the current solution, otherwise another neighbour is selected and compared for its cost value.

(iii) The algorithm terminates when a solution is obtained whose cost is no worse than any of its neighbours.

For example, in the travelling salesman problem, one local search approach is the following. Beginning from a valid tour, its neighbourhood is defined as the set of all tours that are constructed by removing k edges from the tour and then substituting them with k others. This is called the k-opt strategy. Lin (1965) has empirically found that a 3-opt strategy yields results very near the optimum. Lin has also demonstrated that randomly chosen initial tours are very effective. His results have initiated the application of local search on a variety of other problems (eg graph partitioning, job-shop scheduling, etc).

Methods for local search remain problem specific. The two extremes in searching are the *first-improvement* method, where the first favourable transition is accepted, and the *steepest-descent* method, where the entire neighbourhood is searched and the solution with the lowest cost is chosen.

The main disadvantages of local search are its inability to escape local minima of the cost function, the dependence of the obtained local minimum on the initial solution, and the problem of having no upper bound for the computation time.

Two approaches may be used to avoid these disadvantages without loss of the generality of the algorithm:

• execution of the algorithm for a large number of initial solutions
• acceptance of transitions which correspond to an increase in the cost function in a limited way, violating the descending procedure of local search.

The first approach is a traditional way to solve combinatorial optimisation problems approximately. Lin's 2-opt strategy is a well-known example following this method (1965). The last approach was used by the algorithm Kirkpatrick *et al* introduced in 1983, called simulated annealing, which will be described next.

Metropolis *et al* (1953) introduced a simple algorithm that can be used to provide an efficient simulation of a collection of particles in equilibrium at a given temperature. In each step of this algorithm, a randomly chosen particle is given a small random displacement and the resulting change, ΔE, in the energy of the system is computed. If $\Delta E \leq 0$, the displacement is accepted, and the configuration with the displaced particle is used as the starting point of the next step. The case $\Delta E > 0$ is treated probabilistically: the probability that the configuration is accepted is $P(\Delta E) = exp(-\Delta E/K_B T)$. By repeating the basic step many times, one simulates the thermal motion of particles in thermal contact with a heat bath at temperature T. This choice of $P(\Delta E)$ has the consequence that the system evolves into a Boltzmann distribution.

The Metropolis algorithm can be used to generate sequences of solutions of a combinatorial optimisation problem. In that case, the solutions represent the configurations of the ensemble of particles while the cost function and a control parameter T represent the energy and temperature, respectively. The simulated annealing algorithm can now be viewed as a sequence of Metropolis algorithms evaluated at a sequence of decreasing values of the control parameter. It can thus be described as follows:

- initially the control parameter is given a high value and a sequence of solutions of the combinatorial optimisation problem is generated according to a generation mechanism, as in local search algorithms
- the control parameter is lowered in steps, with the system being allowed to approach equilibrium for each step
- the algorithm terminates for some small value of T, for which virtually no deteriorations are accepted any more. The final *frozen* solution is then taken as the solution to the problem at hand.

Simulated annealing can be considered as a generalisation of local search in that it accepts, with non-zero but gradually decreasing probability, deteriorations in the cost function. When the control parameter in simulated annealing is 0, the algorithm corresponds to a version of local search.

One intuitive way of seeing why thermal *noise* is helpful, is to consider the energy landscape in Figure 6.2. Let us suppose that a ball starts at a randomly chosen point on the landscape. If it always goes downhill (and has no inertia), it will have an even chance of ending up at A or B, because both minima have the same width and so the initial random point is equally likely to lie in either minimum. That would be the behaviour of a local search algorithm. If we shake the whole system, we are more likely to shake the ball from A to B than *vice versa* because the energy barrier is lower from the A side. If the shaking is gentle, a transition from A to B will be many times as probable as a transition from B to A, but both transitions will be very rare. On the other hand with violent shaking we make both transitions equally probable. A good compromise is to start by shaking hard and gradually shake more and more gently. At the end, the ball is most probably to be found at the bottom of the global minimum.

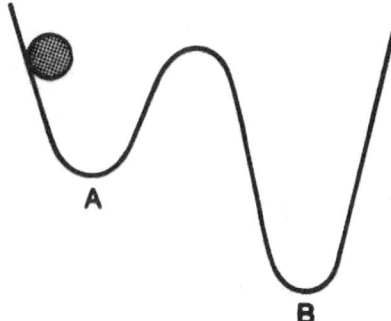

Fig 6.2 An energy landscape

Solutions obtained by simulated annealing do not depend on the initial solution and usually approximate the optimal solution. Furthermore, it is possible to give a polynomial upper bound for the computation time for some implementations of the algorithm (Aarts and van Laarhoven, 1985; Lundy and Mees, 1986).

Thus simulated annealing can be seen as an algorithm that does not exhibit the disadvantages of local search, and remains as generally applicable as local search. However, the following points must be noted:

- the gain of general applicability is sometimes undone by the computational effort, since the simulated annealing algorithm is slower than local search algorithms
- it is not clear whether simulated annealing performs better than repeated applications of local search (for a number of different initial solutions)
- the *cooling schedule* (the way temperature gradually decreases) is an important factor for the performance of the algorithm. Usually, choosing an efficient cooling schedule depends strongly on the structure of the problem under consideration.

6.6   Neural Network Methods in Combinatorial Optimisation

6.6.1   Hopfield Networks
The Hopfield network, introduced by John Hopfield (1982) consists of a set of n processing units. The architecture of Hopfield networks can be described as a strongly connected, undirected graph, with nodes interpreted as neurons and edges as connections (synapses). Connections (i,j) are characterised by symmetric connection weights w_{ij}. Additionally, neurons i carry thresholds Θ_i.

Each neuron is considered to be a binary unit, that is, it can be either in state 0 or in state 1. Thus, the composite state of the network can be described by an n-

dimensional binary vector, while the state space of the network is the set of corners of
the n-dimensional hypercube, namely $\{0,1\}^n$.

The behaviour of the network is described by the following recurrence equation:

$$x_i(t+1) = f(\sum_{j=1}^{n} w_{ij} x_j(t) + \Theta_i) \qquad (6.7)$$

where

$x_i(t)$ is the state of unit i in time t
w_{ij} is the weight of the connection (i, j)
Θ_i is the threshold of unit i
f is the binary step threshold function.

The neuron that is to be updated is randomly chosen. Hopfield showed that there is
a Liapunov energy function (named in analogy with the spin-glass model) for this
system defined in the following equation:

$$E = -\tfrac{1}{2}\sum_{i=1}^{n}\sum_{j=1}^{n} x_i x_j w_{ij} - \sum_{j=1}^{n} x_j \Theta_j \qquad (6.8)$$

where

x_i is the state of unit i
w_{ij} is the weight of the connection (i, j)
Θ_i is the threshold of unit i.

It can be shown that at each step of the computation, the energy of the network
decreases. Thus, Hopfield networks are guaranteed to converge to a stable state, where
the energy function has reached a local minimum.

The preceding paragraphs describe the so-called *discrete* Hopfield networks. The
characterisation comes from the fact that the state of neurons is a binary variable. In
contrast, in the *continuous* Hopfield networks, the state is a continuous variable in the
interval $(0,1)$. In this case the dynamics are described as follows:

$$du_i / dt = \sum_{j=1}^{n} w_{ij} x_{ij} - u_i + \Theta_j$$
$$x_i = f(u_i) \qquad (6.9)$$

where f in this case is the sigmoid function.

The preceding differential system admits the same Liapunov energy function (as in
the discrete case). The network converges to some stable equilibrium point, which is a
solution of the system:

$$u_i = \sum_{j=1}^{n} w_{ij} x_j + \Theta_i$$

$$x_i = f(u_i) \tag{6.10}$$

6.6.2 Noisy Hopfield Networks

One property of Hopfield networks is their robustness to computational noise. The effect of noise to the asymptotic dynamics of the network has been extensively studied (Mitra et al, 1986).

Consider the discrete case. The hint given by neural analogy suggests the noise will interfere mainly with the threshold function. Thus, the noisy dynamics of the network become

$$x_i(t+1) = \left(\sum_{j=1}^{n} w_{ij} x_j(t) + \Theta_i - kB(t) \right) \tag{6.11}$$

where

$x_i(t)$ is the state of unit i in time t
w_{ij} is the weight of the connection (i, j)
Θ_i is the threshold of unit i
$B(t)$ is the white noise in time t
k is the noise factor
f is the binary step threshold function.

The white noise is modelled as a sequence of normal, independent, centred, reduced random variables.

Consider the formula:

$$Erf(x) = \frac{2}{\sqrt{2\pi}} \int_0^x e^{-x^2/2} dx \text{ if } x > 0 \tag{6.12}$$

$$Erf(x) = -Erf(-x) \text{ if } x < 0 \tag{6.13}$$

Let

$$g_k(x) = \tfrac{1}{2}\left[1 + Erf\left(\tfrac{x}{k}\right)\right]$$

be the repartition function of the Gauss probability law of standard deviation k. We then derive the equation for the probability of the state of the updated neuron:

$$P(x_i(t+1) = 1) = P\left(B(t) \le \frac{\sum_{j=1}^{n} w_{ij} x_j(t) + \Theta_i}{k}\right) = \tfrac{1}{2} + Erf\left(\frac{\sum_{j=1}^{n} w_{ij} + \Theta_i}{k}\right) \tag{6.14}$$

and we get

$$P(x_i(t+1) = 1) = g_k \left(\sum_{j=1}^{n} w_{ij} x_j(t) + \Theta_i \right)$$

(6.15)

A Monte-Carlo algorithm simulating the behaviour of noisy Hopfield networks consists of the following loop:

- choose randomly with uniform probability to update the unit
- compute $g_k \left(\sum_{j=1}^{n} w_{ij} x_j(t) + \Theta_i \right)$ (notice that the computation is local)
- perform the random choice of the unit state according to the computed probability.

From the sigmoid shape of g_k shown in Figure 6.3, one can see that for high values of the noise factor the network's evolution is completely erratic (random walk along the edges of the hypercube). On the contrary, for low values of the noise factor, the noisy dynamics follows the main line of the deterministic dynamics.

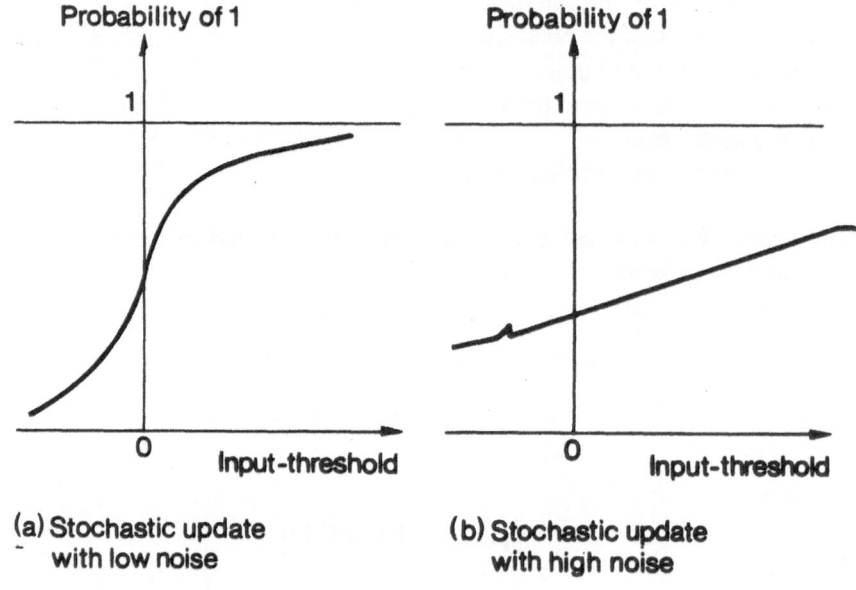

(a) Stochastic update with low noise

(b) Stochastic update with high noise

Fig 6.3 The shape of function g_k

In the general case, the random process defined by the previous stochastic dynamics on the state space of the network is an ergodic Markov chain. Thus, the evolution equations of the average quantities can be derived (Peretto *et al*, 1986). These equations, being highly non-linear, are approximated using mean-field theory which, in order to compute the average state of a unit, neglects the influence of fluctuations of other units. Thus we get:

$$< x_i > = < g_k(\sum_{j=1}^{n} w_{ij}x_j + \Theta_i) \tag{6.16}$$

The average is taken for the stationary distribution of the Markov chain.

One can compare this equation with the equations describing the stable equilibrium point of analogue Hopfield networks defined previously.

6.6.3 Thermodynamics and Hopfield Networks

Let us now consider the random process obtained when the function g_k is replaced by the following function:

$$h_T(x) = \frac{1}{1 + e^{-(x/T)}} \tag{6.17}$$

A new dynamic is defined, namely the Glauber dynamic, formulated by Glauber as a model for isothermic relaxation. The graph of h_T has the familiar sigmoid shape, with T playing the role of the parameter k. The transition probability of the associated Markov chain can be expressed in terms of the energy function.

Consider the network at times t and $t + 1$. Assume that the only change was the transition of a unit i from state 0 to state 1. It is easy to see that this action caused a change in the energy of the network expressed by:

$$\Delta E = -\sum_{j=1}^{n} w_{ij}x_j - \Theta_i \tag{6.18}$$

If the transition of unit i was from state 1 to state 0, then we get:

$$\Delta E = \sum_{j=1}^{n} w_{ij}x_j + \Theta_i \tag{6.19}$$

Let us denote by V the n-dimensional binary vector describing the composite state of the network. We can then derive the transition probability:

$$P(V'/V) = \frac{1}{n} h_T(\Delta E) \tag{6.20}$$

Of course V and V' differ only in one component.

Another equivalent transition matrix commonly used in simulations, is obtained by replacing h_T by h_T' defined as follows:

$$h_T'(x) = \min(1, e^{-(x/T)}) \tag{6.21}$$

This dynamic is called Metropolis dynamic, and was proposed in the 1950s to mimic atomic relaxation.

The stationary probability is the same for both the Glauber and Metropolis dynamics, equal to the Gibbs distribution associated to the energy function:

$$P_T(V) = \frac{e^{-E(V)/T}}{\sum_{\nu} e^{-E(V)/T}}$$ (6.22)

This distribution has a physical interpretation: it is the distribution of maximum entropy among all the distributions of the same average energy.

Based on the Gibbs probability, it is easy to infer the asymptotic dynamics of the network when driven either by the Metropolis or the Glauber dynamics. At high temperatures, the network wanders in state space with almost uniform probability. At low temperatures, the low energy states are far more frequently visited than other states. But one has to take into account that, in low temperatures, escape from local minima (climbing up the hills) has a very low probability of occurrence in a reasonable simulation time.

Another kind of dynamics is proposed in (Szu *et al*), based on the sigmoid function tan^{-1} which is the repartition function of the Cauchy distribution. This distribution has infinite variance, which is supposed to allow the network to achieve long-range jumps at high temperatures and to speed the approach to the ground state.

6.6.4   Boltzmann Optimisers

The purpose of this section is to describe the use of the dynamics described in the previous section in solving optimisation problems. The dynamics is based on statistical physics where a macroscopic system is driven towards an equilibrium state at a given temperature. Subsequent cooling drives the system towards a low energy state. However, due to the existence of a lot of metastable states (local minima of the energy function), low temperature can drive the system towards these states, from which it cannot escape. Analysis of the spin-glass model (similar to the Hopfield network with random connection matrix) has shown that at low temperatures a lot of metastable states exist, separated by high energy barriers. The energy of these states is near the minimum energy. The effect of high temperatures (or equivalently noise) is to smooth the *energy landscape* by diminishing the height of energy barriers separating the metastable states, thus destroying their stability.

The basic idea of simulated annealing described in (Kirkpatrick *et al*, 1983) is to combine the average of high temperature dynamics (free evolution in the state space looking for the minimal state) and low temperature dynamics (slow and constrained evolution trapped in energy wells), by slowly cooling the system. The proposed algorithm is approximately as follows:

- let the system evolve according to the Metropolis or Glauber dynamics for a sufficiently long time (sufficient number of moves or attempts), depending on the size of the network
- slowly reduce the temperature in a geometric way (ratio 0.9 in Kirkpatrick *et al*, 1983)

- terminate when the system is completely frozen (two successive temperatures leaving the state of the system unchanged).

The efficiency of the simulated annealing algorithm depends critically on the *cooling schedule*. In most applications cooling is done in a step by step fashion. However, a mathematical description of the convergence of simulated annealing has been produced, based on a continuous cooling schedule, reported in (Mitra *et al*, 1986). Define the radius r of the network to be the maximum distance (in terms of edges) between any two nodes of the network. Let us denote by L (the discrete Lipschitz bound) the maximum energy difference between energies of neighbouring composite states. It can be shown that the cooling schedule described in the following equation:

$$T_t = \frac{\gamma}{\log(t + t_o)} \qquad (6.23)$$

defines a strongly inhomogeneous Markov chain if $\gamma \geq rL$.

Practically, the ergodicity guarantees the convergence towards the global energy minimum (if it is unique). The result states the relation between the allowed speed of the cooling schedule and the variation of the energy function.

6.6.5   The Travelling Salesman Problem

The main idea of the Hopfield and Tank approach to optimisation problems is the construction of an energy function, whose minima correspond, through an appropriate mapping, to the solutions of the problem in question. This energy function combines both the cost function and the constraints of the problem in such a way that satisfaction of the constraints and minimisation of the cost function leads to minimisation of the energy function. The terms of the energy function are then expressed as weights and thresholds.

This approach has been used to solve many optimisation problems. A representative NP-complete optimisation problem that has been used as a generic problem in the area is the travelling salesman problem (TSP) described in section 6.4 above. Since there are (n - 1)!/2 possible tours whose lengths must be compared, an optimal solution is unlikely to be found by an algorithm bound by a polynomial function of the size of the problem.

Hopfield and Tank (1985), suggested that suboptimal but reasonably 'good' solutions can be obtained using a neural network, whose stable states represent valid tours of the TSP. The mapping used to solve the TSP for n cities, consists of an $n \times n$ nodes network. The nodes are organised in a square matrix where the rows represent the city number and the columns represent their order in the solution path, so a node is described by two indices: one for the city it corresponds to and the other for the visiting order in the final tour.

For example, for five cities numbered A to E, the tour BCAED would be described by a 5 x 5 matrix:

	1	2	3	4	5
A	0	0	1	0	0
B	1	0	0	0	0
C	0	1	0	0	0
D	0	0	0	0	1
E	0	0	0	1	0

For the general case, let x_{ij} denote the state of unit ij. Obviously we wish a composite state of the network to represent a salesman's tour. This is valid under the following constraints:

$$\sum_{j=1}^{n} x_{ij} = 1 \quad i = 1...n \tag{6.24}$$

$$\sum_{i=1}^{n} x_{ij} = 1 \quad j = 1...n \tag{6.25}$$

The length of the tour is a quadratic function of the composite state:

$$\sum_{j=1}^{n-1} \sum_{i_1=2}^{n} \sum_{i_2=1}^{n} d_{i_1 i_2} x_{i_1 j} x_{i_2 j+1} \tag{6.26}$$

The network is synthesised from its energy function. The energy function in turn is a composition of the cost function and terms interpreted as penalised constraints. In Hopfield and Tank (1985) the following formulation is proposed:

$$E = E_1 + E_2 + E_3 + E_4 \tag{6.27}$$

where:

$$E_1 : \frac{A}{2} \sum_{i=1}^{n} \sum_{j_1=1}^{n} \sum_{j_2=1}^{n} V_{ij_1} V_{ij_2} \tag{6.28}$$
$$\scriptstyle j_2 \neq j_1$$

'The salesman can visit a city only once'

$$E_2 : \frac{B}{2} \sum_{j=1}^{n} \sum_{i_1=1}^{n} \sum_{i_2=1}^{n} V_{i_1 j} V_{i_2 j} \tag{6.29}$$
$$\scriptstyle i_2 \neq i_1$$

'The salesman cannot visit more than one city at the same time'

$$E_3 : \frac{C}{2} \sum_{i=1}^{n} \sum_{j=1}^{n} \left(V_{ij} - n\right)^2 \tag{6.30}$$

'The salesman must visit all the cities'

$$E_4 \ : \ \frac{D}{2} \sum_{\substack{i_1=1 \\ i_2 \neq i_1}}^{n} \sum_{i_2=1}^{n} \sum_{j=1}^{n} d_{i_1 i_2} V_{i_1 j} \left(V_{i_2, j+1} + V_{i_2, j-1} \right) \quad (6.31)$$

'The tour's length must be minimal'

Violation of any of the above constraints increases E, whereas a valid TSP tour corresponds to a local minimum of E, with $E_1 = E_2 = E_3 = 0$ and $E = E_4$.

A simpler implementation is proposed by van Laarhoven and Aarts (1988). The energy function is obtained by a weighted sum of the cost function and auxiliary cost functions which are penalised constraints:

$$E_2 = \sum_{i=1}^{n} (\sum_{j=1}^{n} x_{ij} - 1)^2 \qquad\qquad (6.32)$$

$$E_3 = \sum_{j=1}^{n} (\sum_{i=1}^{n} x_{ij} - 1)^2 \qquad\qquad (6.33)$$

This implementation, although simpler, leads to similar results with the previous one. Though the approach seems extendable to a wide range of optimisation problems, difficulties in obtaining regular valid solutions have been reported (Wilson et al, 1988; Hegde et al, 1988). The main problem arises from the fact that the construction of E cannot avoid the existence of local minima which are not valid solutions to the problem. Additionally, since the terms $E_1, ..., E_4$ are combined in E through the parameters A, B, C, D which express the relative weight of one factor over another, the choice of these parameters is not yet formalised. Effectively, the problem of finding a minimal tour through n cities, has been replaced with the problem of finding the best values for A, B, C, D. Wilson et al (1988) have reported that the parameter range for achieving valid solutions becomes increasingly restricted as n increases to approach values of interest.

Other applications to problems in this area include job-shop scheduling (Foo et al, 1988), A/D conversion (Tank et al, 1986), tree searching (Saylor et al, 1986), map and graph colouring (Dahl, 1987).

6.6.6   A General Methodology for Implementing Optimisation Problems

The problem of penalising constraints in combinatorial optimisation is by no means a trivial one. If the penalties are too weak, the energy function minimum may not fulfil the constraints. On the other hand, if penalties are too strong, then the motion is quickly restricted to the set of states fulfilling the constraints, and the algorithm loses its main strength, which allows it to explore the whole of the state space. Increasing the penalty as the temperature is decreasing allows the exploration of the full state space in the beginning, and restricts the motion to feasible states at the end. The main disadvantage of this procedure is that the tuning is problem specific.

A better solution is provided if it is possible to construct a feasible and order-preserving energy function (Aarts and Korst, 1989). Let E be the state space endowed with the neighbour structure defined by its implementation as the configuration space of the neural net. Let f be the cost function and F the subset of states fulfilling the constraints. An energy function g defined on E is said to be feasible if the local minima of g (in terms of the neighbour structure) belong to F. Additionally g is said to be order preserving if its restriction to F induces on F the same order as the original cost function f. The existence of a feasible and order-preserving energy function transforms a constrained optimisation problem into a free optimisation problem for which simulated annealing gives nearly optimal solutions. It turns out that such energy functions have been constructed for classical combinatorial optimisation problems like the TSP and other graph theoretical problems. An example for the crew scheduling problem is given later in this chapter.

6.6.7   Elastic Nets

Durbin and Willshaw (1987) described a parallel analogue algorithm to solve the TSP.

Since a tour of the TSP can be viewed as a mapping from a circle to the plane, so that each city in the plane corresponds to some point in the circle, the problem can be seen as a special case of the general problem of best preserving neighbourhood relationships when mapping between different geometrical spaces.

The algorithm begins from a small circle of nodes centred on the centroid of the distribution of cities. Each node is represented by the coordinates of the related point in the plane. By successive recalculation of the positions of nodes in the plane, the ring is gradually deformed, and finally it describes a tour around the cities (Figure 6.4).

The adjustment of the position of each node depends on two competitive forces. The first pulls that node towards those cities to which it is nearest. The second pulls the node towards its immediate neighbours on the ring, so that nodes tend to be evenly distributed along the tour, thus minimising the total path length.

Both forces are controlled by a *length parameter K* that determines the extent to which a neighbouring node city will affect the currently examined node. At the beginning, K is set to a relatively high value so that all cities have a roughly equal influence on each node whereas the ring is rather 'stiff', in the sense that nodes tend to approach each other. As K is gradually decreased, the ring is elongated and each city becomes associated with a particular section of the ring.

The update rule is:

$$y_j^{new} = y_j^{old} + \alpha \sum_{i=1}^{N} w_{ij}(x_i - y_i) + \beta K \left(y_{j+1} - 2y_j + y_{j-1}\right) \qquad (6.34)$$

where

     N is the number of cities
     M is the number of nodes, of order N
     α, β are constants
     K is the length parameter
     x_i is the vector of the coordinates of city i
     y_j is the vector of the coordinates of node j

and

$$w_{ij} = \frac{\varphi\left(|x_i - y_j|, K\right)}{\sum\limits_{k=1}^{M} \varphi\left(|x_i - y_k|, K\right)} \tag{6.35}$$

where $\varphi(d, K)$ is a positive bounded decreasing function of d, that approaches zero for $d > K$ (ie $\varphi(d, K) = exp\ (-d^2/2K^2)$).

Burr (1988) proposed an enhanced version of the above method leading to faster convergence.

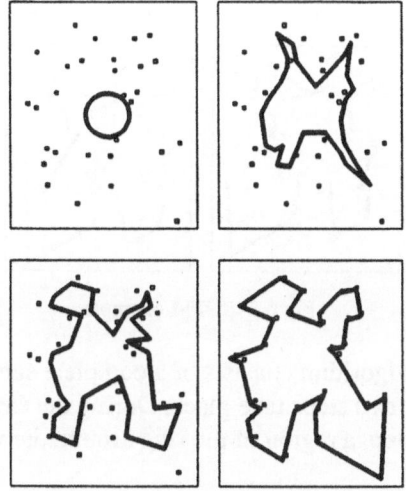

Fig 6.4 Elastic net example

6.6.8   Self-organising Feature Maps

Angeniol *et al* (1988) used a similar method to solve the TSP problem, by exploiting the topology preserving properties of SOFM.

SOFM is an unsupervised, feedforward network that learns through a process that includes a competitive stage of deciding the *winner* node which is closer to the presented input, and an adaptive stage where the winner and its topological neighbours (who are found by a vicinity function) are adjusted, in order to approach the presented input.

The representation of the problem is similar to the elastic-net method: an approximate tour is given by a set of nodes joined together in a ring. Each node is characterised by the two coordinates of the associated point in the plane. The process begins with only one node at the point (0,0) in the plane and the number of nodes grows subsequently according to a node creation-deletion process.

The ring evolves to the ultimate solution path, by continuous recalculations of the positions of nodes (Figure 6.5). These recalculations, being embedded into the two stages of the learning process of the network, are formulated in a rather different way than the previous algorithm.

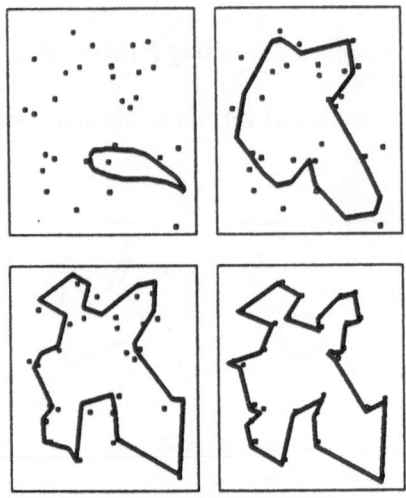

Fig 6.5 SOFM example

Each iteration of the algorithm consists of a complete survey over the N cities. When a city is presented, two steps take place. During the first the node closest to that city *wins*. In the second step, a region of the ring around the winner is pulled towards the presented city.

A vicinity function, that varies inversely with the distance (measured along the ring) of a node from the winner, controls this motion. Thus the winner node is attracted more strongly by the related city, its neighbours being affected to a smaller degree. This vicinity function depends on a *gain parameter G*, whose qualitative characteristics are similar to those of the length parameter K of the elastic net method. When $G \to \infty$ all nodes move towards the presented city with the same strength, while when $G \to 0$ only the winner node moves towards the city in question.

These two steps are formulated as follows:

Step 1: find node j_c, such that

$$V_{jc} = \min_j V_j = \min_j \left(\left| x_i - y_i \right| \right), \qquad j = 1 \dots M \qquad (6.36)$$

where

N is the number of cities
M is the number of nodes, of order N
x_i is the vector of the coordinates of city i
y_j is the vector of the coordinates of node j

Step 2: For all nodes j, calculate

$$y_j^{new} = y_j^{old} + f(n,G)(x_i - y_j) \qquad (6.37)$$

where

$$n = \inf(j - j_c \bmod M, j_c - j \bmod M)$$
$$f(n,G) = \tfrac{1}{\sqrt{2}}\exp(-n^2 / G^2)$$

After an iteration is completed, let $G = (1 - \alpha) G$, where α is a gain parameter.

When a node is found to win two cities during the same iteration it is duplicated, that is, a node is inserted as its immediate neighbour with the same coordinates in the plane. A node is removed from the ring if it does not win a city during three subsequent iterations. The authors claim that this creation-deletion process is critical for the behaviour of the algorithm.

The specific solution of the TSP, which seems easily implementable on an SOFM network, is less demanding as far as the number of nodes and links is concerned, as compared to Hopfield's implementation ($O(n)$ versus n^2 nodes, $O(n^2)$ versus $O(n^4)$ links). Also, the operation of the network is more easily tuned since it is affected by only one parameter, the gain parameter α.

6.7   The Crew Scheduling Problem

Scheduling problems form an important class of combinatorial optimisation problems. They are concerned with optimally satisfying a given set of *tasks*, by utilising several *resources* subject to certain constraints such as priority *constraints* of a task over another, deadlines or immediate service of some tasks, and so on. The goal is to minimise some *cost function* (total service time, cost of that service etc). For example, the scheduling of classes to classrooms, courses to professors, patients to beds and so forth are all described by the general term *job-shop scheduling*. The scheduling of telephone operators, restaurant and similar establishments' personnel, airline crews, etc, are all covered by *manpower scheduling*.

In job-shop scheduling a task may be subdivided into several operations that should be completed sequentially. As long as precedence constraints are not violated, the scheduler is free to arrange operations or whole tasks in time and to assign them to resources that are qualified to handle them and complete them. The most often occurring goal in job-shop scheduling is the minimisation of the total time needed to complete all tasks.

In manpower scheduling the resource, namely the people involved, is important due to its cost. Furthermore, the constraints are usually very complicated.

Focusing on transportation systems, manpower scheduling problems arise in two main contexts:

(i) *Scheduling of vehicle operations and in-vehicle crews*. For example, airline, train and urban or intercity bus networks.

A major constraint is the vehicle schedule which is given and must be respected. The difficulty of the problem arises from many complex constraints such as union rules, government regulations, company policies, etc and the great number of variables.

Another factor that adds to the complexity of the problem is that crews may need to travel to a location other than the crew base to take a new assignment. This procedure is called *deadheading* and since it may be costly, it should be avoided. For example, in an airline company, a deadheading crew occupies seats that would otherwise attract fares.

In the case of *transit crew scheduling*, several models and heuristic approaches have been used. These include decomposition approaches such as the RUCUS (Wilhelm, 1975) and the HASTUS system (Rousseau, 1968) and a matching based approach that solves concurrently the scheduling of vehicles and crews (Bodin *et al*, 1983).

(ii) *Scheduling terminal manpower*. The problem has been described mostly for servicing vehicles, passengers, or freight at terminals (airports, intercity bus terminals, bus stations, subway stations, freight terminals and transfer points, etc). Other applications include telephone operators, supermarket staff, traffic office staff etc. Apart from straight shifts, which is the most common form of employment, split shifts (workdays with large unpaid breaks) may be allowed.

This class of problems is generally simpler because all scheduling is done in the same location. The main difficulty arises from the variations in demand throughout the course of a day or week.

Approaches to solving the problem include a network-flow formulation (Segal, 1974), LP based formulations (Bodin *et al*, 1983), and several heuristics (Couvrey *et al*, 1977; Nicoletti *et al*, 1976).

6.7.1   Planning and Scheduling Problems in Airlines

In the past, many optimisation approaches have been developed to reduce the planning expenditure. Major problems, though, were always present:

- many restrictions could not be mapped as clearly definable parameters, for example legal, political or third party regulations
- the applied programming methods were very expensive due to the enormous number of aircraft, routes and staff. Conventional data processing systems failed to provide quality solutions.

Planning and scheduling problems in airlines can be distinguished at three levels:

(i) decisions regarding position of depots, maintenance halls, purchase of aircraft fleets or route networks are considered as strategic planning
(ii) optimal handling of a given fleet belongs to tactical planning

(iii) scheduling of aircrafts and the assignment of a crew (monthly planning) and short-time changes of the plan (daily planning) are considered as operational planning.

A *leg* is the smallest planning unit and determines the link between two airports. The scheduling of several legs determines a *route*. Including flight and transition time, the route results in aircraft *rotation*.

The flight planning receives the legs required from market surveys and empirical data generated from the routing management. Flight planning must refine them and generate the route planning and aircraft rotation planning.

The next step is the splitting of each rotation into several legs and their reassembly in such a way that an anonymous crew can be assigned to it, including all legal and third party regulations or other constraints. The outcome of the anonymous crew scheduling phase is a set of crew rotations. Rostering them assigns a special crew to each crew rotation forming the service plan for all crew members.

In order to limit the size of the problem, initially we chose to focus our work in the crew scheduling domain, with the additional assumption of a unique aircraft type. Airlines, in manual planning, accept more or less the same assumption. The flight planning area depends more on market surveys and political, economic, social and legal factors. Therefore, putting more effort in market studies and statistical regression analysis would have been outside the scope of this project.

6.7.2   Anonymous Crew Scheduling

The anonymous crew scheduling is concerned with decomposing a set of given aircraft rotations into sequences of flight legs that can be assigned to a crew (an aircraft rotation by itself is usually too long to be flown by a single crew). Aircraft changes are permitted as long as the aircraft type remains the same. Regarding legal restrictions, a crew rotation can last several days. The goal is to minimise the amount of working days.

The construction of aircraft rotations has already assured that for each aircraft type sufficient personnel are available. Then a number of regulations and constraints regarding flight personnel must be considered:

Temporal constraints

It is obvious that the flight leg assigned to a crew must begin after the crew has completed a previous assignment. Furthermore, timing must allow for some other procedures to take place, such as checking in and out of passengers, aircraft inspection, handling of freight etc. An aircraft change further increases the time between flights by an amount specific to each airport.

Local constraints

The arrival airport of a flight leg must be the same as the departure airport of the succeeding leg. Deadhead flights may be permissible to avoid stopping overnight.

Flight duty time (FDT)
FDT is the sum of the flight time of a rotation, the time spent on the ground in between two succeeding flights (ground time) and the time required for briefing and debriefing of the crew. The maximum FDT of a particular crew rotation depends on various factors such as the number of landings, the time of day (or night) the rotation begins, etc.

Number of legs per working day and number of legs per rotation
Upper bounds exist for both these factors, in order to control the number of take-offs and landings of a crew.

Maximum working time
This factor is the sum of all FDTs and has a specific upper bound, for instance, a week.

Rest time
The rest time has a lower bound which is a complicated function of the hours worked up to that point, including the flight legs covered, the time-zone differences, the number of working days that preceded the current one in the same rotation, the type of the flights (long distance or not) etc.

To summarise the basic terms:
Aircraft rotation: the route which an aircraft is flying (contains the legs in turn including the transition time between successive legs), starting and ending at the same airport
Flight leg: a flight between two airports (the smallest planning unit)
Crew: members of flying personnel on a special type of aircraft
Pairing or rotation: a feasible sequence of flight legs to be covered by one crew
Duty period: a continuous block of time when the crew is on duty
Roster: assignment of pairings (or duties in general) to crews together with the rest days, for a period of time (week, month or any other period)
Cost: number of working days.

Starting from the result of the flight planning (set of aircraft rotations) each rotation is considered for an aircraft type. A crew cannot take the same rotation as an aircraft, because the duty period will then be too long. Therefore, the aircraft rotation has to be split into several legs and put together in such a way (regarding the restrictions), that an anonymous crew can fly this crew rotation. The following example will demonstrate the impacts of restrictions to the anonymous crew scheduling:

Example
A set of aircraft rotations is given for the special aircraft type A300 (Figure 6.6). Each day three aircraft start from Frankfurt, Hamburg and Düsseldorf respectively. These three aircraft cover the flight programmes shown in the figure, in a circular

manner, ie the aircraft initially doing rotation 1 will do rotation 2 the next day, and so on.

In this example the following restrictions have to be considered:

- the flight duty time should not be longer than 14 hours. It is to be calculated from the flight time, the ground time and two hours for briefing and debriefing of the crew
- every working day can include up to five flight legs
- between two working days, the rest period for the crew has to be taken into account. In this example the rest period is calculated as the total of the FDT plus two hours, but at minimum ten hours
- the rotation should not be longer than five days.

Aircraft rotation no 1

	242	245	214	217
FRA	0815-0915 MUC	1015-1115 FRA	1215-1300 STR	1345-1430 FRA

	1606	1609	058	
(FRA)	1535-1705 LHR	1755-1920 FRA	2025-2125 HAM	

Aircraft rotation no 2

	066	084	085	120
HAM	0620-0725 FRA	0825-0915 HAJ	1000-1050 FRA	1205-1250 DUS

	212	1746	1749	136
(DUS)	1340-1425 FRA	1555-1705 CDG	1755-1905 FRA	2020-2105 DUS

Aircraft rotation no 3

	109	114	117	034
DUS	0630-0720 FRA	0835-0920 DUS	1005-1055 FRA	1205-1310 HAM

	043	050	057	
(HAM)	1410-1500 FRA	1600-1700 HAM	1750-1850 FRA	

Fig 6.6 Example of flight plan of aircraft A300. Below each flight number is a code for the city (FRA for Frankfurt etc) and the departure and arrival times of the flight

A possible solution is shown in Figure 6.7. The results proposed there contain two crew rotations spanning over a period of six days. Since the routes occur every day, 7x6 = 42 working days per week are needed to cover the given aircraft rotations. The first crew rotation lasts one day. The second rotation lasts five days. On the third day of this rotation the crew rests one day in Düsseldorf. That is because the only possibility of taking off from Düsseldorf is at 6.30, but if this flight were chosen the minimum required rest time (supposedly 10.55 hours) could not be fulfilled.

First rotation
```
        No 1 242        No 1 245        No 1606         No 1 1609
FRA  0815-0915 MUC 1015-1115 FRA 1535-1705 LHR 1755-1920 FRA
```

Second rotation
```
        No 1 214        No 1 217        No 3 050        No 3 057        No 1 058
FRA  1215-1300 STR 1345-1430 FRA 1600-1700 HAM 1750-1850 FRA 2025-2125 HAM
```

```
        No 3 043        No 2 1746       No 2 1749       No 2 136
HAM1410-1500  FRA  1555-1705 CDG 1755-1905 FRA 2020-2105 DUS
```

Rest day in Dusseldorf

```
        No 3 109        No 3 114        No 3 117        No 3 034
DUS  0630-0720 FRA  0835-0920 DUS 1005-1055 FRA 1205-1310 HAM
```

```
        No 2 066        No 2 084        No 2 085        No 2 120        No 2 212
HAM 0620-1725 FRA 0825-0915 HAJ 1000-1050 FRA 1205-1250 DUS 1340-1425 FRA
```

Fig 6.7: A possible solution for crew rotation. Above each leg the aircraft rotation number and flight number are given (see Figure 6.6)

An anonymous rotation is determined by the following main restrictions:

- the number of legs planned per shift
- the number of legs planned per shift in three consecutive shifts of a duty day
- the number of legs planned per duty period
- the maximum Flight Duty Time (FDT)
- the rest period between two shifts
- the transition time between two legs; it will be longer if an aircraft change is planned and it depends on the airport.

The objective is to create an anonymous crew scheduling with minimal shifts for the crew.

6.7.3   Conventional Approaches

The airline crew scheduling problem is typically formulated as a set covering (or set partitioning) problem. This formulation first appeared in Spitzer (1961). we are given a set of objects which we number as the set $S = \{1,2,...,m\}$. We are also given a class $\S$ of subsets of S which we number as the set $\Im = \{S_1,S_2,...,S_n\}$, each of which has a cost C_j associated with it. The problem is to cover all the members of S at minimum cost, using members of $\Im$.

Mathematically, the problem can be seen as an integer programming formulation:

$$\min C\,X \qquad\qquad (6.38)$$

subject to

$$AX \ge e, X_j = 0 \text{ or } 1, \text{ for } j \in \{1,...,n\}$$

where

> m is the number of members of S
> n is the number of members of $\mathfrak{I}$
> C is an n-vector, C_j is the cost of subset j
> X is an n-vector of 0 and 1
> > with X_j = 1 if subset j is used in the solution
> > = 0 otherwise
> A is an m x n matrix of 0 and 1
> > with $A_{i,j}$ = 1 if member i of S, is included in subset j
> > = 0 otherwise
> $e = (1,...,1)$ is an m-vector.

The set partitioning has the same formulation with the exception that the inequality ($\geq$) is replaced by equality (=).

In the context of crew scheduling, S is the set of flight legs that must be covered, and $\mathfrak{I}$ is the set of all feasible pairings (combinations of flight legs assigned to one crew). The above formulation allows the covering of a flight leg by more than one pairing, so two crews can be on the same plane, one deadheading to another location to take a new assignment. When deadheading is not allowed, set partitioning is used.

Apart from crew scheduling, a great variety of scheduling problems can be formulated as set covering (or partitioning) as follows. The usefulness and wide applicability of the set covering model follows from the simple observation that in most cases a problem of the above form can be solved to a satisfactory degree of approximation by the following two-stage procedure:

Stage 1. Using the given constraint set, generate explicitly a subset $\overline{F} \subset F$, such that the probability of an optimal solution being contained in $\overline{F}$ is sufficiently high.

Stage 2. Replace the constraint set by a list of the members of $\overline{F}$ and solve the resulting set covering problem.

Example
Let us consider the case of an airline company that must cover the following flight legs:

Flight	Origin	Destination	Flying time	Time of day
1	Athens	Munich	120	10.00
2	Athens	Paris	175	16.00
3	Munich	London	185	17.00
4	Munich	Paris	180	17.30
5	London	Athens	200	21.00
6	Paris	Athens	210	21.30

The cost of each pairing will be considered proportional to its duration, with the addition of a penalty cost P when a tour does not end in the city of origin. The set of pairings that does not violate any time constraints and constitutes one or more subsequent flight legs in a 24-hour period is:

Pairing	Flight legs	Cost
1	1	$120 + P$
2	2	$175 + P$
3	3	$185 + P$
4	4	$180 + P$
5	5	$200 + P$
6	6	$210 + P$
7	2,6	385
8	3,5	$385 + P$
9	5,1	$320 + P$
10	1,3,5	505
11	1,4,6	510

Using the set covering formulation, the matrix A is as follows;

Legs	Pair 1	2	3	4	5	6	7	8	9	10	11
1	1								1	1	1
2		1					1				
3			1					1		1	
4				1							1
5					1			1	1	1	
6						1	1				1

6.7.4   The Work Done so Far in Airline Crew Scheduling

Probably the most used system by large airlines, is Rubins' algorithm (1973), which has been integrated into a comprehensive user oriented package (IBM and American Airlines). The scheduler provides an initial solution, which is subsequently improved by the algorithm by considering subsets of pairings which are small and easy to solve optimally. Both the quality of the initial solution and the strategy for constructing subsets of pairings are especially important for the success of the algorithm.

Baker and Fisher (1981) describe the implementation of a heuristic set covering crew scheduling systems at Federal Express Corporation of Memphis, Tennessee. The model seems to succeed in a wide variety of large-scale crew scheduling applications. It contains three steps:

(i)   a pairing generator used to enumerate candidate crew schedules
(ii) a heuristic set covering algorithm, used to obtain an initial covering solution

(iii) several solution improvement techniques used to reduce the total cost of the
covering solution.

Crainic (1978) proposed a column generation algorithm which, after constructing a
set of pairings, proceeds with solving the set covering problem with a linear
programming approach (relaxation). Dual variables are used to generate a set of other
pairings with a negative reduced cost that are added to the original problem. This is
pursued until no significant improvement of the cost function is achieved. Then
Salkin's (1975) heuristic is used to obtain an integer solution.

In order to avoid the generation of high order matrices (which can reach the size of
500,000 rows by 100,000,000 columns) the *small matrix* method is used. This
essentially takes only a small subset of the problem into consideration at a time, so the
small matrices generated are solved iteratively. After this is done a few thousand
times, a good solution to the problem exists. Though this approach has been
effectively used by many American airlines, it has not been successful for European
companies which have a star-shaped network with a few points in common. If too
small a subproblem is considered, there can be no improvement in this part of the
problem.

6.8   A Specific Airline Case

In the previous section the airline crew scheduling problem was described informally.
We now give a description in formal terms with the methodology chosen to tackle the
problem. The method does not aim at delivering the optimal solution. The main reason
for that lies in the structural *hardness* of the crew scheduling problem. If we were to
build a neural network delivering the optimal solution we would have to *absorb* the
complexity of the problem into either the size of the network or to the convergence
time. Either would be prohibitive. Thus, the first objective is to deliver a *near to
optimal* solution.

In order to implement a neural network solving an optimisation problem one
essentially transforms the constraints of the problem into the energy functions. In
section 6.8.1 the set of constraints of the problem is described in a formal way. It can
be easily seen that their number is quite large. Thus, a direct mapping of all the
constraints into an energy function seems to be impossible. Furthermore, even if it
were possible, this procedure would yield a large number of penalising variables to be
tuned in a later stage, a very difficult task in every respect. The chosen solution to
these difficulties is to use two separate steps.

First, the aircraft rotations given are transformed to a set of *valid* crew rotations.
The module that performs this task is called the *pairing generator*. The pairings
produced are valid in the sense that they obey all the constraints. Thus, an anonymous
crew could undertake the execution of each pairing. For reasons that will become clear
later, we chose to implement the pairing generator by purely conventional methods. In
the second step, the set of pairings is fed to a neural network (the architecture is
described in section 6.11). The network's task is to select a subset of pairings with near

to optimal cost, covering all the flight legs. This approach has two important characteristics.

First, the constraints of the problem are applied in a phase independent of the neural network. This is due to the fact that the constraints involved are far too many to be described by a neural network architecture. Furthermore, they are dependent on the company involved and other factors mentioned earlier. Embedding the constraints in the network's architecture, even if it were feasible, would result in a case specific network, that for every minor or major alteration of the constraints would require recomputation of its energy function and retuning of its functional parameters. Instead, the separation of the two procedures, namely the application of the constraints and the solution of the set partitioning problem permits easy handling of the data and constraints of a different airline, since only the pairing generator module has to be adapted to the new case.

Second, the set partitioning formulation is used to describe a great number of combinatorial optimisation problems. Since the neural network module is kept independent from the specific details of crew scheduling, by solving the set partitioning problem, its application to a variety of other problems is straightforward.

Also, the problem of tuning the whole system is simplified. Tuning, in terms of our methodology, means rejecting unwanted pairings in favour of the more preferable ones, a task that can be incorporated into the pairing generator.

6.8.1   Problem Formalisation

This section describes the constraints for the crew scheduling problem in a formal way. The constraints were supplied by a major European airline company. They apply only to their A300 fleet.

The input to the problem is a set of aircraft rotations. The solution comprises a set of crew rotations covering all flight legs in the set of aircraft rotations. Each one of the crew rotations must be valid (in the sense that it obeys all the constraints).

We assume that the crew rotation is given as a sequence of flight legs

$$1, 2,..., i,..., L$$
of length L

and for each leg i we assume that the following is known: the departure and arrival airports, the departure and arrival date and time, the aircraft that executes the leg. For each airport x occurring as an arrival or departure point of some leg in the crew rotation, we assume that the ground time and the time zone are given.

Notation
We give below a list of symbols used to represent the constraints:

L :   number of legs in the rotation
i :   the ith leg of the rotation, $i = 1...L$
x_i :   departure airport of flight-leg i
y_i :   arrival airport of flight-leg i

g_x :   ground time of airport x

Z_x :   time zone of airport x, with respect to GMT

d_i :   date of leg i

a_i :   departure time of leg i

b_i :   arrival time of leg i

c_i :   aircraft of leg i

f_i :   flight time of leg i

T_i :   transition time between legs i and $i+1$. Transition time is the time elapsing between the arrival time of leg i and the departure time of leg $i+1$

U :   number of calendar days of the rotation

j :   the jth calendar day of the rotation, $j = 1...U$

K :   number of jobs in the rotation. By the term job we mean a working period between two rest periods. As a rest period we identify any period of time longer than 10 hours between flight legs

k :   the kth job of the rotation, $k = 1...K$

N_k :   the number of legs in job k

n :   the nth leg in a job k, $n - 1...N_k$

FDT_k :   flight duty time of job k. FDT_k is the sum of flight and transition times of the legs in job k, plus 1 hour for briefing and 1 hour for debriefing

R_k :   rest time after job k. R_k is defined to be the period of time between the first leg of the following job $k+1$ and the last leg of current job k

TZ_k :   time zones covered during job k. The time zones covered are calculated as the difference of time zones between the arrival airport of the last leg and the departure airport of the first leg of a job k

CWD_k :   consecutive work days by the end of job k. By this, we refer to the number of days since the last calendar day with no duties, including the day job k finishes

WT_k :   working time by the end of job k. Working time is defined to be the sum of flight duty times of all jobs that took place during a 7-day period. WT_k is defined on the 7 day period ending in the day job k ends.

The following 0-1 variables may now be defined:

$$l_{ij} = \begin{cases} 1 & \text{if leg } i \text{ takes place in } j\text{th day of the rotation} \\ 0 & \text{otherwise} \end{cases}$$

$$t_{ink} = \begin{cases} 1 & \text{if leg } i \text{ is the } n\text{th leg of job } k \\ 0 & \text{otherwise} \end{cases}$$

t_{ink} can be recursively defined as follows:

$$
t_{ink} = \begin{cases}
1 & \text{if } (i = n = k = 1) \vee \\
 & (t_{(i-1)(n-1)k} = 1 \wedge a_i - b_{i-1} < 10h) \vee \\
 & (t_{i-1)(N_k-1)(k-1)} = 1 \wedge \\
 & a_i - b_{i-1} \geq 10 \wedge n = 1) \\
\\
0 & \text{otherwise}
\end{cases}
$$

Constraints 1: Local and temporal sequence
(i) The rotation must start and end in the home base airport

$$x_1 = y_L = \text{home base}$$

(ii) The departure airport of leg $(i + 1)$ and the arrival airport of leg i must be the same

$$y_i = x_{i+1}$$

(iii) Departure date and time of leg $(i + 1)$ must follow arrival date and time of leg i

$$d_{i+1} \geq d_i$$
$$a_{i+1} \geq b_i \text{ if } d_{i+1} = d_i$$

(iv) Transition time T_i between legs i and $i + 1$ must satisfy

$$
T_i > \begin{cases}
|a_{i+1} - b_i| & \text{if } a_{i+1} > b_i \\
|a_{i+1} - b_i + 24| & \text{if } a_{i+1} \leq b_i
\end{cases}
$$

(v) Transition time must be adequate

$$
T_i > \begin{cases}
40\,\text{min} & \text{if } c_{i+1} = c_i \\
g_{x_i} & \text{otherwise}
\end{cases}
$$

ie if the crew does not change aircraft, provision for 40 min must be made. In the case of aircraft change the transition time must be greater than the ground time at the airport.

Constraints 2: Duration
(i) The maximum number of calendar days per rotation is 20

$$U = d_L - d_1 \leq 20$$

(ii) The maximum number of legs per job is 5

$$n_k = \sum_{i=1}^{L}\sum_{n=1}^{N_k} t_{ink} \le 5, \qquad \forall k = 1...K$$

(iii) In three consecutive calendar days the maximum number of legs is 14

$$\sum_{i=1}^{L} l_{ij} + \sum_{i=1}^{L} l_{i(j+1)} + \sum_{i=1}^{L} l_{i(j+2)} \le 14, \quad \forall j = 1...(U-2)$$

(iv) A four-day rotation must not exceed 14 legs

$$\sum_{i=1}^{L}\sum_{j=1}^{U} l_{ij} \le 14, \qquad \text{if } U = 4$$

(v) A five-day rotation must not exceed 20 legs

$$\sum_{i=1}^{L}\sum_{j=1}^{U} l_{ij} \le 20, \qquad \text{if } U = 5$$

Constraints 3: Flight Duty Time

(i) Flight time f_i of leg i

$$f_i = \begin{cases} |a_i - b_i| & \text{if } a_i \le b_i \\ |a_i - b_i - 24| & \text{if } a_i > b_i \end{cases}$$

(ii) Flight Duty Time FDT_k of job k

$$FDT_k = \sum_{i=1}^{L}\sum_{n=1}^{N_k-1} t_{ink}(f_i + T_i) + \sum_{i=1}^{L} t_{iN_k k} f_i + 2$$

(iii) Upper bounds for FDT_k are defined in Table 6.1, ie

$$FDT_k \le \begin{cases} 13.5 \text{ hours} & \text{if } N_k = 2,\, 07{:}01 \le a_{i'} \le 15{:}00 \text{ where } t_{i'1k} = 1 \\ \vdots & \\ 10 \text{ hours} & \text{if } N_k = 5,\, 06{:}01 \le a_{i'} \le 07{:}00 \text{ where } t_{i'1k} = 1 \end{cases}$$

Constraints 4: Rest Time

(i) Rest time R_k following job k

$$R_k = \begin{cases} \left| a_{i'} - b_{i''} \right| - 2 + 24(d_{i''} - d_{i'}) & \text{if } a_{i'} > b_{i''} \\ \left| a_{i'} - b_{i''} + 24 \right| - 2 + 24(d_{i''} - d_{i'}) & \text{if } a_{i'} \geq b_{i''} \end{cases}$$

where

$$i', i'' : t_{i''N_k k} = t_{i'1(k+1)} = 1$$

(ii) Time zones TZ_k, covered during job k

$$TZ_k = Z_{y_{i''}} - Z_{x_{i'}}, t_{i'1k} = t_{i''N_k k} = 1$$

(iii) Consecutive work days CWD_k, by the end of job k

$$CWD_k = w, \text{ if } \sum_{i=1}^{L} l_{ij} \geq 1, \forall j = d_{i'} - (w-1), \ldots, d_{i'}$$

where

$$i' : t_{i'N_k k} = 1$$

(iv) Minimum rest time $R_{k,min}$, following job k. The rest time after completion of job k, must last at least $R_{k,min}$ hours

$$R_{k,\min} = \max(FDT_k + 2, 10)$$
$$R_k \geq R_{k,\min}$$

(v) Rest time R_k must be adequate

Table 6.1 Upper bounds for Flight Duty Time (in hours)

Landings	Departure time				
	0701-1500	1501-1800	1801-2300	2301-0600	0601-0700
2	13.5	13.5	12	11	13
3	13.25	12.25	11.25	10.25	12.25
4	12.5	11.5	10.5	9.5	11.5
5	11	10	9	0	10

Table 6.2 Extra rest time if $TZ \leq 4$ and $FDT \leq 8$

Rest time begins	Extra hours to add to $R_{k,\,min}$
18:01-24:00	0
00:01-04:00	1
04:01-15:00	2
15:01-18:00	1

Table 6.3 Rest time when FDT > 8

Time zones	Rest time
TZ < 4	36 h
TZ > 4	48 h
TZ > 6	72 h
TZ > 8	3 nights

(vi) 144 hours constraint. If a crew had been assigned a job q, with more than three consecutive work days ($CWD_q \geq 3$), and the time zones covered were over 4 ($|TZ_q| \geq 4$), it cannot be assigned a job k, that will cover four time zones or more ($|TZ_k| \geq 4$), in the opposite direction, unless a period of 144 hours (6 days) has passed between jobs q and k. For the comparison of directions, the signs of TZ_q, TZ_k are used.

$$|TZ_k| \geq 4$$

$$\text{if } \not\exists q : q < k \wedge |TZ_q| \geq 6 \wedge CWD_q \geq 3 \wedge sgn(TZ_k) \neq sgn(TZ_q) \wedge d_{i'} - d_{i''} < 6$$

where

$$i', i'' : t_{i'1k} = t_{i''N_q q} = 1$$

Constraints 5: Working time

(i) Working time period WT_k, by the end of job k

$$WT_k = \sum_k FDT_k, \forall k : t_{i1k} = 1, d_i = d_{i'} - 6, ..., d_{i'}$$

where

$$i' : t_{i'N_k k} = 1$$

(ii) Upper bounds for WT_k. If the last leg of job k ends in the Home Base, the Working Time WT_k may extend up to 60.5 hours. Otherwise it must be below 55 hours

$$WT_k \leq \begin{cases} 60.5 & \text{if } y_{i'} = \text{ Home Base, } i' : t_{i'N_k k} = 1 \\ 55 & \text{otherwise} \end{cases}$$

6.9   The Pairing Generator

The solution to the crew scheduling problem was handled by three independent modules:

(i)   the *data preparation* module that takes as input the flight legs data provided by the airline company and extracts from it the data of the flight legs included in the period to be scheduled, in a suitable format

(ii) the *pairing generator* module that accepts as input all the flight legs of the period to be scheduled, and produces as output the pairings that are valid with respect to all the constraints

(iii) the *neural network* module that solves the set partitioning problem on the data produced by the pairing generator.

The data preparation module copies all the details concerning the format of the data supplied by an airline company. In our case, the flight legs were described on a weekly basis. This description was repeated to cover the scheduling period.

The output data from the preparation module are of the form:

unique flight number	:	a number identifying uniquely the flight over the entire scheduling period
departure airport	)	
departure time	)	Time and place of departure and arrival of the flight
arrival time	)	
arrival airport	)	
date	:	date the flight takes place
following flight number	:	the number of the flight that follows after the current flight, in the aircraft rotation.

The pairing generator module has the task of constructing all the valid pairings. This is done by an exhaustive search of the flight legs set, during which a pairing at a time is constructed and the constraints are applied to it. Of course, as a simple optimisation of the search, when a pairing is rejected as illegal, no efforts are made to construct any pairings that are produced by appending flight legs to it.

The whole set is treated as a graph where each airport is a node and each flight leg connecting two airports is an edge connecting the respective nodes. A path on this graph, beginning from and ending with the node corresponding to the crew base and constituted from edges (ie flights) that respect the temporal constraints, is considered as a candidate pairing that is checked for validity. Having applied by construction the local and temporal constraints, the constraints that remain to be checked are the ones concerning duration, flight duty time, rest time and working time.

The output of the pairing generator is an ensemble of pairings. The neural network module takes this ensemble as input and in a specified way constructs a Boltzmann-type neural network whose weights and thresholds are calculated to describe this set. More details on how this is done are given later.

6.10   Conventional Methods for Set Covering Problems

We present here briefly several conventional methods for solving the set partitioning and set covering problems. Both methods are included for two reasons:

(i)   since set partitioning is easily brought to a set covering form, as it will be shown, both solution methods can be used to solve it. So we can compare our neural network solution using either a conventional set partitioning or a convention set covering solution

(ii)   the neural network solution can be extended to cope with the set covering problem as well.

In general, no definite statement can be made about the difficulty of set partitioning and set covering problems. In view of the close similarity between them it is rather surprising that the set covering problem has some important differences. At a closer look, however, we can intuitively understand a reason for this: set partitioning is a *tightly constrained* problem, in the sense that each constraint requires exactly one of many variables to be 1, whereas set covering is a *loosely constrained* problem meaning that at least one of the many variables is required to be 1.

As we have already mentioned it is possible to transform the set partitioning problem to a set covering problem, whereas the reverse transformation is not generally possible.

We give now this transformation, repeating the mathematical formulation of the set partitioning problem:

$$\min C\,X \tag{6.39}$$

subject to

$$AX = e, \ X_j = 0 \text{ or } 1, \qquad \forall j \in \{1,...,n\}$$

where A is an $m \times n$ matrix of zeros and ones, C is an arbitrary n-vector and $e = (1,...,1)$ is an m-vector. The set covering has the same formulation with the exception that $A\,X \geq e$.

We can bring the previous set partitioning problem to the set covering equivalent form by writing

$$\min C X + \theta eY \tag{6.40}$$

subject to

$$AX - Y = e, \quad Y \geq 0, \ x_j = 0 \text{ or } 1, \quad \forall j \in \{1,...,n\}$$

and using $Y = AX - e$

$$\min - \theta e^2 + C'X \tag{6.41}$$

subject to

$$AX - Y = e, \quad Y \geq 0, x_j = 0 \text{ or } 1, \quad \forall j \in \{1,...,n\}$$

where $C' = C + \theta eA$. For sufficiently large θ this problem has the same set of optimal solutions as the corresponding set partitioning problem. A lower bound for 0 is θ is $\theta \geq \sum_{j=1}^{n} C_j$.

The set partitioning and set covering problems are not easy to solve. Karp (1972) has shown the set covering problem to be NP-hard.

This implies that is very unlikely that any efficient algorithm can be guaranteed to find optimal solutions. There is thus a trade-off: we can have an algorithm that runs quickly or one that finds optimal solutions, but not one that does both simultaneously.

6.10.1 Set Partitioning Problems

Several algorithms for the set partitioning problems have been reported in the literature (Balas *et al*, 1976; Bodin *et al*). Most of these algorithms begin by relaxing the binary condition on X_j, ie allowing $0 \leq X_j \leq 1$, and solving the resulting linear programming problem. Several methods are used to get the discrete optimum from the continuous optimum. Among them enumeration (tree-search), cutting planes, and group theory are the principals.

We chose to describe below one of the most used tree-search methods for solving the set partitioning problem, proposed by Garfinkel and Nemhauser (1969). In this enumeration algorithm for the set partitioning problem the solution space is systematically searched by generating partial solutions (assigning 0-1 values to variables taken one at a time) and exploring efficiently the logical implications of these value assignments.

At the start of the process the columns of A are partitioned into t non-empty subsets, called *blocks*, B_j, $j = 1,...,t$, such that block B_j satisfies:

$$A_{i,j} = \begin{cases} 1 & \forall k \in B_j \\ 0 & \forall k \in \bigcup_{l=j+1}^{t} B_l \end{cases} \tag{6.42}$$

for some row i of the matrix A. Thus block k will comprise exactly those subsets (columns) which contain k but do not contain any of $1,...,k$-1 (remember that we have to cover the index set $S = \{1,...,n\}$). Generally $t \le m$ and if $t < m$ this means that some blocks may be non-existent in this particular instance of the problem.

After these permutations of columns the matrix A has the following form:

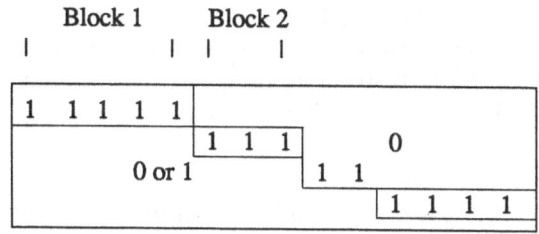

Within each block the columns are usually ordered by increasing costs.

During the course of the algorithm, blocks are searched sequentially, with block k not being searched unless every element i, $1 \le i \le k - 1$, has already been covered in the partial solution. Thus, if any set in block k were to contain elements less than k, the set would have to be discarded (at this stage) due to the no-overcovering requirement.

The tree search proceeds in the standard implicit enumeration fashion (Balas, 1965) with the important advantage that the no-overcovering restriction forms a very powerful exclusion test.

This algorithm performs better on high density than on low density problems (where density means the fraction of non-zero entries in A).

A similar algorithm incorporating a simple lower bound for the remaining subproblem, was also proposed by Pierce (1968) and further improved by Pierce and Lasky (1973). This algorithm also performs better on high density than on low density problems.

Marsten (1974) proposed a somewhat different algorithm using also the block structure. Marsten's algorithm differs from the other enumerative algorithms in that it does not proceed in terms of decisions about individual variables. That is, the basic choice involved is not whether a certain variable should be used or not, but the variables are grouped together in classes and the basic choice involved is which class should be responsible for covering a particular row. The status of individual variables is then determined automatically. The very large number of variables present in any realistic partitioning problem makes this approach quite attractive. This algorithm works better on sparse matrices (ie with small density) and so is appropriate for the set partitioning problems arising in airline crew scheduling problems where the resulting matrice are super-sparse (1-3%).

Apart from enumeration methods, cutting plane methods for solving the set partitioning problem exist. The fact that there is a high frequency of integer solutions among the basic solutions of the linear problem that we obtain by relaxing the set partitioning problem (1976) makes the application of cutting plane methods relatively easy. Most cutting plane methods (Délorme,1974; Salkin and Koncal,1973), use the traditional cutting planes introduced by Gomory and they are general-purpose algorithms which take little advantage of the problem structure.

A group theoretic algorithm was developed by Thiriez (1969).

Finally we mention another method for solving set partitioning via node covering (Balas *et al*, 1973). It is known that set partitioning is transformable to an equivalent set packing problem. On the other hand for each set packing problem there is always an equivalent weighted node packing problem which is equivalent to the weighted node covering problem. Thus one way of solving set packing and set partitioning problems is via solving the corresponding weighted node packing or node covering problems. This method is also used in the neural network solution we proposed.

6.10.2 Set Covering Problems

For the set covering problems the same holds as for the set partitioning, ie algorithms for solving the set covering problems typically rely on branch and bound, implicit enumeration and some other integer programming procedures. In the work of Christofides *et al* (1975) a computational survey of these solution methods can be found as well as an improved tree search method.

Cutting plane heuristics for the set covering problem was studied by Balas (1980) and Balas and Ho (1980). The writers report that algorithms based on cutting planes are efficient for solving large sparse set covering problems and for finding good approximate solutions to problems that are too hard to be solved exactly.

An approach, similar to Marsten's approach for the set partitioning problem mentioned above, is the one proposed by Etcheberry (1977). It is an implicit enumeration algorithm where the branching strategy is similar to the two partitioning of Marsten's algorithm whereas the bounds are obtained by Lagrangian relaxation.

A rather straightforward, greedy heuristic for the set covering problem was proposed by Chvatal (1979). This approach is based on the observation that the desirability of including subset X_j in an optimal cover increases with the ration $|X_j| / c_j$ which counts the number of elements covered by X_j per unit cost. This suggests a recursive procedure for finding near optimal solutions: in each step the subset that satisfies the above requirement is added to the solution until no subset exists. Obviously this is a heuristic of *greedy* nature. Chvatal proved an upper bound for the solutions found using this algorithm. Ho (1982) also gives a worst case analysis for a class of set covering heuristics including the greedy heuristic of Chvatal.

Most of the above mentioned heuristics for the set covering problem are constructive in nature; that is, initially all $X_j = 0$ and then, based on some criterion, an index u is determined such that $X_u = 1$. Typically the same criterion is used repeatedly until a feasible solution is obtained, then redundant elements are usually removed and several methods are used to obtain a local optimum.

In the work done by Vasko and Wilson (1984) and Vasko and Wolf (1988), the basic algorithm called SCHEURI also starts by initialising all X_j to zero and, based on some criterion, which is used repeatedly until a feasible solution is found, an index u is determined such that $X_u = 1$. Redundant elements are then usually removed and several methods are used to obtain a local optimum.

The same algorithm with several criteria has been tested. The choice of the proper variable, X_q, such that $X_r = 1$, depends on the choice of a function, f, to be minimised.

Seven basic functions have been tested. The greedy heuristic was one of them, with very poor worst-case behaviour. A modification of the algorithm, called SCFUNC1TO7, is described in (Vasko and Wilson, 1984), in which each time a variable is to be selected, a random number from 1-7 is generated and that number function is used as the criterion for selection.

The same approach was used (Vasko and Wilson, 1986) to solve the minimum cardinality set covering problem, ie the set covering problem with all costs equal to the same positive number (typically one).

Recently, Vasko and Wolf (1988) have given modifications of the previous SCFUNC1TO7 algorithm, for solving the set covering problem on a personal computer.

6.11   The Neural Network Approach

In the crew scheduling problem the combinatorial optimisation problem to solved can be mathematically formulated as a well-known NP-hard problem, the set partitioning problem (see section 6.8).

In the following sections we will discuss possible neural approaches to the set partitioning problem, and we will focus on the approach based on a Boltzmann machine. The representation chosen and the architectural aspects of the network's operation will be discussed. Initial results have been obtained for real, though small, sets of data given by an airline company. These results are encouraging, though also indicative of further developments needed before the model will handle efficiently datasets with sizes of more practical interest.

For convenience, we initially restate the set partitioning formulation of the crew scheduling problem.

6.11.1   Crew Scheduling - a Set Partitioning Problem
A generally stated scheduling problem is as follows. Given

- a finite set M
- a constraint set defining a family S^* of *acceptable* subsets of M and
- a cost (real number) associated with each member of S^*

find a minimum-cost collection of members of S^* which is a partition of M.

In the airline crew scheduling problem M corresponds to the set of flight legs (from city A to city B, at time t) to be covered during a planning period (usually a few days), while each subset S_j belonging to S^* stands for a possible tour (sequence of flight legs with the same initial and terminal point) for a crew. In order to be acceptable, a tour must satisfy certain regulations (Christofides, 1975).

To set up the problem, one starts with a given set (usually several thousand) of acceptable tours, with their respective costs, and then generates a subset S of S^* such that the probability of an optimal solution being contained in S is sufficiently high. Next, replace the constrained set S^* by S and solve the resulting set partitioning problem (SPP). If the attempt is successful, the solution yields a minimum cost collection of acceptable tours such that each flight leg is included in exactly one tour of the collection.

6.11.2 Hopfield Approach to the Set Partitioning Problem (SPP)

One possible neural approach we could utilise in the set partitioning problem is the one Hopfield first introduced in (1985), to solve the TSP problem. This method, known also as the method of *penalisation of constraints*, is described below.

The network is fully described by its energy function. One function capturing the characteristics of the SPP (its costs and constraints) is the following:

$$E \quad = \quad + \quad A \sum_{i=1}^{m} (\sum_{j=1}^{n} x_{ij} - 1)^2 \qquad\qquad \text{constraints}$$

$$+ \quad B \sum_{j=1}^{n} (\sum_{i=1}^{m} x_{ij} - |S_j|)^2 \quad \text{elements of each subset}$$

$$+ \quad C \sum_{j=1}^{n} (\sum_{i=1}^{m} \frac{C_j}{|S_j|} x_{ij}) \qquad\qquad\qquad \text{costs} \quad (6.43)$$

where m is the number of elements (flights) of the basic set M, n is the number of subsets of M (pairings), x is an $m \times n$, 0-1 matrix such that $x_{ij} = a_{ij} x_{ij}$ and $a_{ij} = 1$ only if element i belongs to subset j, C_j is the cost associated with each subset S_j, $| S_j |$ is the length of S_j and A, B, C are the penalisation parameters.

A network implementing this function is built as m horizontal layers, each layer corresponding to an element (flight). The number of units in each layer is equal to the number of possible pairings, which have to be chosen in an appropriate way and moreover inferior to an upper bound. Thus, for each element i and each subset j a unit u_{ij} is defined. The state of unit u_{ij} determines the value of the variable x_{ij}. In an attempt to diminish the size of the network, it is useful to deal only with units corresponding to variables x_{ij} for which $a_{ij} = 1$.

The connections will include:

- inhibitory connections between units within the same layer in order to guarantee that the same element (flight) is not assigned to more than one subset (pairing)
- excitatory connections between units within the same column to guarantee that all the elements of each subset are accounted for
- inhibitory connections between units of different columns and different layers for minimising the number of subsets
- external connections or bias connections expressing desirability unit u_{ij} is 'ON'.

This approach has some serious disadvantages:

- tuning the penalisation parameters is a difficult task. It is also necessary to increase A dynamically during relaxation
- the size of the network is very large especially in the cases of the set covering and set partitioning problems involved in real applications such as the airline crew scheduling problem.

However, a significant advantage should also be noted. When parameters are well adjusted, the method obtains feasible solutions for both the set partitioning (if such exists) and the set covering problems.

In the following sections, we will focus on another method, which is based on a feasible and order-preserving energy function. This approach requires a network with fewer units and interconnections than the previous one, and leads to approximative solutions of the set partitioning problem.

6.11.3 Boltzmann Machine Approach to the Set Partitioning Problem

The approach we will describe here is based on a transformation of the set partitioning problem (SPP) to the weighted node packing problem (WNP). The latter will then be solved on a Boltzmann machine neural network.

The transformation of the SPP to the WNP involves two steps. Beginning with the SPP, we will first show its equivalence to the set packing problem (SP). The resulting SP is shown to be equivalent to the WNP.

Equivalence between set partitioning - set packing

The set packing problem can be described as follows: we are given a set S and a collection $\mathfrak{I}$ of subsets of S. For each $S_j \in \mathfrak{I}$ there exists a 'cost' $C_j > 0$ related to it. We are asked to find a subcollection of $\mathfrak{I}$ of pairwise disjoint subsets, with maximal cumulative cost. Formally, the set packing problem can be described as follows:

$$\max C\,X \qquad\qquad (6.44)$$

subject to

$$AX \leq e, \qquad X_j = 0 \text{ or } 1, \qquad \forall j \in \{1,...,n\}$$

where A is an $m \times n$ matrix of zeros and ones having the same interpretation as in SPP, C is an arbitrary n-vector and $e = (1,...,1)$ is an m-vector.

If we now consider, for the initial SPP, a sufficiently large number Θ, for example

$$\Theta \geq \sum_{j=1}^{n} C_j \qquad\qquad (6.45)$$

and set $\tilde{C}^T = \Theta e^T A - C^T$, then the following relation holds:

$$\min C^T X \qquad \Theta m - \max \tilde{C}^T X$$
$$\Leftrightarrow$$
$$AX = e \qquad AX \le e$$

where

$$X_j = 0 \text{ or } 1 \quad \forall_j = 1,...,n$$

Thus, the set partitioning problem is equivalent to a set packing problem. Consequently, we can generate a set packing problem, which has the same set of optimal solutions as the set partitioning problem, whenever the latter is feasible (Balas et al, 1976; Salkin, 1975). A proof of this is supplied by Garfinkel and Nemhaucher in (1969).

Equivalence between set packing - weighted node packing problems
The weighted node packing problem (WNP) can be described as follows: given an undirected graph $G(V, E)$, we are asked to find a subset of edges such that no two edges share a node and the total weight of the subset is maximal. A mathematical formulation of WNP is the following:

$$\max C X$$

subject to

$$AX \le e, \quad X_j = 0 \text{ or } 1, \quad \forall_j \in \{1,...,n\}$$

where A is the m x n node-edge incidence matrix of G, C is an arbitrary n-vector of weights and $e = (1,...,1)$ is an m-vector. If $C_j = 1 \quad \forall i = 1,..., n$
the above problem is to find in $G(V, E)$ a maximal subset of nodes which are mutually non-adjacent.

The set packing problem acquired from the original set partitioning problem, is also equivalent to a weighted node packing problem whose costs $\hat{C}^T = (C_1,...,C_n)^T$ are the same for each node as those of set packing, ie

$$\hat{C} = \left(C^T e |A_1| - C_1, C^T e |A_2| - C_2, ... C^T e |A_n| - C_n \right) \qquad (6.46)$$

or

$$\left(\hat{C} = C^T e_n e_m^T A - C^T \right)$$

where $| A_j | =$ the number of 1s in column j.
Actually, the associated WNP is constructed as follows

(i)   denote by A_j the jth column of the matrix A of the generated set packing problem

(ii) the intersection graph $G_A = (N,E)$ of A has one node for every column of A, and one edge for every pair of non-orthogonal columns of A (ie $(i,j) \in E$ if and only if $A_i A_j \geq 1$)

(iii) let A_G be the node-edge incidence matrix of G_A, and denote by WNP the weighted node packing problem whose weights $\hat{C}_j$ are the same for each node as those of set packing

(iv) then, WNP is defined as follows:

$$\max \hat{C}^T X$$
$$A_G^T X \leq e_q \tag{6.47}$$

where

$$X^T = [X_1, X_2, ..., X_n]$$
$$X_j = 0 \text{ or } 1, \quad \forall_j \in \{1, ..., n\}$$

and e_q a column 1-vector with q elements, $q = |E|$.

We conclude that the solution X is a feasible (optimal) solution to set packing if and only if it is a feasible (optimal) solution to the weighted node packing problem (Lemke et al, 1971). That is, a subset of vertices is independent (contains no edges) and is therefore a solution to the node packing problem if and only if the corresponding columns of A do not intersect.

Now, having shown the equivalence between the SPP, the SP and the WNP, it is obvious that one way of solving the set partitioning problem as well as the set packing problem is to solve the associated weighted node packing problem.

6.11.4 The Neural Network Architecture

In order to solve the set partitioning problem with a neural network, the equivalent WNP will be solved instead. This will be done with a Boltzmann machine with n units, each unit corresponding to a subset S_j. Thus, for each subset S_j the state of a unit u_j determines the value of the variable X_j.

There are two kinds of connections between the computing elements, excitatory connections which make it more likely that the neuron receiving them will be firing, and inhibitory connections, which make the neuron receiving them less likely to be firing. There are excitatory connections on each unit (external connection) and between units, which represent disjoint subsets. There are inhibitory connections between units, which represent no disjoint subsets. Information is represented by a set of connection strengths *a priori* specified. These strengths are symmetrical, ie two computing elements have the same inhibitory or excitatory effect on each other. Strengths over excitatory connections have non-negative real values, while strengths, over inhibitory connections have negative real values. More precisely, the connection strengths are defined as follows:

$$W = (w_{ij}) = \begin{cases} \Theta|S_i| - c_i & \text{if } i = j \\ -\{\max\{\Theta|S_i| - c_i, \Theta|S_j| - c_j\} + \varepsilon\}\delta_{ij} & \text{if } i \neq j \end{cases} \quad (6.48)$$

where $i = 1,...,M$, $j = 1,...,N$ and $|S_i|$ is the cardinality of the subset represented by the unit u_i, c_i is the cost associated to the subset S_i, $\Theta = \sum_{i=1}^{N} c_i$. ε is a positive, very small real and $\delta_{ij} = 1,(0)$ if pairs S_i, S_j are not disjoint (otherwise).

The computing elements are assumed binary, that is, can only be 'on' (firing) or 'off' (no firing). The set of states of all computing elements at a moment in time is represented by a state vector u, corresponding to the solution vector, x^u.

Figure 6.8 shows an instance of the SPP in a set formulation and also the corresponding bipartite graph. The Boltzmann machine network associated with this example is given in Figure 6.9 and the connection strengths, the matrix W, is given in Figure 6.10.

SS elem	S_1	S_2	S_3	S_4	S_5
a	1	0	1	0	0
b	0	1	0	1	0
c	1	0	1	1	0
d	0	1	0	0	0
e	1	0	0	0	1
c^T	1	1	1	1	1

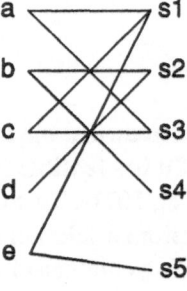

Fig 6.8 An instance of the SPP : (i) in a set-formulation and (ii) the corresponding bipartite graph. The cost vector equals the unit vector. There are two feasible solutions, $x^1 = (1,1,0,0,0)$ and $x^2 = (0,1,1,0,1)$

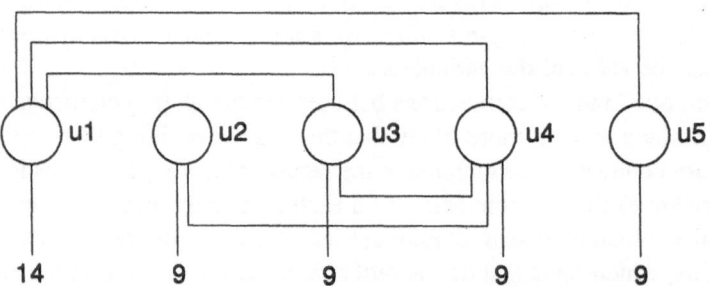

Fig 6.9 The Boltzmann machine network associated with Figure 6.8. The connections with strengths of zero, are removed

$$W = (w_{ij}) = \begin{bmatrix} 14 & 0 & -14 -\varepsilon & -14 -\varepsilon & -14 -\varepsilon \\ 0 & 9 & 0 & -9 -\varepsilon & 0 \\ -14 -\varepsilon & 0 & 9 & -9 -\varepsilon & 0 \\ -14 -\varepsilon & -9 -\varepsilon & -9 -\varepsilon & 9 & 0 \\ -14 -\varepsilon & 0 & 0 & 0 & 4 \end{bmatrix}$$

Fig 6.10 The connection strengths for Figure 6.9 including information about total cost, number of elements to cover, and number of elements and its costs for each subset

Once defined, the network operates in a probabilistic way and in a sequential mode. The computing elements are allowed to change their state only once at a time. Each computing element i, is chosen randomly, and it is allowed to change its state u_i with probability

$$p(u_i) = \varphi(\Delta C / T) \tag{6.49}$$

where

$$\varphi(x) = \frac{1}{1+e^{-x}}, \quad \Delta C = (1-2u_i)(\sum_j w_{ij}u_j + w_{ii})$$

and T is a positive real number which controls the behaviour of the system.

Now since the connection strengths are assumed symmetrical the network will always settle into a local maximum (Vasko et al, 1984; Poliac, 1987; Aarts, 1987). Moreover, if the connection strengths are well chosen for the undersolving optimisation problem, that is, they properly capture information about satisfaction of the constraints and optimality of the objective function, then the above local maximum would represent an appropriate solution of the problem. A good measure of the appropriateness of the connection strengths is the satisfaction, by the synthesised global consensus function, of the two following properties (Aarts, 1987):

(i) each local optimum gives rise to a solution which satisfies all the problem's constraints

(ii) higher local optima give rise to lower cost solutions.

The significance of the previous properties is obviously of great importance. The first implies that the network always provides a feasible solution. The second implies that the higher the consensus of the network, the better the solution provided.

Concerning the proposed network, which implements the SPP with the connection strengths W defined in (i), it can be demonstrated that it holds the second property. However, the first property is not always true. In fact, the set U of all local maxima is composed of two subsets U^1 and U^2, where U^2 includes only vector states which are feasible solutions of SPP, and U^1 includes only vector states which are partial solutions of the SPP. It can be proved that property (i) holds in a general sense, by considering the valid partial solutions.

6.11.5 The Choice of Model Feature Parameters

The parameter ε, included in the definition of the connection strengths W, assures valid partial solutions (mutual exclusion), while also being significant for the overall behaviour of the network. In fact, small values attributed to it define a weak inhibition between the connected units and thus more freedom to all units to pass by the firing state, according to the sigmoid shape of equation 6.49. The capture, therefore, of high consensus state vectors, is more likely to happen. In contrast, when large values of the parameter ε are allowed, then strong inhibitions are defined, and thus the possibilities for some units to fire are reduced. The result is that the network gets stuck quickly in a local maximum, satisfying problem constraints and less optimality of the objective function. We think that it would be advantageous at the beginning of the method to let the parameter ε have values near zero, and later during the evolution of the network, as it has captured more and more information about the objective function, to let it increase progressively. However, this idea has not yet been tested: in all of the work we have carried out, ε was set equal to 0.1, following empirical trials of a range of constant values.

The parameter T included in equation 6.49, which is analogous to the temperature in physical applications of the Boltzmann machine, is crucial for both solution quality and computation time.

Several decrement laws were tested:

- a geometric decrement (used also in Kirkpatrick *et al*, 1983), in which the nth Metropolis search method is applied with temperature $T_n = (0.9)^n T_0$, where T_0 is the initial temperature
- a logistic type decrement with $T_n = \dfrac{T_{n-1}}{1 + \log(t_n)}$ where $t_n = t_{n-1} \times (1 + rate)$ and $t_0 = 1.0$, $n = 1, 2, \ldots$, and *rate* is a small adjustable parameter, the role of which is discussed in the section on computational results
- $T_n = \dfrac{To}{1 + \log(t_n)}$ where parameters are as previously
- a polynomial time cooling schedule, which was presented by Aarts (1987) with

$$T_n = \frac{T_{n-1}}{1 + \dfrac{T_{n-1}\log(1+rate)}{3\sigma_{T_{n-1}}}}, \quad n = 1, 2, 3, \ldots \tag{6.50}$$

where *rate* determines the amount by which the value of T is decreased.

6.11.6 Comments

Extensive testing of the Boltzmann machine simulator has been carried out for both artificial and real airline crew datasets of various dimensions, using neural network software running on an IBM PC. The aims were to investigate:

(i) the effectiveness of the method for finding near-optimal solutions, and
(ii) the dependence of performance on dataset size, ie the scaling behaviour.

The results indicate first of all that the method is sound, and produces good solutions reliably. Computing times inhibited efficient exploration of the scaling behaviour, but nevertheless permitted the initial conclusions that there is likely to be a size limitation (independent of computing power) which would fall short of commercial airline requirements. It was therefore more fruitful to investigate methods for improving network architecture and performance, rather than pursue a detailed analysis and presentation. Hence the interim comments on performance are presented in summary form below, and are followed by accounts of two new developments which show promise for solution of large-scale problems.

The structure and the connection matrix for a Boltzmann machine-type neural network seem to be appropriate for solving set partitioning problems. The synthesised consensus function has suitable features, which produce good results when the parameters involved in the model operation are well adjusted. When the network is lying in a stable state, a valid partial solution with low cost is always obtained.

Furthermore, the higher the consensus of a stable state the better the partial solution, with a higher probability for a feasible solution. Additionally, near-optimum consensus stable states give rise to feasible solutions, while optimum stable states provide the optimum solution of the SPP.

We point out that there are still many possible improvements to this method, for both solutions quality requirements and computational time. Particularly, the use of an improved simulated annealing in our Boltzmann machine network will improve the solutions obtained and it will also considerably reduce the computing cost. Moreover, adopting the parallel updating rule (at least on the computing elements with zero connection strengths), we can expect considerable improvements in speeding up the process.

Some of the more important features of the model are summarised as follows;

- it always gives rise to valid partial solutions with few elements not covered, when the instances are not feasible
- it performs well for instances of super-sparse matrices (density 1-3%)
- it deals with the cardinality cases of the SPP, where costs are not associated with the subsets
- the network would benefit from the parallel updating rule, in terms of improving the speed of convergence.

Furthermore, the model needs more tests on large real-world applications in order to be able to conclude its effectiveness. However, it has the ability to accept more constraints such as precedence constraints, in order to provide valid rotations when we deal with real-world applications such as the airline crew scheduling problem. Such information should be captured by the connections among the computing elements as penalties incorporated in the connection strengths.

6.12   Improving the Performance of the Network

The project has successfully demonstrated a method for mapping the aircrew scheduling problem (via the set partitioning problem) onto a Boltzmann machine. However, it is clear that the performance of such a system on a full scale problem would not be acceptable. Two other approaches to the network architecture (both modifications of the Boltzmann machine) have been investigated.

The first of these is the so-called *guard network* system (Adorf *et al*, 1990). An implementation of this method has confirmed the reported performance improvement although it has yet to be tested on full scale aircrew data.

The second technique, the *multiscale* approach, has also been investigated. An implementation of the method has been integrated into the existing Boltzmann machine simulator and is presently being evaluated.

Both these methods show significant improvements over the simple Boltzmann architecture. It is also apparent that they provide new ways of approaching the overall design of a constraint-based optimisation system, in that it may prove possible to improve performance further by alternative ways of partitioning the problem.

6.12.1   Guard Networks

This approach follows the work of Adorf *et al* (1990). The problem with the naïve Hopfield net is that the evolution tends to settle into local minima with fewer than N neurons active. In this condition no further network updates are allowed and the net cannot escape into the deeper global minima. This problem has long been recognised with symmetric, zero diagonal Hopfield networks. Adorf and Johnston's solution is to employ a second network to monitor the performance of the main net. They call the second network a guard network.

The idea here is to provide a guard which monitors each row and enforces the desirable condition that at least one neuron in each row is active. Figure 6.10 shows the additional architecture of the guards.

The guard bias γ is set to a value $< \theta$, the guard input weight. The guard output weight is set to ϕ, a large positive value, large enough to overcome all inhibitory weights connected to neurons in this row. The guard neuron only fires if the whole row is inactive, otherwise it does nothing.

A guard neuron is assigned to each row of the main network. The guard is asymmetrically coupled to the row and its input weights and threshold are set so that it only fires when none of the neurons in the row that it guards are active. When the guard fires it provides a large positive stimulus to *all* the neurons in its row, the size of the stimulus is large enough to overcome all the potential inhibitory stimuli that any neuron in the row could be receiving: hence all the neurons are now encouraged to fire. The price to be paid for this desirable behaviour is that the network viewed as a whole is now asymmetric, so monotonic reduction of the total energy function is no longer guaranteed.

Adorf and Johnston assert that the best update scheme for the guards is to make them operate on a much faster timescale that the rest of the net: after a neuron is updated, its guard should also be updated to ensure that the state of the guards always reflects the state of the main net.

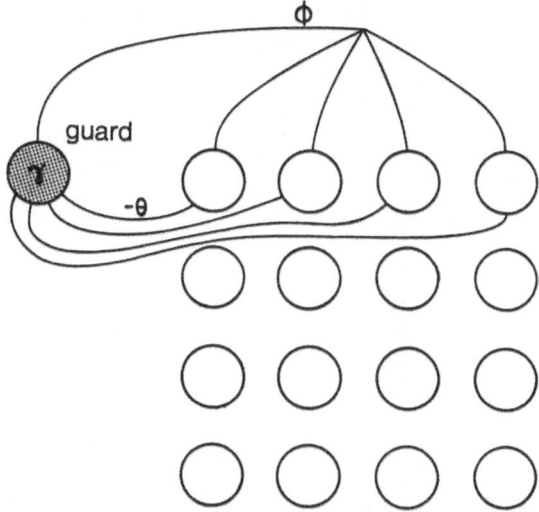

Fig 6.10 Additional guard neurons enforce the condition that at least one neuron in each row remains active

On its own, we found that this scheme tends to lead to oscillation of the energy function: random selection of the next neuron to update is not appropriate.

The paper suggests a heuristic for choosing which neuron to update. First, select a row at random and update its guard, then choose the neuron in the row which is *most inconsistent* with the current state of its input and update this neuron. That is, select the neuron which corresponds to

$$\text{neuron } ij \text{ with max} \begin{cases} x_{ij} \text{ if } y_{ij} = 0 \text{ and } x_{ij} \geq 0 \\ -x_{ij} \text{ if } y_{ij} = 1 \text{ and } x_{ij} < 0 \end{cases}$$

If more than one neuron is *most inconsistent* then choose at random which to update.

Another important implementation issue is the storage of the weight matrix $W_{ij,mn}$ which is not only sparse, but often its elements are not needed for calculation. Elements of the weight matrix need only be calculated when the appropriate neuron is active.

Running on a 16MHz 386 PC, we implemented the above scheme for solving the *N*-Queens problem in C. For comparison we also solved the problem using Prolog, a logic programming language with built-in backtracking facilities. Table 6.4 shows the results of the simulation.

The guard network is certainly effective for this problem. For larger sized problems we were unable to find any solutions using the conventional approach, but the guard network successfully found solutions. We expect that the execution times could be reduced by a factor of five with optimised coding of the algorithm.

Table 6.4 Comparison of execution times for compiled Prolog and Guard
Network solutions to the N-Queens problem. The network times are approximate
times to reach a solution with 50% probability

Number of Queens	Prolog solution	Guard net solution
8	1 s	8 s
10	1 s	15 s
12	2 s	30 s
16	71 s	40 s
20	1,929 s	60 s
32	>8 h	300 s

6.12.2 The Multiscale Approach

The *multiscale method* originates from related work in the field of numerical analysis
(Chatelin and Miranker, 1982; Miranker, 1981). A general description of the
multiscale approach is presented in (Mjolsness *et al*, 1990 and 1991), but it is quite
difficult to apply the method directly to the set partitioning problem. Our approach,
although following the general guidelines, is quite different and leads to a better
insight into the operation of Hopfield-type networks.

The basic principle is that we can speed up convergence in a large optimisation
problem by introducing a smaller approximate version at a coarser scale and
alternating the relaxation steps between the fine scale and the coarse scale instances of
the problem.

The development of a multiscale approach suitable for the Boltzmann machine is
not straightforward, because a transformation of special nature must be applied to the
original fine scale variables.

Suppose that we have a Hopfield-type neural network with n binary units (n-net)
operating as a Boltzmann machine. The connection strengths of the network are w_{ij},
with $w_{ij} = w_{ji}$ and $w_{ii} = 0$ ($i = 1,...,n$, $j = 1,...,n$). Also the threshold values are θ_i, ($i = 1,...n$). Suppose now that at the moment we wish to switch to a smaller network with p
nodes (which we shall call the p-net) the state vector of the network is $\mathbf{y}^* = (y_1^*,...,y_n^*)$,
where the components $y_1^*, i = 1,...,n$ assume binary values.

We denote by $\xi = (\xi_1, ..., \xi_p)$ the state of the p-net. Since the p-net must also
operate as a Boltzmann machine, the components, ξ_i, $i=1,..., p$ must be binary. The
mapping from the n-net to the p-net is based on partitioning the nodes of the n-net into
p disjoint groups, so that each group corresponds to a node of the p-net. Hence, in
order to construct the p-net, we must specify how the nodes of the n-net are grouped
together.

We shall use the notation $s(k)$ to specify the number of nodes corresponding to
group k ($k=1,...,p$) (ie the size of the group). Then the correspondence between groups

of nodes in the n-net and nodes in the p-net can be expressed by means of the mapping function $v(k,k')$ ($k=1,...,p, k'=1,...s(k)$) as follows. The equation
$v(k,k') = i$, ($i=1,...,n$) means that node i of the original network corresponds to the k'th element of group k. Thus, the mapping function represents each node i of the n-net in terms of the index k of the group to which it belongs and the respective index k' within that group.

In the case that the state of the n-net is $\mathbf{y}$ the corresponding consensus function $C(\mathbf{y}^*)$ can be written:

$$
\begin{aligned}
C(\mathbf{y}^*) &= \tfrac{1}{2}\sum_{i=1}^{n}\sum_{j=1}^{n} y_i^* y_j^* w_{ij} + \sum_{i=1}^{n}\theta_i y_i^* \\
&= \tfrac{1}{2}\sum_{k=1}^{p}\sum_{k'=1}^{s(k)} y_{v(k,k')}^* \sum_{l=1}^{p}\sum_{l'=1}^{s(l)} y_{v(l,l')}^* w_{v(k,k')v(l,l')} \\
&\quad + \sum_{k=1}^{p}\sum_{k'=1}^{s(k)} y_{v(k,k')}^* \theta_{v(k,k')}
\end{aligned}
\tag{6.51}
$$

Suppose that while being in state $\mathbf{y}$ we decide to switch to the small network. Suppose also that we have a way of constructing the p-net and, after operation of that network, its final state is $\xi = (\xi_1,...,\xi_p)$. Then the transformation for obtaining the new state $\mathbf{y} = (y_1,...,y_n)$ of the original network has the following form:

$$
y_{v(k,k')} = y_{v(k,k')}^* + u_{v(k,k')}\xi_k
\tag{6.52}
$$

The quantities $u_{v(k,k')}$ (or u_i) are defined as

$$
u_{v(k,k')} = 1 - 2 y_{v(k,k')}^*, \quad k=1,...,p; \quad k'=1,...,s(k)
\tag{6.53}
$$

This means that if $y_i^* = 0$ then $u_i = 1$ and if $y_i^* = 1$ then $u_i = -1$. The advantage of the above transformation (6.52) is that it can assure that the new state vector $\mathbf{y}$ of the n-net contains binary components. The physical interpretation of the transformation is very interesting: for each group k, $k = 1,...,p$, if the final state of unit k in the p-net is 1 (ie $\xi_k = 1$), all nodes belonging to group k change their state (with respect to the state vector $\mathbf{y}$). Inversely, if the resulting value ξ_k is 0 (ie $\xi_k = 0$), then the nodes belonging to group k retain the state they had before switching to the p-net.

The connection weights x_{kl}, ($k = 1,...,p$; $l =1,...,p$) of the p-net can be computed by means of the following argument. The consensus of the new state $\mathbf{y}$ of the n-net is:

$$
\begin{aligned}
C(\mathbf{y}) &= \tfrac{1}{2}\sum_{k=1}^{p}\sum_{k'=1}^{s(k)}\sum_{l=1}^{p}\sum_{l'=1}^{s(l)} y_{v(k,k')} y_{v(l,l')} w_{v(k,k')v(k,l')} \\
&\quad + \sum_{k=1}^{p}\sum_{k'=1}^{s(k)} y_{v(k,k')}\theta_{v(k,k')}
\end{aligned}
\tag{6.54}
$$

Using the transformation equation (6.52) and the consensus definition we can show that

$$C(\mathbf{y}) = C(\mathbf{y}^*) + C(\xi) \tag{6.55}$$

where the quantity $C(\mathbf{E})$ is given from the following:

$$C(\xi) = \tfrac{1}{2} \sum_{k=1}^{p} \sum_{l=1}^{p} x_{kl} \xi_k \xi_l + \sum_{k=1}^{p} w_k \xi_k \tag{6.56}$$

where

$$x_{kl} = \begin{cases} \sum_{k'=1}^{s(k)} \sum_{l'=1}^{s(l)} u_{v(k,k')} W_{v(k,k')v(l,l')} u_{v(l,l')} & \text{if } k \neq l \\ 0 & \text{if } k = l \end{cases} \tag{6.58}$$

and the threshold w_k of each node k ($k = 1,...,p$) of the small network is:

$$w_k = \sum_{k'=1}^{s(k)} \theta_{v(k,k')} u_{v(k,k')} + \sum_{k'=1}^{s(k)} \sum_{l=1}^{p} \sum_{l'=1}^{s(l)} u_{v(k,k')} W_{v(k,k')v(l,l')} y^*_{v(l,l')}$$

$$+ \tfrac{1}{2} \sum_{k'=1}^{s(k)} \sum_{l'=1}^{s(k)} u_{v(k,k')} W_{v(k,k')v(l,l')} u_{v(l,l')} \tag{6.57}$$

It is important to note that, since the original network has symmetrical weights (ie $w_{ij} = w_{ji}$ for all $i = 1,...,n; j = 1,...,n$), we can easily observe from equation (6.57) that the p-net has also symmetrical weights (ie $x_{ki} = x_{lk}$ for all $k = 1,...,p;\ l = 1,...,p$). This symmetric behaviour ensures convergence of the p-net when used in Hopfield-type operations. Another important observation concerns the computation of the values x_{kl}, using equation (6.57). The importance lies in the fact that in order to compute the weight x_{kl} only the weights w_{ij} connecting the nodes belonging to groups k and l are involved. This fact implies on one hand a reduced complexity and on the other hand that there is an inherent parallelism in the transformation task, ie the weights of the small network can be computed in parallel, and each computation requires a different portion of the connection matrix W.

As a conclusion we can state that once being in a state $\mathbf{y}$ of the original network, we can create a small p-net as described above. Then we can perform iterations on this network until it reaches a large positive value of the energy consensus. Using equation (6.52) we can return to the original network whose consensus is now given by equation (6.55), ie it is increased by an amount equal to the small network's consensus.

An important point in developing the overall strategy is to specify the conditions for switching between the small and large nets. This must be considered in conjunction with the specification of the cooling schedule. There are several choices. For example, each network could have its own cooling schedule, ie we could start the original network from an initial temperature and perform some iterations (temperature decrements) in that network, and then switch to a small network where we perform the cooling procedure from a suitable initial temperature until we reach a low temperature.

Then we could return to the original network where we continue the cooling schedule from the point it was previously interrupted. Another possible strategy considers a unified cooling schedule for the entire system. This means that we keep a global temperature variable which is successively decremented either in the n-net or in the p-net. We chose to experiment with the second approach, which seems to be more elegant, and to keep one cooling schedule independent of the network alternations.

The basic element that should be kept in mind when designing the switching strategy is the overall execution time of the strategy. There are three main benefits when operating on the coarse-scale network:

(i) Since the size of the p-net is considerably smaller than the size of the original network, the computation time required to perform an iteration is considerably reduced

(ii) The state-space of the p-net is logarithmically reduced with respect to the state space of the n-net and the number of iterations required to perform a search of the state-space of the p-net is very low. Thus, the number of iterations that should be performed at each temperature on the p-net is significantly lower.

(iii) As stated above, the notion of group update that is incorporated in the multiscale approach assists the original network in escaping from energy traps and can lead to higher consensus maxima.

Due to (i) and (ii) above, the execution time associated with performing iterations on the p-net is a small fraction of the overall execution time.

On the other hand, there is an overhead associated with the cost of the transformation, ie with the cost of creating the p-net and computing its weights. Since this computation depends on the current state of the original network, it must be performed every time we wish to switch from the original network to the coarse-scale network. This is a fact that imposes a restriction on the number of switches between the networks. It should be noted that the computational cost of the switch from the coarse-scale to the fine-scale network is negligible. Therefore, in order to achieve acceleration it is important that the benefit from increasing the consensus using the p-net exceeds the transformation overhead.

The strategy that we have selected for our experiments considers a unified cooling schedule, ie there is a global temperature updated by all networks. We start from a suitable initial temperature and perform iterations on the original network. When the temperature reaches a specific value (ie after some temperature decrements), we construct the corresponding coarse-scale p-net and continue the execution (iterations and decrements of the global temperature) using this network. When the temperature reaches a certain value again we return to the n-net where the operation continues. This procedure is repeated until the temperature reaches a very low value (near zero). The rate at which the temperature decreases is the same for all networks.

In order for the overall computation time to be reduced, the number of switches must be properly specified. Also the temperature values at which switching occurs must be suitably adjusted. These issues are discussed in the next section.

6.12.3 Implementation and Preliminary Experimental Results

In order to experiment with the multiscale approach, the code of the Boltzmann machine simulator has been adapted. Attention has been paid in order to provide a flexible implementation, which allows several alternative switching strategies with slight modifications in the source code.

As already stated, we have chosen to experiment with a unified cooling strategy, during which the global temperature is suitably decremented on the coarse-scale and the fine-scale networks. In order to achieve acceleration, we have to adjust the number of switches and the temperature values at which switching should occur.

The overhead associated with the creation of the coarse scale p-net, ie with the computation of the connection weights x_{ki} and the threshold values w_k, $k = 1,..., p$, appears to play a critical role in the above decisions. As already stated, we wish the computational benefit obtained by performing iterations on the p-net (instead of using the n-net) to exceed the construction overhead. It is easy to estimate the complexity of that overhead using equations 6.57 and 6.58. We consider that the sizes of all groups k, $k = 1,...,p$ are equal (or almost equal, ie $s(k) = n/p$ or $s(k) = n/p + 1$ (which is the experimental case). In this case, it is easy to observe that the computational cost of creating the p-net turns out to be (approximately) equivalent to the computational cost of performing $1.5n$ iterations of the fine-scale n-net. This computational cost is not too high, since it is equal to (or smaller than) the time needed for the original network to perform a search of the state-space, while being at a specific temperature. Thus, if we use the p-net to perform iterations for a sufficient number of temperature values (instead of performing those iterations on the n-net), the expected benefit will be substantial.

In experiments, we have tried to evaluate the multiscale method using several sizes of the coarse-scale network and strategies involving different numbers of switches and different switching temperatures. Real world data provided by the pairing generator are used. The method is tested in the case of large networks consisting of thousands of nodes (typically from 1000 to 5000 nodes), while the size of the p-net does not exceed 200 nodes. In each experiment we are obliged to execute both the original and the modified Boltzmann machine code in order to compare the results. This fact makes the evaluation task time consuming. We are concerned with both the quality of the solution and the speed of convergence.

Several choices concerning the temperature decrement rate, the size of the p-net constructed at each switch and the points of switching have been investigated. Preliminary experiments indicate a substantial reduction of the overall execution time and, in most cases, a better quality of solution. However, it is clear from the present discussion that the issues involved are many and can play a significant role in the performance of the model. Hence, further investigation and more experiments are needed to evaluate the method and establish the best strategy to follow in solving the crew scheduling problem.

References

Aarts E H L and van Laarhoven P J M (1985) Statistical cooling: a general approach to combinatorial optimisation problems. Philips Journal of Research *40*, 193-226

Aarts E H L and Korst J H M (1987) Boltzmann machines and their applications. Proc of Parallel Architectures and Languages, European Conference, Eindhoven, The Netherlands, June 1987

Aarts E H L and Horst J H M (1989) Simulated annealing and Boltzmann machines. Wiley & Sons

Adorf H M and Johnston M D (1990) A discrete stochastic neural network algorithm for constraint satisfaction problems. International Conference on Neural Networks, San Diego CA, USA, June 1990 *3*, 917-924

Angeniol B, De la CroixVaubois G and Le Texier Y (1988) Self-organising feature maps and the TSP. Neural Networks *1*, 289-293

Baker E and Fisher M (1981) Computational results for very large air crew scheduling problems. The International Journal of Management Science *9*, 6, 613-618

Balas E (1965) An additive algorithm for solving linear programs with zero-one variables. Operations Research *13*, 517-546

Balas E and Padberg M W (1976) Set partitioning: A survey. SIAM Review *18*, 4, 710-760

Balas E and Samuelson H (1973) Finding a minimum node cover in an arbitrary graph. MSRR 325, Carnegie-Mellon Univ, Pittsburgh, USA

Balas E (1980) Cutting plans from conditional bounds: A new approach to set covering. Math Prog *12*, 19-36

Balas E and Ho A (1980) Set covering algorithms using cutting plans, heuristics and subgradient optimisation: A computational study. Mathematical Programming *12*, 37-60

Bodin L, Golden B, Assad A and Ball M (1983) Routing and scheduling of vehicles and crews - the state of the art. Computer and Operations Research *10*, 2, Special Issue

Bodin L D, Kydes A S and Rosenfield D B (-) Approximation techniques for automated manpower scheduling. Research paper, UPS/UMTA-1, SUNY, Stony Brook, New York, USA

Burr D J (1988) An improved elastic net method for the TSP. IEEE Int Con on Neural Networks *1*, 69-75, San Diego CA, USA, 1988

Christofides N (1975) Graph theory: An algorithmic approach. Academic Press

Christofides N and Korman S (1975) A computational survey of methods for the set covering problem. Management Science *21*, 5, 591-599

Chvatal V (1979) A greedy heuristic for the set covering problem. Mathematics of Operations Research *4*, 3, 233-235

Couvrey B and Yansouni B (1977) A method of estimating manpower levels at an airport. ACIFORS Symposium Proceedings, Germany, 340-355

Crainic T (1978) The crew scheduling problem of an aviation company. Publication 122, Centre de Recherche sur les Transports, Université de Montreal, Canada

Dahl E (1987) Neural network algorithm for an NP-complete problem: Map and graph colouring. Proc IEEE First Int Conf on Neural Networks, San Diego CA, USA, 1987

Dantzig G B and Wolfe P (1960) Decomposition principle for linear programming. Operations Research *8*, 1, 100-111

Délorme J (1974) Contribution to the resolution of the covering problem: truncating methods. Thèse de Docteur Ingénieur, Université Paris VI

Durbin R and Willshaw D (1987) An analogue approach to the TSP using an elastic net method. Nature, 326, 689-691

Etcheberry J (1977) The set covering problem: A new implicit enumeration algorithm. Operations Research *25*, 5, 760-772

Foo Y and Takefuji Y (1988) Integer linear programming neural networks for job-shop scheduling. Proceedings of the IEEE Int Conf on Neural Networks, San Diego CA, USA, 1988

Garfinkel R S and Nemhauser G L (1969) The set partitioning problem: Set covering with equality constraints. Operations Research *17*, 848-856

Gomory R E (1958) Outline for an algorithm for integer solution to linear programs. Bulletin of the American Mathematical Society *64*, 5

Hegde S U, Sweet J L and Levy W B (1988) Determination of parameters in a Hopfield/tank computational network IEEE International Conference on Neural Networks, II, 291-298, San Diego CA, USA, 1988

Ho A C (1982) Worst case analysis of a class of set covering heuristics. Math Prog *23*, 170-180

Hopfield J J (1982) Neural networks and physical systems with emergent collective computational abilities. USA Proc National Academy of Sciences *79*, 2554-2558

Hopfield J J and Tank D W (1985) Neural computation of decisions optimisation problems. Biological Cybernetics *52*, 141-152

Hu T C (1972) Integer programming and network flows. Addison Wesley

Karmarkar N (1984) A new polynomial time algorithm for linear programming. Combinatorica *4*, 373-395

Karp R (1972) Reducibility among combinatorial problems. Complexity of computer computations, 85-103, Plenum Press, New York, USA

Khachian L G (1979) A polynomial algorithm for linear programming. Dolklady Akad Nauk USSR, 244(5) 1093-1096. Translated in Soviety Math Dolcady *20*, 191-194

Kirkpatrick S, Gelatt C and Vecchi M (1983) Optimisation by simulated annealing. Science, 220:671

Lemke C E, Salkin H M and Spielberg K (1971) Set covering by single-branch enumeration with linear programming subproblems. Operational Research *19*, 998-1022

Lin S (1965) Computer solutions of the TSP. Bell Tech Journal *44*, 2245-2269

Lundy M and Mees A (1986) Convergence of an annealing algorithm. Math Prog, 34, 111-124

Metropolis N, Rosenbluth A, Rosenbluth M, Teller A and Teller E (1953) Equation of state calculations by fast computing machines. J ChemPhysics *21*, 1087-1092

Marsten R E (1974) An algorithm for large set partitioning problems. Management Science *20*, 5, 774-787

Mitra D, Romeo F and Sangiovanni-Vincentelli A (1986) Convergence and fine time behaviour of simulated annealing. Advances in Applied Probability *18*,747

Nicoletti B, Natali L and Scalas M (1976) Manpower staffing at an airport: Generalised set covering problem. 1976 ACIFORS Symposium Proceedings, Miami, Florida, USA, 473-503

Peretto P and Niez J J (1986) Stochastic dynamics of neural networks. IEEE Transaction Systems, Man and Cybernetics *16*, 1

Pierce J C (1968) Applications of combinatorial programming to a class of all-zero-one integer programming problems. Management Science *15*, 191-209

Pierce J F and Lasky J S (1973) Improved combinatorial programming algorithms for a class of all-zero-one integer programing problems. Management Science *19*, 528-543

Poliac M O, Lee E B, Slagle J R and Wick M R (1987) A crew scheduling problem. Proc IEEE First International Conference on Neural Networks, 779-786

Rousseau J M (1984) Crew scheduling methods in the operation and planning of transportation systems. Transportation Planning Models, 439-472

Rubin J (1973) A technique for the solution of massive set covering problems with application to airline crew scheduling. Transp Sci 7, 1 34-48

Salkin H M and Koncal R D (1973) Set covering by an all integer algorithm: Computational experience. J Assoc Comput Math *20*, 189-193

Salkin H (1975) Integer programming. Addison-Wesley

Saylor J and Stork D (1986) Parallel analogue neural networks for tree searching. AIP Conference Proceedings 151: Neural networks for computing

Segal M (1974) The operator-scheduling problem: A network-flow approach. Operations Research 22, 808-823

Spitzer M (1961) Solution to the crew scheduling problem. AGIFORS symposium, October 1961

Szu H and Hartley R L (-) Simulated annealing with Cauchy probability. Optics letter

Thiriez H (1969) Airline crew scheduling: A group theoretic approach. PhD dissertation, MIT, October 1969

Vasko F J and Wilson G R (1984) An efficient heuristic for large set covering problems. Naval Research Logistics Quarterly *31*, 163-171

Vasko F J and Wolf F E (1988) Solving set covering problems on a personal computer. Computer Operations Research 15(2) 115-121

Wilhelm E (1975) Overview of the Rucus package driver run cutting program. Workshop on Automated Techniques for Scheduling of Vehicle Operators for Urban Public Transportation Services, Chicago Ill, USA

Wilson G V and Pawley G S (1988) On the stability of the TSP algorithm of Hopfield and Tank. Biol Cyber, 58, 63-70

Chapter 7

Methodology

7.1 Introduction

7.1.1 What is Methodology?

Methodology is the study of a whole body of methods used in a particular branch of activity. By this definition, we will not attempt to cover the complete methodology of applying neural networks to industrial applications, as we are not going to concern ourselves with the more detailed aspects of the implementation of neural networks. Thus we will not attempt to give more than the barest of statements concerning the detailed structure of nets: the numbers of nodes to use, the best learning algorithms, the most appropriate activation functions etc. We believe this would be unwise for a number of reasons: the detailed discussions derived from 3 years research and many different applications would be far too long; and, more significantly, we feel that any such conclusions would date rather rapidly as new techniques arise in the literature.

Instead we will concentrate on distilling the essence of the ideas which arose from the whole ANNIE exercise. We will extract and recapitulate where appropriate specific points of relevance discussed in the earlier applications chapters, and will also derive more general rules which can be seen only when considering the complete picture. Thus we hope to address the higher level questions such as:

- Why use a neural network at all?
- What are the potential advantages and disadvantages?
- What should you know about your problem before you start?
- Which types of real industrial problem are best suited to network solutions?
- What is important about the data used for training?

We believe this approach will indicate above all else that you must understand your problem well if neural networks are going to prove to be of value: there are rarely simple solutions to complex problems nor is there an easy prescription for how and where to implement a particular approach. You must have a precise idea of what you want to do and must know the best existing methods. Only then will neural networks be seen to be what they are: a useful additional tool in the armoury of data analysis methods which should be used when and where appropriate.

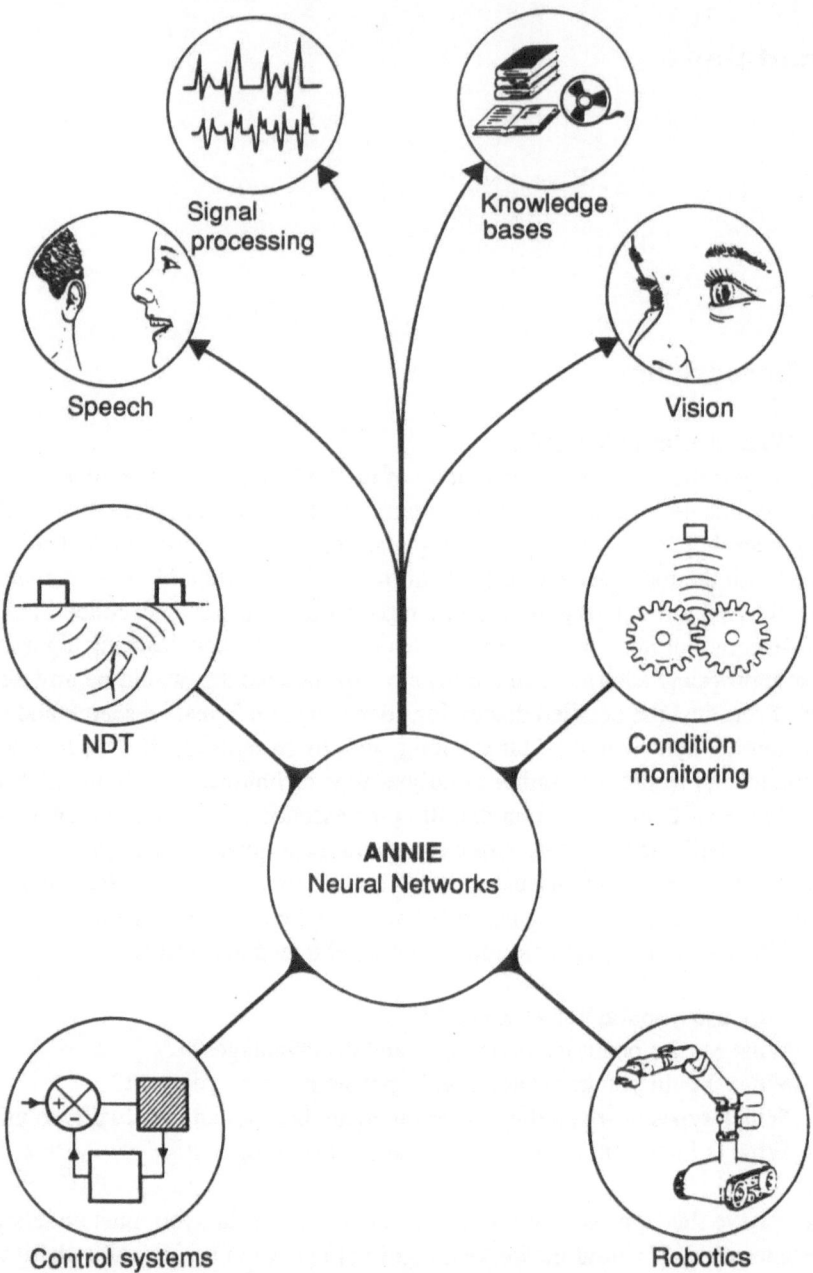

Figure 7.1 The ANNIE project was founded on the immediate needs of manufacturing industry, with a view to spin offs into long-term requirements within IT. The methodology derived allows an assessment of the potential impact of the technology to be determined in a range of different application areas.

7.1.2   Relationship to the ANNIE project

The ANNIE project set itself challenging aims and objectives. One of the main goals was to develop sufficient understanding through generic studies and specific applications that it would become apparent which areas (especially IT areas) are best approached using neural networks. Some of the areas originally conceived of as being of interest are illustrated in Figure 7.1. The methodology which has arisen addresses just this problem of identifying suitable candidates and of determining how best to implement a neural network solution.

The wide range of studies carried out allows us to draw general conclusions. However, as noted above, throughout the duration of the project the field was (and is still) a rapidly evolving one. Even with lessons that are general and which are consequently unlikely to change too rapidly, it should still be remembered that the IT area itself is evolving rapidly. It is therefore likely that the applications themselves will change with time, and that constant review with respect to the *fundamentals* of the methodology will be necessary.

7.1.3   Relationship to the Handbook

In the next section we will compare neural network techniques with related conventional methods which are used to carry out tasks of industrial significance. The relative advantages and disadvantages of these will be noted, and hence some indication of which sort of problem is best suited to which technique should become apparent.

In section 7.3 we will tackle some of the practical problems found when attempting to implement neural network solutions, paying particular attention to the problem of data representation. We shall where appropriate refer back to the implementations discussed in earlier chapters, but the aim will be to note general points believed to be of relevance to a wide range of application areas.

Section 7.4 looks in more depth at the specific application areas covered by ANNIE: pattern recognition; control and optimisation. We will draw on those examples pursued within ANNIE as the source material, but aim to highlight practical aspects concerned with these areas in general. The final section is devoted to a discussion of the whole ANNIE experience, and gives some pointers to the future.

7.2   Conventional and Neural Network Approaches

7.2.1   Introduction

The first stage of deducing a high level methodology for the application of neural network solutions is to determine when they should be implemented in preference to equivalent conventional methods. This section aims to provide useful input to that process by describing the differences between neural network and conventional approaches to the sort of problems of interest to industrial users. It does this in two stages:

- presenting the advantages and disadvantages associated with neural network methods
- outlining other factors which affect and help you in determining your choice of approach.

In each stage, the general aspects are qualified where possible by the experiences obtained within ANNIE. Note that we make no attempt to define explicitly the boundary between neural network and conventional methods, but adopt the rather grey distinction which has been accepted throughout the rest of the handbook, where the methods outlined in Chapter 2 and Appendix 2 are those primarily being considered.

7.2.2   Advantages and Disadvantages

In comparing the differences in conventional and neural network approaches to the solution of industrial problems we will first outline the potential advantages and disadvantages associated with each. Some of these are summarised in Table 7.1, and below we will describe each in more detail and relate them to the work carried out within ANNIE.

Table 7.1 Some potential advantages and disadvantages of neural network solutions

Advantages	Disadvantages
speed of execution	blind application
fault tolerance	no defined degrees of freedom
quality of results	difficult to interpret
ease of implementation and use	no defined limits on applicability
special purpose hardware	difficult to know why decisions are made
acts as a learning machine	subject to over-learning
sensor fusion capability	speed of learning

Speed

One of the most widely quoted potential advantages of neural networks is that they offer substantial advantages over many conventional methods in terms of speed. This arises for two reasons:

(i) the capability to *summarise* a set of data (compared with, say, a nearest neighbour approach) and to avoid an iterative approach when producing a suitable mapping; and

(ii) the enormous potential increase in computing power achievable when networks are implemented in fully parallel hardware.

These benefits apply to the performance of networks in testing or recall mode. During learning quite the reverse is true: for large dimensionality problems with many training examples, many hours of CPU time will often need to be expended before sufficient convergence is achieved. Whether speed can be exploited as an advantage or if it will act as a disadvantage therefore depends on the particular application. Does it need frequent retraining? Is it an on-line application? Is the network of a size that it is practical to implement it in hardware and yet will save significant time? Is it economically sensible or necessary?

It is worth noting that in the real applications studied within ANNIE, fully parallel hardware implementations rarely proved to be a necessity in order for some advantage to be demonstrated. However, there are applications (some real-time operations in control and pattern recognition) where speed will undoubtedly prove to be a distinct advantage for neural networks, and certainly this was the case in several ANNIE applications.

Fault tolerance and sensor fusion
In the context of neural networks, fault tolerance is generally taken to refer to the capability of a parallel distributed system to tolerate failures of specific nodes (or links) without causing catastrophic breakdown of the system. Clearly this quality is only exhibited once a parallel hardware implementation has been effected, but is potentially very important for wafer-scale VLSI . Its value is greatest in those, usually safety-critical, applications where there are the most stringent requirements on real-time fault behaviour. In project ANNIE, these requirements did not arise explicitly, although there was some work carried out to investigate the degradation of performance as a function of damage to the network, for the control application.

Another aspect to fault tolerance with neural networks is their supposed capability to deal with faults to, say, the sensor(s) which provide input. This combines two separate points:

(i) tolerance to noise; and

(ii) the ability to fuse data from a variety of different, possibly independent, sources.

The robustness to noisy data was demonstrated very effectively in the ANNIE pattern recognition applications where the images involved were intrinsically noisy. The ultrasonic scanning application also demonstrated the effectiveness of sensor fusion; combining signals from various angles was found to be essential for optimal performance.

Neural networks, by virtue of their flexibility and non-linear behaviour, are able to perform well in more complex situations than the ultrasonic case where they have to fuse data from sensors with completely different response functions. The solder joint scanning application was an example of this where the 2-D and 3-D images were combined successfully. Additionally, sensor fusion can be a means of providing some form of redundancy into a system to improve reliability. This was demonstrated in the control applications of autonomous vehicle operation and robot position location.

Quality of results

In some instances it is easy to provide simple, meaningful measures of the quality of results achieved, but this is not always the case. For example in pattern classification problems the success rate acts as a useful and simple criterion, but in hard optimisation problems the ideal solution is not always known. This makes simple objective measures more difficult to obtain although criteria such as quality achieved for a given expenditure of CPU time can be used. Even if objective success rates are calculable, subjective aspects to the quality of the results obtained often remain. This is particularly so if there are several different measurable aspects which can be combined only by assigning relative weightings to each. If, however, we concentrate on simple measures like success rates, what evidence is there from the ANNIE project that the quality of results is better with neural networks?

The generic studies carried out comparing neural network and conventional classifiers indicated that there is no simple message as to which does in fact give the highest quality of results. In simple cases where the class structure is known or where the dimensionality of the problem is small, there are usually some conventional methods which offer superior performance. If the problem is well understood, say involving only linear decision boundaries, then the best method can probably be chosen by inspection. If the class distributions are known *a priori,* then it is impossible to do better than the optimal Bayes classifier.

However, the structure of the problem may well be unknown, or else the dimensionality may be large with no obvious way to reduce it (without possibly throwing away important information). In these cases neural networks methods, by virtue of their great flexibility for carrying out non-linear mappings from one space to another, will frequently offer the highest quality. Even then, it should be stressed that a more optimal solution could still probably be obtained if the understanding of the problem was increased to the extent that the structure was known or the dimensionality could be reduced.

Special hardware needs

The implementation of neural networks in parallel hardware may have other advantages over and above the increase it can give in speed of execution. Special requirements might include power consumption and the fault tolerance indicated above. These potential advantages are obviously very much application-specific, and did not play a major role in the considerations behind the topics studied in project ANNIE.

Ease of implementation

Implementing a neural network solution to most problems can be very much simpler than implementing a solution using a comparable conventional approach. This is especially so if little or no attempt is made to preprocess the data or to reduce its dimensionality through feature extraction for instance. The success of applying neural network methods in such a naive fashion is, however, not considered to be very great. During project ANNIE the blind application of, for example, conventional back-propagation networks in pattern recognition and control gave far from optimum results in most cases. Suitable preprocessing and data representation were crucial to success. In the area of optimisation, not only does the requirement to build in constraints reduce the ease of implementation itself, but also the scaling of the problem determined that for realistic problems a more careful and structured approach was necessary.

There are occasions, however, where it is a sufficient rather than an optimal solution which is sought. In these instances the ease with which a neural network solution can be adopted might suggest that some economic or efficiency advantage can be obtained. The reliance on such solutions will be enhanced if sufficient and representative training data are available. If there are no obvious features of the data, then the network solution is likely to show considerable advantages compared with the exhaustive searches or heuristic methods which might then otherwise be needed. Many examples exist outside of project ANNIE where huge savings in time for implementation have been achieved over alternative solutions such as expert system approaches.

Neural networks as adaptive learning machines

Another very valuable feature of neural networks is that they can act as adaptive learning machines. Thus they can be used to control a process automatically and to adapt continuously to variations in the environment in an intelligent fashion - as learning machines they offer the possibility of being retrained (adapting) to cope with changing circumstances.

This learning capability of neural networks (shared with some conventional approaches) could be particularly valuable in, say, a pattern recognition application of quality control. If the criteria for acceptance were to change, it would be possible to define a new training set, learn the new criteria and replace the trained network without changing any existing software or hardware implementation. This particular advantage of neural network solutions was rarely encountered within ANNIE, but is likely to become of more significance once the number of real industrial applications increases.

Blindness in operation

In the sections most points relate to potential advantages of neural networks, depending on the particular applications. The main disadvantage for many applications, apart from speed during learning is the fact that one tends to operate them in a *blind* fashion. This disadvantage can be broken down into a number of separate aspects:

(i) *Neural networks are black boxes.* It is easy to use and consider neural
     networks to be acting merely as *black boxes*. This is one of the reasons that
     they are easy to implement, but it means you can be given some solution with
     little if any indication of its reliability. The temptation is therefore to accept
     the output provided, although you have no real idea as to why it has made the
     decision. Assurance concerning the applicability of the network to any test
     data, and the reasons behind any decision it makes, can be obtained if the
     internal behaviour of the network is deduced. In principle, by taking the
     network apart and examining it internally, its high level representation of the
     data (the features) can be abstracted. It should then be possible to predict its
     behaviour in any new circumstances or operating regimes. In practice this is
     rarely possible, although theoretical work investigating such approaches is
     being pursued.

(ii) *Understanding complexity.* It would be helpful if the user knew precisely the
     number of degrees of freedom possessed by the network. Although upper
     limits to the number can be given, the complex constraints and
     interdependencies of the weights in a network may reduce it in a far from
     obvious fashion. There is theoretical work aimed at clarifying this area, but at
     present it is still difficult to know how complex any network actually is and
     how well matched it is to the dataset on which it is designed to operate.

(iii) *Limits of applicability.* Another problem is to try and understand the limits of
     applicability of a trained network. Once a network has been trained on a given
     set of data, what assurance is there that it will behave sensibly when given
     unseen (test) data? There is good evidence that networks - which can
     frequently be viewed as no more than a means of mapping from one space to
     another - will perform reliably on test data if they fall 'within' successfully
     learned training data. However, this poses the question, "What does 'within'
     mean for high dimensionality datasets?". This is not always obvious: a
     pragmatic approach is to use sufficiently large and representative datasets for
     training that you have no reason to believe that new data will not be of a
     similar nature to some already presented to the network. In real applications,
     however, this can be a practical problem

(iv) *Getting the right training set.* For the ANNIE applications it was frequently
     found that only small-sized datasets existed. In such circumstances the first
     reaction was to attempt to generate further datasets, although this could be
     costly in time and money. If this was not possible, and it was felt that the
     process behind the application was sufficiently understood that it could be
     modelled accurately, then simulated data could be used to supplement the real
     dataset.

The approach to these problems within ANNIE has been a largely pragmatic one:
testing on large and realistic datasets; using visualisation tools to observe the
movement of decision boundaries; adopting an empirical approach to determining the
value of parameters; and defining criteria *a priori* so that sufficient rather than optimal

solutions can be accepted on the basis of a statistically acceptable performance. Although this has convinced us of the usability of neural network solutions in a number of application, the disadvantages associated with the blindness of the approach are real; they necessitate the expenditure of considerable time and energy for validation if the reliability of solutions is to be assured.

7.2.3   Divergence and Choice

Having discussed the advantages and disadvantages associated with neural network and conventional methods we now focus on other ways in which they diverge and how one might derive criteria for making suitable choices.

Firstly it should be noted that there are usually conventional mathematical approaches which offer similar processing possibilities to any neural network methods one may wish to use. Thus in a problem for which an error back-propagation network might be felt to be appropriate, it would probably be possible to use some form of multiple linear regression, similarly an optimisation problem being solved by a Hopfield network or a Boltzmann machine could be approached by a local area search technique or simulated annealing. Why therefore use a network rather than the conventional mathematical equivalent?

The studies within ANNIE would tend to support the argument that in general the answer to this question is, "Don't", qualified by "unless you have a good reason to". We have found that conventional methods should be fully understood and if necessary tried first. Only if these have been fully examined and shown to be inadequate (by suitable criteria such as those suggested below) should network methods be explored. In addition to understanding the analysis methods it is crucial to understand as fully as possible the details of the application.

The understanding should be attempted at the mechanistic level - the physics of the processes which underlie the data - and at the statistical or ensemble level. This means that suitable tools should be used to assess the complexity of the problem - including simple statistical measures and more complex techniques such as principal component analysis. Measures should be defined and evaluated which determine the nature of any training set ensuring, for example, that it is truly representative.

The requirements of any solution should ideally be defined beforehand. What success rates are required? What level of precision will be satisfactory? What is the range of data values to be covered? Will criteria for classification be subject to frequent alteration? Only with these questions clearly answered will it be possible to use the possible advantages and disadvantages already discussed as a basis for making the choice on the way forward. Once this is clear, the task of implementation begins. This is discussed in the next section.

7.3   Implementing Solutions

In this section we will review aspects concerned with implementation of network methods. This section covers the lowest level of our discussions into the methodology

behind the application of neural networks to industrial problems. We will still be
attempting to make general comments, not restricting them to particular architectures
or paradigms, but will concentrate on two key points: how to implement solutions in
particular systems; and how data representation should be handled.

7.3.1   Breaking down the Problem

By identifying in the previous section the potential advantages and disadvantages of
neural network approaches, we have essentially dealt with the inputs needed to answer
the questions of when and where they should be implemented. Even knowing these
points, it can still be difficult to make the necessary assessment of the best approach.
For the same application, there may be subjective criteria, or aspects such as time and
cost which will lead to different assessments on different occasions. Whatever the
basis of the criteria you choose to set, it will often be the case that the preferred option
will involve networks as only part (or parts) of a complete system broken into
component modules.

    In order to determine how and in which part of the system to use neural networks,
it is of course necessary to break it down in some logical fashion. This once more
involves the requirement to understand your application sufficiently well that you can
do this. The networks can then be employed in successive filtering operations or in
carrying out that bit of the problem for which no clear features are available. To
discover which parts these are demands a flexible approach. In ANNIE we have found
the use of software simulation to be of great value even if the final solution requires a
hardware implementation. If possible the environment which is developed should be
able to cope with a variety of different paradigms and preprocessing techniques. Once
the problem is better understood the advantages can be evaluated and, possibly using
suitable benchmarks, the best way forward can be deduced.

7.3.2   Data Representation

How the data is represented prior to presentation to a network is crucial to any success
which is going to be achieved. In this respect it is no different from any other
approach. It is only because of the flexibility of neural networks and their ability to
cope with highly complex multivariate data that we may tend to forget that wherever
intelligence and prior knowledge can be built into the solution it should be done so
explicitly.

Data encoding

The most fundamental aspect to data representation is the coding of the data
themselves. If the distance between variables is important, then some coding should
be chosen for which this quantity is conserved. An obvious example here is the binary
representation of numbers which very poorly conserves Hamming distance. Thus the
Hamming distance between 15 and 16 is very much larger than that between 15 and
14. This obvious example has some more subtle variants. In a three-class problem, if
all are of equal *a priori* likelihood a network output should not have a single output
node target of -1,0 or 1 for each of the three classes, as this would impose a bias

towards class 0, the distance between -1 and 1 being further. Instead three node outputs are required. Another example is when using angular measures. When working with such measures it should be remembered that the underlying arithmetic is modulo 2π. This once more can impose unwanted biases. Because the same target output can have more than one valid value given to it, such representations can give rise to conflicting and inconsistent training datasets in which two examples with identical inputs and target outputs can have two apparently different target values. This will give rise to a compromise being learnt for the response to that particular input, which in this case is completely wrong. Such possible conflicts and the need to ensure single-valued representation are not of course restricted merely to the angular measures cited here.

The scaling of the data is also important in ensuring that undue weight is not given to any one parameter within a dataset. Suitable methods to choose the scaling are dependent on the detail of the problem, stressing the importance of understanding the statistics of the datasets.

Preprocessing

By preprocessing we refer here to any processing of the raw data prior to its presentation to a neural network. We can split this into two categories:

(i) data preparation (including the data encoding discussed above) which is applied both to input and target data; and

(ii) possible fixed transformations which are applied to the input data.

Concentrating on the second of these two, we find that in practice there are usually several stages of preprocessing. These are typically carried out by applying simple statistical methods (averaging, taking moments etc) but can extend to more complex filters including other neural networks.

It can be argued that in a classification problem, sufficient preprocessing will reduce the input data to such a level that the classification itself is trivial. The difficulty is knowing what preprocessing is required, and the goal is to remove unnecessary information from the dataset whilst retaining that needed to make any discrimination. It is important always to try and understand what that information is if possible. For example in pattern recognition the position of the image may be unimportant, so centring is useful preprocessing to carry out. In this way knowledge in a very real sense is being built into the solution. Similarly the preprocessing carried out in optimisation tasks can reduce the search space whilst building in constraints.

In ANNIE the level of preprocessing was determined by an understanding of the process supplemented occasionally by heuristic or empirical methods. In the solder joint classification problem different preprocessors were applied and their effect on the data could be compared. In optimisation, heuristics can be used to cut down the huge space being explored to one of a more manageable size through appropriate preprocessing.

7.4   ANNIE Applications

Above we cited some instances of where the ANNIE project results support the
general assertions made regarding the advantages and implementation of neural
network solutions. In the sections which follow we will concentrate on examining the
methodology behind each of the application domains covered by the project. Once
more by citing specific examples of the experiences of the project, we hope to draw
out general messages about the application areas as a whole.

7.4.1   Pattern Recognition

In the pattern recognition area we studied supervised learning methods in some depth.
Fair comparisons indicated that in terms of accuracy of classification it was not
possible to determine any single best approach to adopt. In varying circumstances
different classifiers performed the best, depending on the class structure, the training
data, and the degree of class separation. In the absence of more detailed knowledge,
neural network methods offered a good and unbiased attempt at solution. When this is
combined with their many other possible advantages it can frequently make them
appear to be an attractive option.

The need for and the role of preprocessing was very obvious in the real
applications studied. Suitable choices were able to compensate for the rather sparse
data which made up the training sets, and allowed some *a priori* knowledge to be built
into the solution. This meant that the real gains possible from processing image data -
primarily speed and ease of implementation - could be realised. Building in invariance
to various transformations could also often be achieved through appropriate
preprocessing, although sometimes important aspect was satified through use of an
appropriately defined and adjusted training set.

The requirement for preprocessing was one of the many indications that networks
rarely provide the complete solution. They do however act as useful additional tools
to use when necessary. Thus we showed instances where combinations of networks
(hybrid) systems proved to be very powerful (particularly with the ALOC ultrasonic
data) or when the use of networks in cascaded form (ie as a filtering mechanism, such
as in the acoustic emission studies) was the most effective.

The choice or definition of the class structure proved also to be important, in
particular with the solder joint application where far better success was achieved if an
attempt to distinguish and classify all possible defect types was not attempted. Rather,
it was better to use one set of methods to distinguish the more simply identified defects
and to use a neural network as a novelty filter to distinguish between the remaining
good joints and the more difficult 'blow out' defects.

The pattern recognition experiences also indicated the importance of getting the
right training set. The lack of sufficient quantities of well-characterised data was often
the barrier to initial successes. However, this lack of data is often the case in real
applications, and so efforts have to be made to generate more, and to allow the
networks to learn suitable variances for example. The small training sets also quite

frequently led to overfitting of the data. Ways to overcome it were to use smaller or different nets, as well as enhancing the size of the training set etc.

Finally, real improvements without special hardware were demonstrated, but the speed in allowing real-time classification offered distinct advantages in all of the inspection applications studied.

7.4.2   Control

Some of the findings of the pattern recognition applications are equally valid for the control area. In particular the dependence on preprocessing, obtaining the best data representation, and composing a sufficient and representative training set. The coding of information in binary and the angular information discussed above were particularly relevant for the case of the autonomous robot. In the case of vehicle location, the choice of data presentation as velocities or acceleration information made significant differences to performance. The need to understand and problem was also clearly demonstrated in both applications and acted as an aid to the choice of networks, a breakdown of the problem into simpler components, and a guide to the most appropriate data representation to adopt.

In order to test out different preprocessing methods and representations, it proved to be very valuable for a specially-tailored software environment to be defined and implemented. (This proved particularly valuable in the pattern recognition area where it was possible to make it sufficiently general that several different applications could be studied using the same software.) This also brings out another feature of the work in the control area: the need to be prepared to test variations in different networks, and to employ different learning strategies. This empirical approach to the problem was needed in order to obtain improved performance and to define limitations of the applicability of any one approach.

The role of sensor fusion was very important in this area, with the localisation of a robot being carried out effectively and easily using a simple neural network architecture. The comparative ease with which this solution could be implemented goes towards demonstrating that use of a neural network approach could well be justified on efficiency grounds.

Finally, the practical application of a control system to a mobile robot makes special demands on the hardware used, in particular low power consumption and weight considerations. Control applications can also be safety critical, demanding fault tolerance. A network solution, implemented on special hardware, can provide for both these demands in a way which some more conventional processing system might not.

7.4.3   Optimisation

The work on optimisation showed that a neural network (the Boltzmann machine) could be used as part of the overall solution to a hard problem, although it is unlikely on its own to be able to solve realistically-sized problems. This is due to its scaling behaviour, and suggests that for NP-hard optimisation problems it has no immediately significant advantage compared with other methods. However, there were indications that once the problem was broken down appropriately, neural network techniques could then be used to advantage in providing improved solutions to these difficult problems.

It should be recognised that in this field what is sought frequently is sufficiently good rather than optimal solutions. It is important therefore to adopt a pragmatic stance and once more to seek empirical solutions where necessary. Recognising the heuristics should be used where possible, it was found that the best point for their application was early in the processing stage. Thus the best approach was to use conventional methods (a pair generator) to cut down the search space dramatically, and to take into account the constraints of the problem. This leaves the network (of a reduced size) to deal with the complex part which remains.

An obvious advantage of this approach is that by being smaller the network is able to converge on a solution more rapidly. In addition, by taking into account all (or nearly all) the constraints at a preprocessing stage it is not necessary to build in a constraint-satisfying mechanism at the network stage. This is advantageous since it can be difficult to express the more complex constraints in the network cost function without involving the somewhat arbitrary tuning of parameters in order to prevent either constraints from being violated or far from optimal solutions being produced. A final advantage is that the network part itself can be used in a variety of different applications. If the constraints are varied it is not necessary to make major changes to its parameters or structure; this can all be done at the preprocessing stage.

The structured approach described above also allows better control to be exercised over the nature of the solutions which are generated. Thus the nets can be guided towards the likely areas of best solution. Successive use of networks as filter mechanisms is therefore demonstrated clearly. Extensions to this breakdown have been investigated with recent advances in multiscaling techniques allowing the structure of the problem to be segmented and treated successfully with a number of linked networks. Another feature, the use of networks to guard others so that local minima problems are avoided, also appears to offer great promise for the future.

7.5   Discussion

7.5.1   The ANNIE Experience

It might be argued that many of the points made in the previous sections are well established and, indeed, have been known for quite some time. However, we feel that the ANNIE project has played an important part in supporting many of these ideas and in developing them further when considering real applications. There are some notable differences from some of the claims which were being made at the start of the project. For example, it was felt that neural networks would show a distinct improvement over conventional classifiers in terms of success rates and that this was likely to lead to a whole area in which networks would be the default option. The experience within ANNIE is that matters are not quite so simple. On the other hand, the project has revealed that distinct advantages can be had in some circumstances without the need for special or exotic hardware which some also have suggested is the principle source of advantage for this new technology.

So what are the main messages from the entire experience? Repeatedly it was found within the project that blind application of any methods is unlikely to give the best and certainly not the most reliable results. The better understood the problem is, the better the methods which can be employed to solve it. Therefore you must understand the physics or systematics of the process to which the analysis is being applied, and in order to gain this understanding you must be prepared to use a whole variety of different tools. This inevitably means that when deciding the best approach, you should not use a neural network unless it is going to offer advantage for some definable reason. However, in deciding the possible advantages, you should not forget that the full picture must be taken into account. When all important factors affecting the acceptability of a particular approach are included - economic considerations, speed of implementation, robustness, reliability, and efficiency etc - the neural network methods become more favoured.

You should not expect that the neural network will provide the solution on its own. Any problem will normally benefit from being split into components, each of which should be solved by the method most suitable to it. Often use of a network as a filtering component of a complete system can be its best role, and the only way to determine just how and where this should take place will frequently be by adopting an empirical approach. To allow rapid experimentation and a variety of different options to be exercised, we found that it rapidly becomes necessary to develop your own software. During the ANNIE project each application area ended up developing its own software environments, with considerable success. Standard packages can often provide a useful first insight into analysis of data, but the details of your particular application, and in particular the needs of data representation demand that some specialist software is almost certain to be necessary.

7.5.2   Pointers to the Future

We can only guess at what the future will hold in this field. The potential still appears to be large and we would expect that increasingly there will be real-world applications in which neural networks will play a part. In some cases we believe they will provide a significant technical edge over other methods.

Practically we would expect to see an increased use and understanding of the role of hybrid networks where filtering processes and "guard" monitoring networks are used as parts of larger systems. We would expect that the individual components of networks might become complex, embedding Kalman filters or being driven by maximising entropy rather than minimising training errors.

We would hope that theoretical progress will lead to a better understanding of the details of network implementation. Limits to the size of usable networks should be established, and formal information theory should be developed which gives objective measures of the suitability of a specific training set, and what reliability can be attached to any solutions generated. The topology and structure of networks will be determined in part by optimising methods such as genetic algorithms, which will reduce the problems associated with over-fitting.

The development of dedicated hardware chips for parallel processing is going to continue to make a significant impact for real implementations, since in time-critical applications a possibly unique advantage of neural networks will be exploited.

Because of the pace of neural network research during the duration of the ANNIE project, it has inevitably meant that some of the techniques used early on have become out-dated. We see no reason why this should not continue to be the case in this field, particularly in the area of hardware implementations of parallel architectures. The message therefore is clear: that one should stay in touch with developments, and ensure that any installed systems are designed with sufficient flexibility that new improvements can be integrated easily if needed. However, we do not expect that progress will ever be such as to negate the need for users to input their own knowledge of the problem. We would certainly not wish to suggest that there is any reason for holding back now from implementing a neural network solution on the grounds that some miraculous solution may be just around the corner!

Appendix 1:   Partners in the ANNIE Consortium and Project Staff

AEA Technology
Harwell Laboratory
Oxon OX11 0RA   UK
Contact: **Dr John Collingwood**
Tel: +44 235 435222
Fax: +44 235 432726

CETIM
Service Essais Mesures Controles
52 Avenue Felix-Louat
BP67, 60304 Senlis Cedex
FRANCE
Contact: **Dr Mohammed Cherfaoui**
Tel: +33 44 583359
Fax: +33 44 583400

IBP Pietzsch GmbH
Hertzstrasse 32-34
D-7505 Ettlingen
GERMANY
Contact: **Mr Rigobert Opitz**
Tel: +49 7243 709134
Fax: +49 7243 709191

Siemens AG
Hammerbachstrasse 12-14
Abt U9 321   Postfach 3220
D-8520 Erlangen
GERMANY
Contact: **Mr Rainer Meier**
Tel: +49 9131 183881
Fax: +49 9131 187547
and
Otto Hahn Ring 6
D-8000 Munich 83
GERMANY
Contact: **Dr Peter Mengel**
Tel: +49 89 836 48756
Fax: +49 89 636 46192

**National Technical University
of Athens**
Athens
GREECE
Contact: **Mr Andreas Stafylopatis**
Tel: +30 1 775 7401
Fax: +30 1 775 7501

British Aerospace Plc
Sowerby Research Centre
FPC 267 PO Box 5
Filton, Bristol BS12 7QW
UK
Contact: **Dr W Andrew Wright**
Tel: +44 272 363538
Fax: +44 272 363733

Alpha SAI
72-74 Salaminos St
Kalithea, Athens
GREECE
Contact: **Mr George Panayotopoulos**
Tel: +30 1 958 2506
Fax: +30 1 958 5079

Artificial Intelligence Ltd
Greycaine Road
Waford, Hers WD2 4JP
UK
(now ceased trading)

KPMG Peat Marwick GmbH
Gross Gallusstrasse 10-14
6000 Frankfurt am Main 1
GERMANY
Contact: **Mr Hans Naumann**
Tel: +49 69 21 64345
Fax: +49 69 21 64610

Contributors to the ANNIE project

Project Manager

Dr A T Chadwick (1988-90)
Dr J C Collingwood (1990-91)

Technical

AEA Technology

Ms C Baxter
Dr C M Bishop
Dr S F Burch
Mr D J Clayworth
Mr I F Croall
Dr A H Harker
Mr G B Hesketh
Mr K D Horton
Mr I Kirk
Dr J P Mason
Dr T Mathews
Ms S Peach
Mr M Raymond
Mr C Robertson
Prof C G Windsor

Alpha SAI

Mr E Antippas
Mr N Ioannidis
Mr G Karagiorgos
Mr A Koumasis
Dr G Panayatopoulos
Mr N Verikios
Mr V Zissimopoulos

Artificial Intelligence Ltd

Dr M Gittins
Dr P Williams
Dr P Thornton

British Aerospace plc

Mr M Brown
Mr A C Crowe
Mr P Edwards
Mr J Hubbard
Mr G Lamont
Mr S Wadey
Dr W A Wright

CETIM

Dr M Cherfaoui
Mr P Pelletier
Mr Y Perrin
Dr J Roget

IBP Pietzsch GmbH

Mr M Frondorf
Mr A Gayer
Mr H T Kazmeier
Mr R Opitz
Dr K Overlach
Mrs U Schley
Mr T Weber

KPMG Peat Marwick GmbH

Miss G Hantschel
Miss U Hantschel
Mr H Naumann
Mr J Richter

NTUA

Mr S Karamanlakis
Prof S Kollias
Ms D Kitsiou
Mr D Kontravdis
Mr A Likas
Mr A Tyrakis
Prof G Papageorgiou
Prof A Stafylopatis

Siemens AG Erlangen

Mr B Gentner
Mr R Meier
Mr A Rudert

Siemens AG Munich

Mr L Listl
Dr P Mengel

Technische Hochschule Darmstadt

Prof M Glesner
Mr M Huch
Mr W Poechmueller

Project Consultant

Prof T Kohonen

Project Administration

Mrs J Clark
Mr L Cousins
Mrs E Davison
Miss S Harrison

ESPRIT Project Officer

Mr J-J Lauture

Referees

Dr F Fogelman
Dr S Garth
Dr R Linggard

Handbook sub-editor, design and typesetting

Mrs N Hutchins

Appendix 2: Networks Used in the Project

A2.1 Introduction

This appendix presents in more detail the theoretical background to the networks used (or tested) in this project. It is provided to assist in understanding the applications.

A2.2 Associative Networks

Neural network models are widely used as *associative memory* paradigms. The concept of memory usually involves a storage mechanism, and a recall mechanism based upon associations encoded within the memory. In a simple form of associative recall, a spatial input pattern, the key, is transformed into a corresponding output pattern, the recollection. Two types of transformation should be distinguished, the *auto-associative* recall, whereby an incomplete input pattern recalls its complete version, and the *hetero-associative* recall, whereby pairs of different data are associated. Many results concerning associative memories are independent of implementation, others are based upon particular network realisations.

Early work on associative memories concerned mainly *linear networks*. In such networks, associations are encoded in the connection matrix W by using correlation techniques, see for instance (Kohonen, 1972, 1988). In most cases, linear networks operate as hetero-associative memories following a simple 'one-shot' feedforward procedure. An input pattern X is presented to W and an output pattern Y is produced, which represents the interpolative recollection corresponding to the input pattern. The *Optimal Linear Associative Mapping (OLAM)* introduced by Kohonen is one of the most elaborate systems of this type.

In traditional *nonlinear associative networks*, vector patterns A_i, i -1,..., m are stored in a matrix memory W. An input pattern A is presented to the network and an output pattern A' is produced after performing multiplication of A with the connection matrix W and some subsequent nonlinear operation such as thresholding. A' is either accepted as the desired recollection or fed back to produce A'' and so on. A stable network will eventually produce a fixed output. This feedback procedure behaves as if input A was unidirectionally fed through a sequence of matrices identical to W. Memories of this type are auto-associative. The most popular paradigm of 'unidirectional' auto-associative memory is the *Hopfield network* (Hopfield, 1982).

In the general case, associative memories are hetero-associative. The accuracy of the recollection can be considerably increased if an iterative forward and backward procedure is performed instead of the 'one-shot' feedforward operation described above. This idea leads to a class of two-layer nonlinear feedback networks known as *bidirectional associative memories*. This type of network, introduced by Kosko (1988a), is shown to be a minimal feedback architecture that ensures stable hetero-

A common feature of the above discussed categories of associative memory is their fixed *encoding* procedure, which consists in the direct off-line computation of the connection matrix W. We will now describe in some detail the major paradigm of each one of the three memory types.

A2.3   Linear Associative Networks

A2.3.1 Optimal Linear Associative Mapping (OLAM)

This technique, developed by Kohonen, aims to form a transformation between spatial patterns or equivalent vectorial representations, which can be designed optimally. Optimality is understood in the sense that we can have a solution to the paired-data association problem which minimises the effect of noise or other imperfections present in the inputs.

Kohonen (1988) defines the associative recall as a mapping in which a finite number of input vectors is transformed into given output vectors. Formally, the problem can be put as follows; if we have m input pattern vectors $x_k \in R^n$ and m output vectors $y^k \in R^p$ to find the p-by-n matrix operator W, which satisfies simultaneously the m equations:

$$y_k = Wx_k \qquad\qquad (A2.1)$$

ie to find the solution to the matrix equation

$$Y = WX \qquad\qquad (A2.2)$$

where Y, X are the matrices with the y_k and x_k as their columns, respectively.

The general approximate solution to the above equation is obtained from the theorem of the pseudo-inverse (Albert, 1972) and guarantees the minimisation of the error $(WX - Y)$. The solution is given by

$$W = YX* \qquad\qquad (A4.3)$$

where $X*$ is the pseudo-inverse of the matrix X. A sufficient condition that an exact matrix solution exists is $X*X = I$, ie the x_k are linearly independent and so the number of input vectors is less than their dimensionality ($m < n$). There exist various procedures for the computation of $YX*$. Among these we mention: several linear corrective algorithms, the method of gradient projection or best exact solution when $m < n$, the method of linear regression or best approximate solution when $m > n$, and the recursive solution in the general case (Kohonen, 1988).

It is clear that so far we have described the hetero-associative recall procedure. In the auto-associative recall, where an item is retrieved by its fraction, the optimal solution has the form

$$W = XX* \qquad\qquad (A2.4)$$

But we know that $X X^*$ is the orthogonal projection operation on the subspace of the x_k vectors, k - 1 ... m and so, given an input vector x, the recollection according to (A2.1) is

$$XX^* x = x'$$ (A2.5)

where x' is the orthogonal projection of x in that subspace. Indeed x' is the best linear combination of the x_k that approximates x in the sense of least squares and thus the mapping is optimal in this sense. Although we extracted the previous result as a special case of the general solution, it can be obtained independently (Kohonen *et al*, 1976), and the fact that it is possible to implement auto-associative mappings by the orthogonal projection operation is interesting in the context of associative memory.

A2.3.2 Correlation Matrix Memory
We often adopt as a solution for the matrix equation problem (A2.2) the matrix product

$$W = YX^\mathsf{T}$$ (A2.6)

using the correlation matrix formalism, where X, Y as above. This is the *correlation matrix memory* also referred to as *linear associative memory* (LAM) introduced by Anderson (1968) and improved by Anderson (1972) and Kohonen (1988). In the case where the vectors x_k are orthonormal the pseudo-inverse X^* of X is equal to its transpose X^T, and thus the optimal solution (A2.3) is

$$W = YX^* = (Y(X^\mathsf{T}X)^{-1}X^\mathsf{T}) = YX^\mathsf{T}$$ (A2.7)

ie the correlation matrix and the OLAM are equivalent. Indeed, this is the only case to the LAM approach that perfect recall is guaranteed.

A2.3.3 Novelty Filter
If in the auto-associative recall instead of extracting x' as a result we extract the component $x - x' = x''$ of the input vector x, we have a system which is called a novelty filter. x'' represents the 'maximally new' information in the vector x and so the novelty filter provides a component-wise measure of the difference between the presented pattern and all the known patterns.

A2.3.4 Optimal Linear Identification
If the purpose is only to identify a stored pattern and not to recollect the pattern or an approximation of it, we can use the previous optimal hetero-associative mapping with some identification tags in the place of y_k vectors. The unit vectors

$$u_1 = (1, 0, 0, ...0)^\mathsf{T} , u_2 = (0, 1, 0, 0, ...0)^\mathsf{T} ,...$$

are the simplest such tags. In that case equations (A2.1) and (A2.2) take the form

$$u_k = Wx_k \tag{A2.8}$$

and

$$U = WX \tag{A2.9}$$

respectively.

A2.3.5 Optical Associative Memories

OLAM have enjoyed optical implementations. The development of optical associative memories comes with the augmented use of optical computing in recent years. We distinguish two categories of optical associative memories: the *holographic memories* and the *non-holographic memories*. In the former we exploit the central characteristic of holograms, ie the spatial distribution of memory traces, for the storage of masses of pictorial information. In the latter the matrix operations for associate recall are performed by multiplying light intensities using some kind of light modulating matrix arrays and summing up convergent light beams by discrete photosensitive elements (Kohonen, 1988). One implementation of OLAM with holographic memories is the system of Giles and Fisher (1985). In general, contemporary optical systems make extensive use of new material. In fact, the lack of proper materials has limited holographic computing at the experimental level.

A2.4   Hopfield Networks

Hopfield (1982) introduced a neural network architecture, which became known as the *Hopfield network*. Originally, the Hopfield network was a single layer network, operating in discrete time. Later on (1984) the discrete time constraint was removed by Hopfield himself, resulting in a continuous time Hopfield network.

The activity of the network is described by an energy function. Hopfield (1984) was able to show that under certain assumptions to be described later, the energy decreases with time, and the network converges to a local minimum, where it stabilises. The energy function describing the activity of the network as a whole is:

$$E = -\tfrac{1}{2}\sum_{i=1}^{n}\sum_{j=1}^{n} y_i y_j w_{ij} - \sum_{j=1}^{n} y_j \theta_j \tag{A2.10}$$

where y_i is the output of unit i, w_{ij} is the strength of the connection (i, j) and θ_i is the threshold of unit i. The connection strengths w_{ij} as well as the unit thresholds θ_i can be positive and/or negative real values. Their values compared to n (the number of processing units) can be large (exponential) or small (polynomial). y_i can take either binary ($\{0,1\}$) or continuous values ($[0,1]$). Last, there is no restriction on the topology of the network, which can vary according to the needs of the problem it is to solve.

It is exactly through this energy function that Hopfield networks make their way into the area of *combinatorial optimisation*. Specifically, it can be shown that if a cost function of an optimisation problem can be manipulated into a form that allows identification with terms of the energy function, then a Hopfield network can be constructed that solves the problem.

Two modes of operation of the Hopfield network are studied: *synchronous* and *asynchronous*. The first is interpreted by means of a global clock common to all processing units. At each time unit *t* all the processing units simultaneously apply the network's update rule, and modify their local state. On the contrary, under the asynchronous mode the processing units are not synchronised in any way. Thus, only one processing unit is active at any time unit. The active processing unit is picked at random under a uniform probability distribution.

A2.4.1 Stability

In order to address stability issues for the Hopfield networks, two architectures must be separately considered: *symmetric* ($w_{ij} = w_{ji}$, i, j - 1 ... *n*) and *asymmetric* networks. In the case of symmetric networks the computation is guaranteed to converge only under the asynchronous mode of operation. Moreover, it can be proved that in each step of the computation the energy of the network decreases, and finally reaches a local minimum. The remarkable fact is that at this point the network reaches a stable state (for all i, x_i remains constant). On the contrary, under the synchronous mode of operation it is possible for the network to cycle and never reach a stable locally optimum solution. As a matter of fact, Poljak and Sura (1983) have shown that if a symmetric Hopfield network operates synchronously, and does cycle, then the cycle is of length two. In an asymmetric network (under both synchronous and asynchronous operation) there is no similar relationship between the local minima of the energy function and the stable states of the network.

The update rule of the operation phase of the Hopfield network is distinct between the discrete and the continuous time case. For the discrete time case the operation of the network can be described as follows:

$$x_i(t+1) = \sum_{j=1}^{n} w_{ij} y_j(t) + \theta_i \tag{A2.11}$$

$$y_i(t+1) = f(x_i(t+1)) \tag{A2.12}$$

where $x_i(t)$ is the activation of unit i at time t, $y_i(t)$ is the output of unit i at time t, w_{ij} is the strength of the connection (i,j), θ_i is the threshold of unit i and f is the binary step threshold function.

For the continuous time case the operation of the network can be described as follows:

$$x_i = \sum_{j=1}^{n} w_{ij} y_j - x_i + \theta_i \tag{A2.13}$$

$$y_i = f(x_i) \tag{A2.14}$$

where x_i is the activation of unit i, y_i is the output of unit i, w_{ij} is the strength of the connection (i,j), θ_i is the threshold of unit i and f is the sigmoid function.

The generalisation of Hopfield networks (generally any kind of feedback networks) to include higher order connections, and the resulting advantages, are discussed by Maxwell *et al* (1986). In particular, high-order associative memories yield a better and better polynomial approximation to a sampled mapping, as the dimensionality of the correlations increases. With appropriate arrangement of the weights, the networks display scale and translation invariance. The dynamics of a unit are

$$\dot{x}_i = \sum_{j=1}^{n} w_{ji} y_j + \sum_{j=1}^{n}\sum_{k=1}^{n} w_{jki} y_j y_k + \sum_{j=1}^{n}\sum_{k=1}^{n}\sum_{l=1}^{n} w_{jkli} y_j y_k y_l + ... - x_i + \theta_i \qquad (A2.15)$$

where x_i is the activation f unit i, y_i is the output of unit i, $w_{jk...i}$ is the additive strength of the indirect connection from unit j to unit i and θ_i is the threshold of unit i. If connectivity matrices $w_{jk...i}$ are invariant under any permutation of their indices, then Hopfield's convergence theorem applies to higher order networks. Their stability can also be shown with an extension to the ABAM theorem (Kosko, 1988b).

Hopfield networks learn off-line, so before entering the operation phase, the connection strengths w_{ij} must be set to appropriate values. In the case where a Hopfield network is used as an associative memory, the problem of learning is described as follows: A set $\{x^{(1)}, x^{(2)},...,x^{(m)}\}$ of m n-dimensional column vectors is given, which are to be 'stored' by the network. Then the connection strengths w_{ij} are computed according to the formula:

$$w_{ij} = \sum_{k=1}^{m} (2x_i^k - 1)(2x_j^k - 1)$$

$$(A2.16)$$

with $w_{ii} = 0$ for $i = 1...n$. It should be noted that the above formula, known as the outerproduct construction, was first presented in another context in Cooper (1973).

In the wider area of optimisation there is no systematic way for computing the connection strengths w_{ij}. This fact is not a surprise, as the following elementary example will show. Consider the problem of the Maximal Independent Set (MIS). We are given an undirected graph $G(V,E)$ (V the set of vertices, E the set of edges), and we are asked to find a collection of vertices M such that no two vertices are adjacent (ie not connected by an edge). Furthermore M must be maximal in the sense that there exists no vertex that can be added to M. Given G we can construct a Hopfield network that solves the MIS problem for G. The network has exactly the same topology with G, the connection strengths $w_{ij} = -1$ for all, $i,j = 1...n$ ($w_{ii} = 0$), and the unit thresholds $\theta_i = 1/2$ for all $i = 1...n$. It is easy to see that the previous network indeed solves the MIS for G, but what could be a systematic way for computing the network?

A2.4.2 Simulated Annealing

One of the major problems of the Hopfield network, is its inability to escape local *et al*, 1983), which is similar to the technique used in *Boltzmann machines*. The idea of an energy surface whose minima correspond to the solutions of the original problem is maintained. The update rule, though, is governed by a Boltzmann probability distribution. A unit is turned on (states take binary values), according to the probability

$$p(i) = \frac{1}{1 + \exp(-\frac{x_i}{T})} \qquad (A2.17)$$

where x_i is the activation of the unit, as described in the discrete time Hopfield network. The temperature factor (T) is introduced to allow one to proceed from an initial stage of high temperature, where the basic features of the solution appear, to a final state of low temperature, where the specific details are to be determined. At the limit $T \rightarrow 0$, this machine becomes non-probabilistic and becomes the conventional Hopfield network.

Performance

The performance measure most commonly used in the Hopfield networks is their *capacity*, of great importance especially when the network is used as an associative memory. The discrete asynchronous mode of operation is assumed in this issue. Information in Hopfield networks is stored as stable states. Suppose that a set $\{x^{(1)}, x^{(2)},...,x^{(m)}\}$ of m n-dimensional column vectors is given, which are to be stored, by an n-neuron network. This means that the vectors must be stable states of the network. The question is what is the maximum number m such that any m vectors of n binary digits can be made stable states in a network of n neurons by a proper choice of the connection strengths w_{ij} and the thresholds θ_i?

Hopfield suggested, from numerical results concerning the outer product construction, that $m \leq 0.15n$. It can be shown (Abu-Mostafa and St Jacques, 1985) that no matter how the connection weights and the thresholds are calculated, n is an upper bound for m ($m \leq n$). A more recent result (McEliece *et al*, 1987) states that, if the connection strengths are calculated according to the outer product construction, then it can be proved that $m \leq n/2log(n)$. Finally, in the case of higher order networks, it can be shown than $m \leq 2^{n-1}$ (Prados, 1988).

A2.5   Bidirectional Associative Memories

The *Bidirectional Associative Memory* (BAM), introduced by Kosko (1988a) is a two layer nonlinear feedback network that behaves as a hetero-associative content addressable memory. Bidirectional, ie forward and backward information flow, produces two-way associative search for stored pattern associations. Information passes forward from one layer to the other through the connection matrix W, and then passes backward through the matrix transpose W^T. It is the choice of the transpose that makes BAM a minimal two-layer feedback network, in the sense that all other two-layer networks require more information in the form of a backward connection matrix different from W^T. When the BAM units are activated, the network quickly evolves to a stable state of two-pattern reverberation or resonance in the spirit of Grossberg's adaptive resonance (1981). The stable reverberation corresponds to a local minimum of the system energy.

A BAM can store and recall a number of pattern pairs (A_k, B_k), $k = 1,2,...,m$, where the kth pattern pair is represented by the vectors $A_k = (a_1^k, a_2^k,...,a_n^k)$ and $B_k = (b_1^k, b_2^k,...,b_p^k)$. Let us accordingly denote by F_A and F_B the two layers (fields) of the network, where F_A is composed of n units and F_B is composed of p units. Although bidirectional stability can be proved for both binary and bipolar valued units, it has been shown that it is better on average to use bipolar valued pattern vectors.

The BAM is a fixed network, in the sense that learning is performed off-line and consists in the straightforward construction of the n-by-p weight matrix W, where w_{ij} is the connection weight from the ith unit of F_A to the jth unit of F_B. The bipolar pattern pairs $(A_1,B_1),...,(A_m,B_m)$ are encoded by means of a simple but general procedure based upon correlation (Hebbian) techniques. The idea is to memorise the association (A_k, B_k) by forming the correlation matrix or vector outer product $A_k^T B_k$. Then the m associations are superimposed by simply adding up the correlation matrices pointwise:

$$W = \sum_{k=1}^{m} A_k^T B_k$$

(A2.18)

Backward connection is then given by

$$W^T = \sum_{k=1}^{m} (A_k^T B_k)^T = \sum_{k=1}^{m} B_k^T A_k$$

(A2.19)

The above encoding formalism is fundamental in linear associative memories and provided an infrastructure for feedforward linear systems, such as Kohonen's Optimal Linear Associative Mapping. However, the BAM uses this principle in a nonlinear multi-iteration procedure to achieve hetero-associative mapping.

The BAM recall is a nonlinear feedback procedure operating in discrete time. Let $A = (a_1,a_2,...,a_n)$ and $B = (b_1,b_2,...,b_p)$ denote the state vectors of layers F_A and F_B respectively, during the operation of the network. Each unit in F_A and each unit in F_A independently examines its input sum from the units in the other layer and changes state or not according to a threshold function. The input sum to the jth F_B unit is the column inner product $AW^j = \sum_{i=1}^{n} a_i w_{ij}$, where W^j is the jth column of W. The input sum to the ith F_A unit is similarly $BW_i^T = \sum_{j=1}^{p} b_j w_{ij}$, where W_i is the ith row of W. By taking 0 as the threshold for all units, the new state values are calculated using

$$a_i = \begin{cases} 1 & \text{if } BW_i^T > 0 \\ -1 & \text{if } BW_i^T < 0 \end{cases}$$

(A2.20)

$$b_i = \begin{cases} 1 & \text{if } AW^j > 0 \\ -1 & \text{if } AW^j < 0 \end{cases} \tag{A2.21}$$

Units maintain their current state if the input sum equals the threshold.

When an input pattern is presented to F_A the units of F_A feed forward their state values through W to F_B. Then the values of the F_B units are calculated and fed back through W^T to F_A. The values of the F_A units are calculated and the procedure is repeated until all F_A and F_B state values cease to change, thus reaching an equilibrium. The procedure is analgous when a pattern is presented to F_B. The operation of the network can be synchronous or asynchronous without any limitation (such as the asynchronous operation restriction in discrete Hopfield networks), since in either case bidirectional stability can be shown for arbitrary matrices.

BAM stability is shown by defining a Lyapunov energy function

$$E(A,B) = -\tfrac{1}{2} AWB^\mathsf{T} - \tfrac{1}{2} BW^\mathsf{T} A^\mathsf{T} = -AWB^\mathsf{T} \tag{A2.22}$$

the last equality following from $BW^\mathsf{T} A^\mathsf{T} = B(AW)^\mathsf{T} = (AWB^\mathsf{T})^\mathsf{T} = AWB^\mathsf{T}$, since the transpose of a scalar equals the scalar. It follows that $E(A,B) = E(B,A)$, which establishes that the BAM energy is a well defined concept. It can be easily shown (Kosko, 1988) that the energy function decreases for any change in the state values of the F_A and F_B units. Hence, the BAM procedure converges to some stable point (A_f, B_f), such that $E(A_f, B_f)$ is a local energy minimum. Moreover, the weight matrix W is an arbitrary (real) matrix having no effect on the stability analysis. This implies that every matrix is bidirectionally stable.

The BAM is a nearest-neighbour pattern associator. Its storage capacity for reliable recall is roughly estimated as $m < min(n,p)$, ie no more pairs can be reliably stored and recalled than the lesser of the dimensions of the pattern spaces.

The BAM concepts discussed so far pass over to the continuous case, if we assume real state values which are transformed by bounded monotonically-increasing signal functions. This results in a dynamic model that constitutes a direct generalisation of the continuous Hopfield model (1984), which is itself a special case of the Cohen-Grossberg theorem (1983). The generalisation of the continuous BAM model is the *Adaptive Bidirectional Associative Memory* (ABAM) (Kosko, 1987), which extends global bidirectional stability to real-time unsupervised learning. An extension to the ABAM model is the *Competitive ABAM* (CABAM), which includes intra-layer connections (Kosko 1988b). Apart from the above extensions, Kosko also introduced the *Temporal Associative Memory* (TAM) model (1988a), which applies the discrete BAM concepts to temporal sequence encoding and recall.

A2.6   The Boltzmann Machine

The Boltzmann machine consists of a network of simple computing elements. These neurons can have two discrete states 'on' or 'off'. They are connected by synapses with different (real) weights, which represent a local quantitative measure for the desirability that the two connected neurons are on. See (Hinton *et al*, 1984 and 1985) for a general overview of the Boltzmann machine.

As in the simulated annealing algorithm an energy function is defined over the network. The energy function of the Boltzmann machine is defined as the sum of the product of the activations and the synaptic weights between two active neurons. As in simulated annealing the algorithm searches for the global minimum of the system by reducing the energy. Often only a local minimum is reached. To give the system a possibility to escape from such a minimum in the Boltzmann machine the state of an individual neuron is iteratively adjusted by a stochastic function. The applications of the Boltzmann machine are optimisation and pattern recognition.

A2.6.1 Mathematical Model

The network of the Boltzmann machine, shown in Figure A2.1, normally consists of input units, hidden units and output units. In special cases, especially in some optimisation problems, all neurons can be used as input and output units.

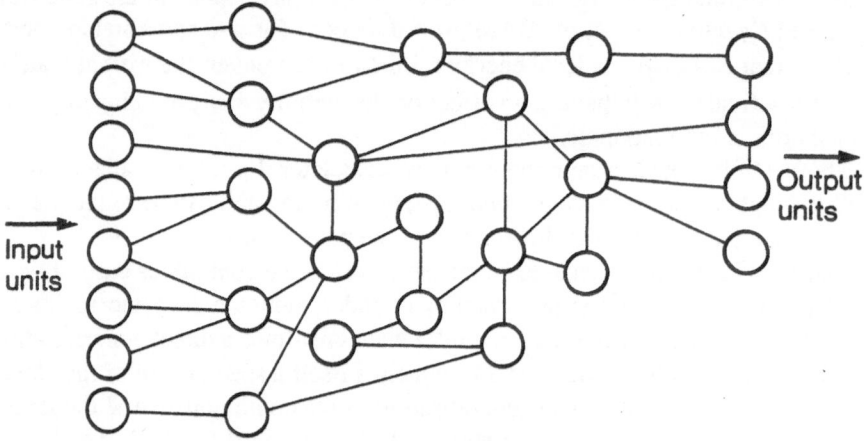

Fig A2.1 The architecture of a Boltzmann machine

The neurons are binary, that means they have only the states 0 and 1 ('off' or 'on'). The synaptic weights are real and may be positive or negative. The energy function is defined as:

$$E_k = \sum_{i=1}^{N} \sum_{j=1}^{N} w_{ij} u_i^k u_j^k$$

(A2.23)

where w_{ij} is the synaptic weight and u_i is the state of unit i in step k. If the state of neuron i is changed (from step k to k'), the change of system energy will be:

$$\Delta E_{kk'} = E_{k'} - E_k = (u_i^{k'} - u_i^k(\sum_{j=1, j \neq 1}^{N} w_{ij} u_j^k)$$

(A2.24)

The aim of the Boltzmann machine is to minimise the energy of the system by changing the neuron states. If the system proposes to change a neuron state, first the resulting $\Delta E_{kk'}$ is calculated. Then the proposed change of state will be accepted with the probability p_i:

$$p_i = \frac{1}{1 + e^{\frac{\Delta E_{kk'}}{T}}}$$

(A2.25)

T is the control parameter or temperature of the system like the temperature in simulated annealing. In contrast to simulated annealing (where all state transitions which decrease the system energy are always accepted and only state transitions which increase the energy are accepted with a probability less than 1), in the Boltzmann machine all state transitions (including those which decrease energy) are accepted with a probability, given by the equation above.

The probability function of the Boltzmann machine is shown in Figure A2.2. For $\Delta E_{kk'}$ negative the probability to accept is (for low temperature) near 1 and for $\Delta E_{kk'}$ positive it is near 0. In this way the Boltzmann machine reaches a good global minimum in short time. It may be shown that the absolute global minimum is reached in infinite time. A maximum instead of a minimum may be reached if the sign of the probability function is changed.

A2.6.2 The Boltzmann Machine Learning Rule
The Boltzmann machine may be used as an optimisation system. However, the synaptic weights must be hand-crafted by the user for the particular problem to be solved. For pattern recognition, there exist learning rules to teach the network how to assign the weights.

Step 1: In the first step the input vector is presented to the input units and the desired answer to the output units (input and output are binary vectors)

Step 2: Start the system with a high temperature so the units change their state with $p \approx \frac{1}{2}$. Let the network run. Decrease the temperature slowly, so that there are more than N (number of neurons) state changes between each temperature step. Decrease the temperature until the system reaches an equilibrium.

Step 3: Now let the system change many ($>>N$) states of randomly chosen units and generate a statistic about p_{ij}, ie the probability that neuron i and j are on

Step 4: Take away the output vector

Step 5: Decrease temperature again as in step 2

Step 6: As step 3 but generate a statistic about p'_{ij}
Step 7: Now adjust the synaptic weights: $\Delta w_{ij} = \varepsilon(p_{ij} - p_{ij'})$
        (ε is a scale factor)
Step 8: Repeat step 1-7 until Δw_{ij} reaches the desired minimum.

The system has reached a global minimum after learning.

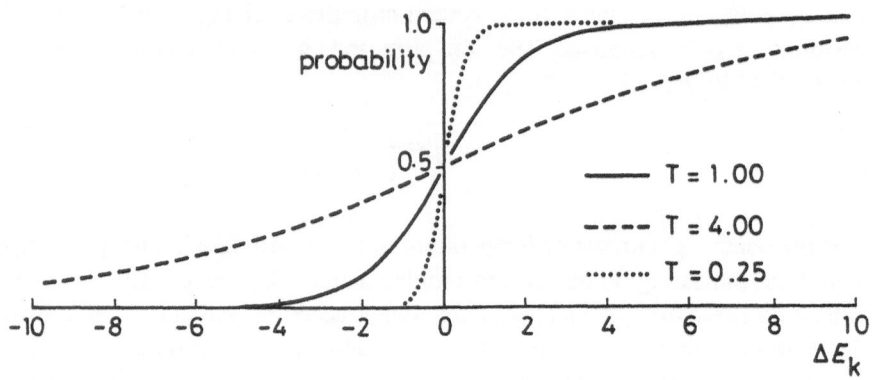

Fig A2.2 The probability function associated with the Boltzmann machine

A2.6.3 Results

The Boltzmann machine solves optimisation problems which are amenable to transformation into (0-1) problems (problem solution can be represented by a binary vector). In principle, all NP-hard problems can be transformed into 0-1 problems, but for most of the problems the transformation is very difficult. If the problem has been transformed then it may be implemented directly on a Boltzmann machine, this means that synaptic weights in network must be selected by the user. Then the Boltzmann machine searches for the optimum (Koarst and Aarts, 1988).

In the implementation on conventional (sequential) computers and even on parallel computers the Boltzmann machine is slower than simulated annealing, because in a Boltzmann machine at each step a probability must be calculated. With a hardware implementation, where calculation of the probability function can be performed extremely quickly, the Boltzmann machine may be faster than simulated annealing.

A2.7   Error Feedback Networks

Error feedback networks have, in common with associative memory, distinct learning and operation phases. In this type of network adjustment of weights is performed so as to minimise the error of the network with respect to its performance upon a set of known input/output vectors. This ability of an error feedback network to generate outputs which are completely different from its inputs enables it to be used as a *classifier*.

Over the past 45 years most neural networks have used the McCulloch and Pitts model (1943) or generalisations thereof as the basic computational unit. According to this model, each neuron performs an analogue summation of its inputs and uses simple thresholding to determine its boolean output. Some of the more popular generalisations of this model include a sigmoidal nonlinear function as a soft threshold.

A2.7.1 Feedforward Multilayer Networks

An important class of neural networks, which generally use such nonlinear functions, are *feedforward multilayered networks*. These networks consist of layers of neurons, which are connected in a feedforward way. The input patterns are presented at the bottom layer of the network, while the network outputs are generated at the top layer. Intermediate, or hidden, layers play an important role in these networks, because they provide the network with the ability to form internal representations and mappings from its input to its output.

Thus the neurons, or units, of hidden layers can be thought of as feature detectors, which capture important underlying regularities of the input information. A crucial property of feedforward networks is their ability to improve their performance by learning new information. This is accomplished by modifying the interconnection strengths among neurons, ie the weights of the network, according to some prespecified rules. The most frequently used technique for the training of such networks is supervised learning and using the error- backpropagation algorithm. Another technique, which can also be used for this purpose is reinforcement learning.

A2.7.2 Recurrent Networks

A more general class of neural networks are recurrent networks, the architecture of which includes feedback connections. The existence of feedback generally improves the computational abilities of the network, because the processing is performed in several steps or iterations, in contrast to the simple step processing of feedforward networks. A technique which can be used for the training of such network is the Boltzmann machine.

A2.7.3 Learning

In all the above-mentioned techniques, an error function is defined, the minimisation of which provides the required modification of the network interconnection weights. In most cases the response of the network to its input signals is used to define this error function. The computed error, which gives a measure of the network's performance, is fed back to the network for updating the interconnection weights. As a consequence, the networks which are trained by these techniques are based on error-feedback.

A2.8   Error Feedback Learning

In supervised training, feedforward networks are presented with pairs of input and desired output patterns. Two sweeps are then performed through all the weights. On the first sweep the input patterns propagate forward through the network and each hidden and output neuron computes its own output value using a prespecified set of weights. In the case of a multilayered network, the desired outputs are presented at the top output layer. On the second sweep the error between these target outputs and the actual outputs of the network is computed first. Then this error is propagated backward to the bottom layer and is used by the learning algorithm to update the network interconnection weights.

Various learning techniques have been developed and used for the supervised training of feedforward neural networks, including the simple least-mean-squared (LMS) and perceptron procedures, as well as the more recently introduced back-propagation algorithm.

A2.8.1   The Error Function

The supervised training of feedforward networks is generally based on the minimisation of a sum of squares error function. For a network with N_L outputs and after the presentation of K pairs of input and desired output patterns, this function is

$$E = \tfrac{1}{2} \sum_{r=1}^{K} \sum_{m=1}^{N_L} [d_m - y_m]_r^2$$

(A2.26)

where the symbols d and y denote the desired and actual output of the network respectively and the subscript r shows the dependence on the rth input presentation.

A2.8.2   The LMS Error-correction Algorithm

The capabilities of single-layered networks, called *perceptrons*, have been studied and analysed in earlier work on neural networks (Rosenblatt 1962, Minsky and Papert 1988). Adaptation of the networks' interconnection weights can be performed using the LMS error-correction algorithm (Widrow and Hoff 1960, Widrow and Winter, 1988). The LMS or Adaline learning rule minimises the error function E, by starting with an initial set of weights and repeatedly changing each weight by an amount proportional to the derivative of E with respect to it

$$\Delta w_{ij} = \mu \frac{\delta E}{\delta w_{ij}}$$

(A2.27)

where the parameter μ is a small positive number controlling the convergence of the method, and w_{ij} denotes the weight connecting the output of the ith neuron to the input of the jth neuron. The computation of the partial derivative in equation (A2.27) requires that a soft threshold, like a sigmoid, is used to determine the neuron's output.

A2.8.3 The Perceptron Rule

In case of binary threshold neurons using a step thresholding function, the LMS rule is replaced by the perceptron rule, which takes into consideration only the sign of the partial derivative of the error function. According to this rule Δw_{ij} is set equal to plus or minus μy_i, where y_i is the output of the ith neuron, depending on whether the jth neuron is off and should be on, or is on and should be off. There is no weight change, when the jth neuron is behaving correctly.

A2.8.4 Multilayer Network Learning

Although the above algorithms are capable of updating the weights of single-layered networks, they cannot provide appropriate solutions to the training of multilayered networks. The fundamental difficulty associated with the adaptation of neurons which belong to hidden layers has been solved using the back-propagation technique (Werbos 1974, Parker 1985, Rumelhart et al, 1986a). This algorithm is examined in some detail in section A2.9.

A2.8.5 Reinforcement Learning

A specific learning procedure, which differs somewhat from supervised learning, in that it does not receive a set of desired outputs, is *reinforcement learning*. In this technique, the network and the environment interact in a closed loop. The environment provides the network with a time-varying vector of inputs and the network responds with a time-varying vector of outputs. The environment receives the network's outputs and computes a scalar, stochastic or deterministic, signal r, which is called reinforcement and measures the network's performance. An appropriate function of the reinforcement signal, such as the expected value of it

$$J(w) = E[r|w] \tag{A2.28}$$

is then minimised by the reinforcement learning algorithm. It is generally assumed that some of the units are stochastic. In Williams (1987, 1987a) a general class of reinforcement learning algorithms is presented, where

$$\Delta w_{ij} = a_{ij}(r - b_{ij})e_{ij} \tag{A2.29}$$

where a_{ij} is a learning rate factor, b_{ij} is a reinforcement baseline and e_{ij} is defined as follows

$$e_{ij} = \frac{\delta ln(g_i)}{\delta w_{ij}} \tag{A2.30}$$

where g_i is the probability density function determining the output y_i as a function of the parameters of unit i and of its input.

A2.9   The Back-propagation Algorithm

Back-propagation is based on the steepest descent optimisation technique and uses the chain rule to calculate the derivatives of the error function (A2.26) with respect to the weights of the hidden or output units of feedforward multilayered networks. The error signals and derivatives are first computed at the top of the network as the difference between the desired and actual outputs and then propagated back to the bottom input layer. The sigmoid function, which is used in each neuron, provides the differentiability required along the network's signal paths. Each weight of the network is then updated using appropriate formulas (Rumelhart *et al*, 1986a).

An important property of back-propagation is the effective distributed updating of the estimated weights, which is amenable to parallel implementation and can be viewed as analogous to biological neural nets. However other properties of this technique, such as the back-propagation of the error derivatives, have been criticised as not biologically plausible. Moreover, despite its effectiveness, the algorithm's rate of convergence is too slow to be used in many practical situations. For this reason, the development of faster versions of back-propagation having superior convergence properties, is currently being investigated. A brief description of some current methods is given in the following section.

A2.9.1 Description of the Method

Let us define a vector $\mathbf{w}$ consisting of all the weights w_i^l of the network. A necessary condition at a global or local minimum of equation (A2.26) is that the derivatives of this function with respect to $\mathbf{w}$ be zero. A system of nonlinear equations is derived in this way, the dimension of which depends on the number of layers L and the number of neurons per layer N_l, for $l = 1,...,L$.

From optimisation theory (Luenberger 1984, Ortega and Rheinbolt 1970) it is known that there exist many iterative techniques for minimising a given error function, such as steepest descent, Newton's conjugate gradient and least squares methods. In general, all these techniques (Dennis and Schnabel, 1983) converge to a local minimum of the error function, which hopefully is a good solution to the minimisation being performed. The basic idea behind all methods is to update the interconnection weights iteratively, so that a new vector position at time $k + 1$ is computed in terms of the position at time k as follows

$$\mathbf{w}_{k+1} = \mathbf{w}_k - \mu P \mathbf{h} \qquad (A2.31)$$

where $\mathbf{h}$ is the direction of minimisation and P is the step length. The parameter μ can be chosen equal to a small positive number, which is less than unity, but it may also be selected using a line search strategy across the minimisation direction.

Steepest descent

Steepest descent uses the local downhill gradient of the error function in the place of $\mathbf{h}$ and sets P equal to unity. It has a low $O(n)$ computational complexity where n is the number of weights in the network. Conventional back-propagation, which is based on this technique, can update the input weights of all neurons belonging to the same layer in parallel, using information only about the specific weight to be updated. As a result, the algorithm is simple, but has only linear convergence, thus it can converge very slowly, especially when a minimum is approached. The momentum variant of the algorithm uses a damping of its behaviour

$$\mathbf{w}_{k+1} = \mathbf{w}_k - \mu\mathbf{h} + \alpha(\mathbf{w}_k - \mathbf{w}_{k-1}) \qquad (A2.32)$$

where the momentum factor α can be either fixed, or iteration-varying. In the latter case, it may vary as a function, for example, of the squared ratio of successive local gradients. The algorithm's performance can be improved in this way, but most of the above-mentioned problems still remain.

Conjugate gradient
Conjugate gradient is a technique which can be used to provide an improved variant of back-propagation. This technique can also be implemented in a form having a computational cost $O(n)$ (Shanno, 1978). Various forms of this method, which can be efficient in terms of convergence and parallel implementation, exist.

Quasi-Newton methods
Quasi-Newton methods use a matrix P as a step size in equation (A2.31). This matrix approximates the inverse of the Hessian matrix H, which contains the second derivative of the error function with respect to the weights (Luenberger, 1984). These methods require the inversion and storage of an $n * n$ matrix, resulting in a rather high $O(n^2)$ computational and storage cost.

The computation cost can be reduced by building up a sequence of matrices, which tend in the limit to the inverse Hessian matrix. Exact or approximate line searches (Dennis and Schnabel, 1983) may be used to compute the most appropriate convergence rate factor n in each step of the method. The application of Quasi-Newton methods as a variant to back-propagation has been examined by Watrous (1986).

An approximate scheme, which implements the Newton method in the form of a second order back-propagation learning algorithm, at a computational cost $O(n)$, is developed in Parker (1987). Another technique, which uses only the diagonal terms of the Hessian matrix to improve conventional back-propagation is presented in Le Cun (1987). Other implementations of the Newton method, which have a storage cost of only $O(n)$ operations have also been proposed in the literature (Werbos, 1988).

Least squares technique
The least-squares technique also uses a matrix P as a step size in equation (A2.31). This technique approximates the Hessian matrix, by keeping only products of first derivatives of the error function. Moreover, it limits the size of the weight updating increments ($\mathbf{w}_{k+1} - \mathbf{w}_k$), in order to implement an effective Taylor-series expansion of

the minimised error function (Luenberger 1984; Ortega and Rheinbolt 1970; Dennis and Schnabel 1983).

This is performed by adding an identity matrix, scaled by a factor μ, to matrix P, instead of multiplying it by the factor μ, as is described in equation (A2.31). The μ factor can be selected using a line search or a trust region strategy (Dennis and Schnabel, 1983; Wang, 1988) so that it is ensured that the error function is minimised in each step of the method. The performance of the method varies between the performance of the steepest descent and Newton's methods, depending on the value of the convergence rate factor n which is used. As a consequence, the method can have quadratic convergence, similar to the Quasi-Newton methods, close to a minimum of the error function.

A2.9.2 Generalisation and Network Design Methodologies

Generalisation may be viewed as the main property that should be sought when designing a network, since it determines the amount of data needed to train the network, so that it provides a correct response when a pattern outside the training set is presented at its input. In the advent of neural networks, it was assumed that there was not much need for remodelling the underlying problem and that an artificial neural network solution could be obtained instead, by training from empirical data, with little or no *a priori* information about the application. Recent studies, however, indicate that the right network architecture is fundamental for a good solution to exist. However, this is in general a difficult task. The lack of modelling, in classification problems, is not restricted to neural computing techniques alone, since the same can be said about statistical pattern recognition (Duda *et al*, 1973) in general, where simple Gaussian mixtures (usually two) often delimit the boundary of the developed theory. Although some moderate sized problems can be solved using general unstructured networks, it is not expected that an unstructured network generalises correctly on every problem. It has also been indicated (Le Cun, 1989a) that some *a priori* knowledge about the task can be built into the network, even in highly regular problems, such as image recognition. Questions, such as whether it is possible just to feed into the learning algorithms the images to be recognised (containing as many features as there are pixels) or first to preprocess and extract relevant features, such as line ends, arise in these areas. It could be mentioned here, that in the ANNIE work on pattern recognition, experimental studies have been performed to investigate the cases arising from the above questions, in the processing of image data.

The problem of what constitutes a sufficiently large training sample and a suitable network architecture has been a subject of recent research (Baum et al, 1989; Blumer et al, 1991; Haussler, 1989; Valiant, 1984). Results indicate that the likelihood of correct generalisation depends on the size of the hypothesis space, ie the set of networks which give good generalisation, and the number of training examples. If good generalisation is required, when the generality of the network is increased, the number of training examples should also be increased.

Some theoretical upper and lower bounds on the number of examples that are necessary to achieve generalisation have been recently developed (Baum, 1990). Let us assume that m random examples are chosen from some fixed, but arbitrary, probability distribution and that we try to load them on a feedforward network with W weights and N units, ie to train the network to learn them correctly. If we can find a set of weights, such that at least a fraction $(1-\varepsilon/2)$ of the training data are correctly loaded, for m sufficiently large ($m > (64W/\varepsilon)$ $\ln$ $(64N/\varepsilon)$), then there is large confidence (at least $1 - 8$ $exp(-\varepsilon m/32)$) that the network will correctly classify all but a fraction ε of future examples belonging to the same distribution. A similar result can be shown for the case where a learning procedure is used, which starts with W' weights, but during learning kills some *useless* synapses, resulting in a network with a smaller number W of weights. The following lower bound has also to be given: Let us train a network with n inputs, which are completely connected to a hidden layer of k units, the latter being connected to an output unit. Any learning algorithm, which uses fewer than (roughly) W/ε training examples, where $W = k(n+1)$ will be fooled by some distributions, ie it will not have good generalisation performance. This implies that using a net with W weights to fit fewer than $O(W/\varepsilon)$ examples is overfitting, so that generalisation cannot be guaranteed. Using various assumptions, it may also be possible to extend these results to the case of networks with more than one hidden layer.

As has already been mentioned, generalisation depends not only on the number of training examples, but also on the size of the solution space, or on the form of the distribution from which examples are taken. In practical situations there are many situations in which high rates of generalisation are obtained using sample sizes that are comparable to or smaller than the number of weights in the network (Denker, 1989). In such cases, the natural distribution may be trivial. For example, in classifying two different patterns, the distributions might roughly be two delta functions with some scatter; in this case it is, therefore, quite easy to classify them with very few training examples.

Since overfitting effects are due to oversized networks, an increase in the likelihood of correct generalisation can be obtained, by minimising the number of free parameters in the network. This, however, should be done without reducing the size of the network to the point where it can no longer compute the desired function. A good compromise becomes possible, when some knowledge about the task is available and can be used, with an increased effort, in the design of the architecture. Some techniques, which can be used to produce a reduced size network are described next.

A2.9.3 Back-propagation and Network Reduction Constraints
There are mainly two techniques for reducing the size of a network. The first is problem-independent and dynamically deletes *useless* connections *on-line*, during training. Various approaches have been proposed for this purpose, which add a constraint term (that is a function of the network complexity) to the minimised error criterion, that penalises big networks with many parameters (Chauvin, 1989; Hanson, 1989; Ishikawa, 1989)). The simplest case is to add the constraint term $\sum_i w_{ij}^2$, which is equivalent to weight decay during minimisation, but other, more complex constraints are also possible. Weight decay has recently been a very popular subject of

research. A drawback of this technique is that it generally slows down the convergence of the method.

The second technique is weight sharing. This technique imposes equality constraints among the connection strengths, in the sense that several interconnections in the network are controlled by a single weight. An interesting property of this technique is that it can be implemented with very little computational overhead. Weight sharing can be used for shift-invariant feature extraction and is generally used in many pattern recognition problems.

Other techniques which can be used for reducing the network size are based on weight-space transformations. These techniques use some transformation to produce another space of parameters which is more suitable for the specific minimisation task. Some problems of these techniques are the choice of the correct transformation and the knowledge of the Jacobian matrix of the transformation, which is required so that the gradients of the error function are computed.

An example which uses constraints in designing a network for recognising handwritten numerals is given in (Le Cun, 1989b). It is based on results of classical work in visual pattern recognition, which indicate the advantage of extracting local features and combining them to form higher-order features.

A2.9.4 Acceleration Techniques for Back-propagation
Back-propagation converges slowly, even for medium-sized network problems. This fact results from the usually large dimension of the weight space and from the particular shape of the error surface. Oscillation between the sides of the deep and narrow valleys, for example, is a well-known case where gradient descent provides poor convergence rates. The search for faster and more robust training methods must meet two requirements: simplicity, in order to reduce the total computational load, and locality, in order to maintain compatibility with distributed hardware architectures.

Back-propagation is generally used to minimise a quadratic function of the error at the output of a multilayered network, updating the network weights in the direction that yields the maximal error reduction. As a result, the weight update is largest among the components with largest derivatives. The following cases may, however, also occur (Sutton, 1986). First, if a certain component of the gradient vector has a small value, this may indicate that the error surface has a gentle slope along that direction and may be far away from the minimum, so that bigger steps in that component can be used for faster convergence. Second, if a certain component of the gradient has a large value, it may correspond to a direction orthogonal to a deep and narrow valley, so that smaller steps in this direction might help to avoid oscillations and to enable the bottom of the ravine to be reached.

A technique that is currently used for speeding up the learning rate lets each synapse have its own learning rate parameter, increasing or decreasing its value according to the number of sign changes observed in the partial derivative of the error function with respect to the corresponding weight (Jacobs, 1988). The basic idea of this technique is that if the sign of a certain component of the gradient remains constant for several iterations, it corresponds to a smooth variation of the error surface;

the learning rate for this component should, therefore, be increased. On the other hand, if the sign of some component changes in several consecutive iterations, the learning rate parameter should be decreased to avoid oscillation.

A simple and effective way of implementing this idea is to adapt the learning rate parameter n_{ij} in the weight update equation

$$w_{ij}(k) = w_{ij}(k-1) + n_{ij}(k)\delta_{ij}E(k) \qquad (2.33)$$

where $n_{ij}(k)$ is the specific learning rate parameter of the ijth synapse at iteration k, according to the following rule

$$n_{ij}(k) = d_l n_{ij}(k-1) \qquad (2.34)$$

for $l = 1,2$, with $l = 1$ corresponding to the case where the derivative of E has the same sign in the k - 1 and k iterations and with $l = 2$ implying the contrary. In general d_1 is slightly greater than unity (between 1.1 and 1.3) and d_2 slightly below $1/d_1$. the above equations imply an exponential increase and decrease of the learning rate parameter. More details about the algorithm can be found in (Almeida *et al*, 1990).

An interesting feature of the method is that it can be combined with other, higher order variants of back-propagation that use an approximation of the Hessian matrix of the error function in the minimisation procedure (Le Cun, 1989b). These variants are not believed to bring a tremendous increase in learning speed, but converge reliably,without requiring extensive adjustments of the learning parameter. We have combined the above-mentioned acceleration technique with a second-order least squares back-propagation variant, based on the Marquardt-Levelberg optimisation technique (Kollias *et al*, 1989).

A2.9.5 Extensions to the Method
Application to recurrent networks

Back-propagation has been developed as a learning algorithm for feedforward networks. An interesting topic of research has been the application of back-propagation to recurrent networks. In Minsky and Papert (1988) and Rumelhart *et al* (1986a) it is shown that for every network with feedback, there is a feedforward one, which has identical behaviour over a finite period of time. This is achieved by adding an identical feedforward network at the top of the original one, as many times as required to cover the whole period of time.

However, this formulation is neither efficient, nor elegant. Various techniques have been developed, which implement the back-propagation learning technique to recurrent networks (Almeida, 1987; Pineda, 1987 and 1988; Rohwer and Renals, 1988). A serious problem in this case is the proof of convergence. Work is being done to develop specific recurrent architectures, which ensure that the above mentioned form of back-propagation converges and can therefore be effectively used.

Combination with unsupervised learning

For the case of accuracy and for better real-time convergence, attempts have been made to combine back-propagation with unsupervised learning (Werbos, 1988; Lippmann, 1987), such as the content-addressable memory scheme developed by Kohonen (1988b). A combination of unsupervised learning with other error-correction learning techniques, such as the LMS algorithm is performed in Moody and Darken (1988).

A2.10 Self-organising Networks

Self-organising networks, as the name implies, have the common characteristic that they are able to recognise similarities in data without being explicitly taught what their output should be. In this section, the self-organising mapping of Kohonen and the Adaptive Resonance Theory of Grossberg are considered.

A2.10.1 The Kohonen Self-organising Feature Map

It is known that there are different regions in the brain either performing different tasks such as speech control, analysis of sensory signals etc, or relating to certain sensory modalities such as vision, hearing etc.

Within these regions there are some fine structures which reveal a kind of order. This order preserves a mapping between the topographical ordering of the sensory origins of the signals and the ordering of the areas where they are received in the brain (for example sensory signals of shoulder, arm and hand, the *somatosensory map*, or the mapping of different frequencies by the hearing, the *tonotopic map*). These localised responses to sensory excitations seem to be achieved by a special kind of lateral feedback. Around each neuron or functional unit of neurons there is a small area of neurons with excitatory connections to this unit surrounded by a wider 'ring' of inhibitorily connected neurons again surrounded by a range of excitatorily connected neurons.

In 1984 Kohonen introduced in his book 'Self-organisation and Associative Memory' (1984), a kind of artificial neural network which he called the self organising feature map. The architecture and functional laws of this network are motivated by the above mentioned features of the brain.

General description

The self organising feature map is a single layer network with (time-variable) lateral feedback which provides local response to external excitations. The processing elements obey an adaptation rule which represents a form of unsupervised learning. Learning occurs in neighbourhoods; upon excitation with an external input, not only the local responding element but a whole neighbourhood will learn to improve the response to this input pattern. This kind of architecture and learning will produce a network which automatically learns to recognise the main topological features of the distribution and ordering of a collection of input patterns.

The network maps an input pattern to a corresponding network unit according to the learned features, preserving some kind of ordering or topology under this mapping (eg input patterns which were neighbours in the input-space will also be neighbours after the mapping) and approximating the input distribution density by the point density of the weight vectors.

Ordered mappings

Assume that we have a set of input patterns $\{x_i : i = 1,...,n\}$ that can be ordered in some metric or topological way such that $x_1Rx_2Rx_3...$, where R stands for a simple ordering relation, eg with respect to a single feature that is implicit in the representations. We define a topology-preserving mapping as:

Definition. A one-level self-organising system, like that of Figure A2.3, is said to produce a one-dimensional topology-preserving mapping if for $i_1 > i_2 > i_3 > ...$

$$y_{i_1}(x_1) = \max_i\{y_i(x_1): i = 1,...,n\}$$

$$y_{i_2}(x_2) = \max_i\{y_i(x_2): i = 1,...,n\}$$

$$y_{i_3}(x_3) = \max_i\{y_i(x_3): i = 1,...,n\} \text{ etc}$$

where $y_i(x_j)$ is the response of unit i to the input pattern x_j. The above definition is readily generalisable to two and higher-dimensional arrays of processing units. In this case the topology of the array is simply defined by the definition of neighbours to each unit.

Feature maps

A critical point of self-organisation is the formation of local response by lateral feedback. We briefly demonstrate this phenomenon by considering a one-dimensional array of processing units, each of which receives the primary input x_i and a great number of lateral connections from the outputs of other units (Figure A2.3). The strength of lateral interaction is a function of distance and is usually described as having the form of a Mexican hat. The array of processing units can also be two-dimensional.

The signal transfer of this network can be expressed in terms of the following equation:

$$y_i(t) = \sigma\left[x_i(t) + \sum_{k=-S}^{S}\gamma_k y_{i+k}(t-1)\right] \tag{A2.33}$$

where σ stands for a sigmoid nonlinearity, $[-S,S]$ is the width of lateral feedback and γ_k are coefficients indicating the strength of lateral feedback.

We can describe the 'clustering' phenomenon due to lateral feedback as follows: the initial activity distribution in the network may be more or less random, but over time the activity develops into *clusters* or *bubbles* of a certain dimension. As the bubble should be formed at the local maximum of activity, it may be completely expedient to define the computation process as consisting of two steps:

(i) find the activity maximum
(ii) define a subset of units in the array around this maximum, corresponding to the bubble.

Learning

Upon excitation with an input signal, due to the lateral feedback the network forms 'bubbles' of activation. That means the elements of a neighbourhood of the unit with the highest input activation produce positive output while all other units have zero output. Input activation for unit u_c is computed by merely summing up the weighted inputs:

$$x_c = \sum_j w_{jc} a_j$$

The size of this bubble or neighbourhood depends on the ratio of positive weights to negative weights.

Once the neighbourhood N_c is formed, learning occurs within this neighbourhood. All units belonging to N_c move their weight vectors W_i a certain amount towards the input pattern A_k. Outside the neighbourhood N_c no learning takes place.

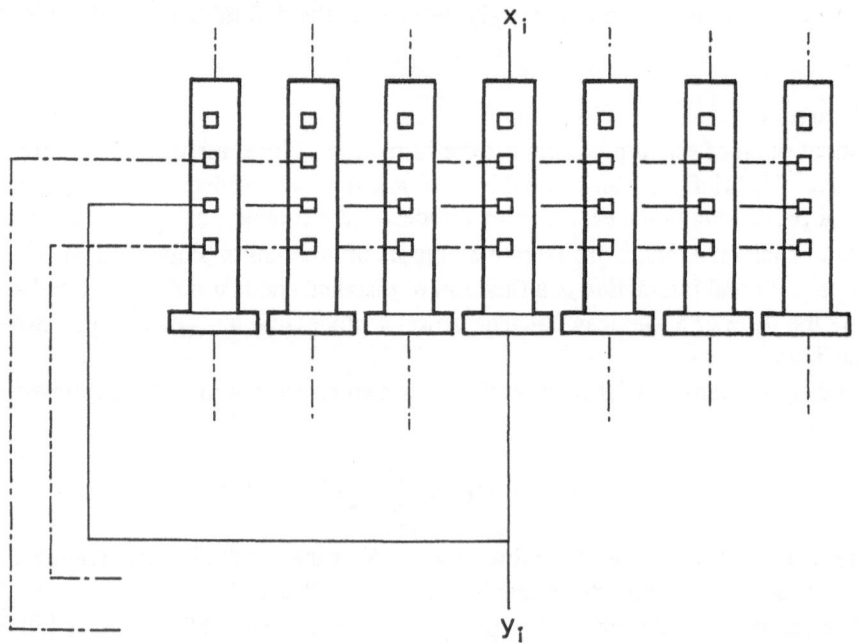

Fig A2.3 Lateral inhibition in an array of elements

A2.10.2 Adaptive Resonance Theory

The ideas in adaptive resonance theory (ART) partly stem from Grossberg's thesis that if a parallel architecture is to offer significant benefits, then the whole system must be organised in a parallel distributed way. Systems akin to back-propagation generally have distinct 'learning subsystem' and 'operational subsystem' components.

ART networks are, therefore, intrinsically more complicated than other architectures: the learning subsystem and operational subsystem interact continuously and no distinction is drawn between the 'learning phase' and 'operational phase' (Carpenter and Grossberg, 1986a).

Architecture of ART1

Figure A2.4 shows a typical architecture of an ART network designed for binary inputs taken from Carpenter and Grossberg (1986b). Two successive stages F_1 and F_2 of the attentional subsystem encode patterns of activation in short term memory (STM). Bottom-up and top-down pathways between F_1 and F_2 contain adaptive long term memory (LTM) traces which multiply the signals in these pathways. The remainder of the circuit modulates these STM and LTM processes.

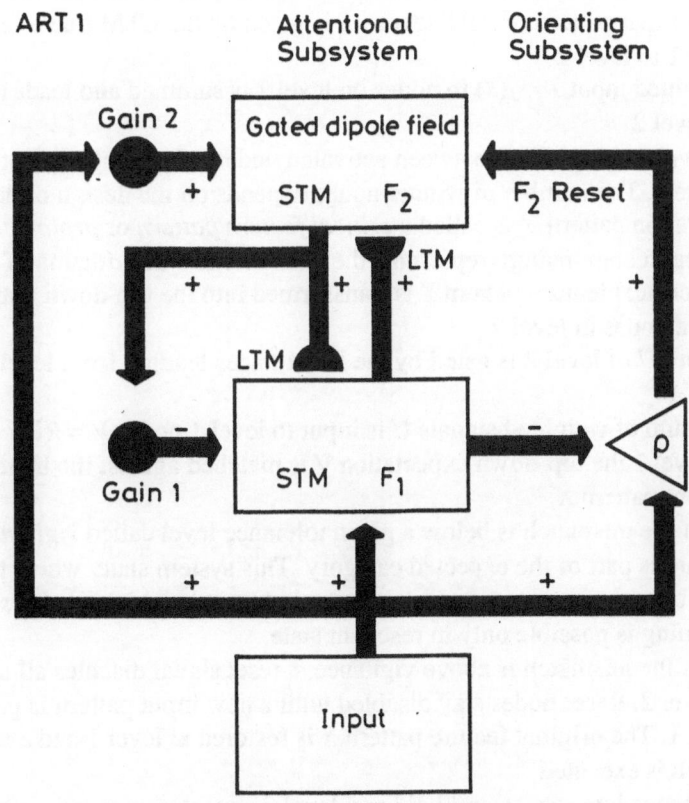

Fig A2.4 The architecture of ART1 for binary inputs

Modulation by gain control enables F_1 to distinguish between bottom-up input patterns and top-down priming, or template, patterns, as well as to match these bottom-up and top-down patterns. Gain control signals also enable F_2 to react supraliminally to signals from F_1 while an input pattern is on.

The orienting subsystem generates a reset wave to F_2 when mismatches between bottom-up and top-down patterns occur at F_1. This reset wave selectively and enduringly inhibits active F_2 cells until the input is shut off.

Operation of the network

The following sequence of operations is accomplished by an ART system when an input signal is presented.

(1) Depending on the problem area the input to ART comes from, the input signals preprocess in an appropriate way

(2) The input signal I is transformed into a feature pattern X by the nodes of ART level 1

(3) The output function of nodes in ART level 1 implies thresholding takes place. So all nodes that have an activation level above the threshold value are firing

(4) The output pattern of ART level 1 S is gated by the LTM traces leading from level 1 to level 2

(5) Weighted input $T = f(S)$ to nodes on level 2 is summed and leads to activation on level 2

(6) In level 2 competition between activated nodes selects the highest winner or winners. The number of winner nodes depends on the design of level 2. This activation pattern Y is called a *critical feature pattern* or *prototype*. The critical feature pattern represents the characteristic information of one category

(7) The critical feature pattern Y is transformed into the top-down output U by the active nodes in level 2

(8) Output U of level 2 is gated by the LTM traces leading from level 2 to level 1

(9) The sum of weighted signals U is input to level 1 nodes $V = f(U)$

(10) At level 1 the top-down expectation V is matched against the bottom up feature pattern X

If the mismatch is below a given tolerance level called *vigilance*, the input feature is part of the expected category. This system state, where the balance is held between bottom-up and top-down inputs at level is called *resonance*. Learning is possible only in resonant state.

If the mismatch is above vigilance, a reset signal disables all active nodes at level 2. Reset nodes stay disabled until a new input pattern is presented to level 1. The original feature pattern X is restored at level 1 and a new category search is executed

(11) The procedure repeats until either a level 2 expectation matches the input feature at level 1 or the continuous disabling of level 2 nodes leads to

uncommitted nodes at level 2. If uncommitted nodes at level 2 are selected, a new category is established and the new critical feature pattern is learned.

(12) If no uncommitted nodes are available at level 2 for creation of a new critical feature pattern, the system stops.

During the search for a matching critical feature pattern, no learning is in effect. Grossberg tells that the update of LTM traces is a very slow process compared to the STM activation initiating the search process.

The gain control parameter is an auxiliary parameter which is used to tell the nodes at level 1 whether top-down input must be processed.

Features of the model

Self-scaling computation units: critical feature patterns. The system is able to adjust its ability to ignore or notice noise in input patterns depending on the complexity of the learned categories. In simple patterns with a few features, a small bit image change in the pattern is significant; this is not so with complex patterns. The system accomplishes this by storing a critical feature pattern which represents a particular class.

Self-adjusting memory search. ART is capable of a parallel memory search that adaptively updates its search order to maintain efficiency as its recognition code becomes arbitrarily complex due to learning.

Direct access to learned codes. In an ART model, as the learned code becomes globally self-consistent and predictively accurate, the search mechanism is automatically disengaged. Subsequently, no matter how large and complex the learned code may become, familiar input patterns *directly access*, or activate, their learned code, or category. Unfamiliar patterns can also activate this category is they have enough in common with learned categories.

Environment as teacher: modulation of attentional vigilance. Although an ART system self-organises its recognition code, the environment can also modulate the learning process and thereby carry out a teaching role. This teaching role allows a system with a fixed set of feature detectors to function successfully in an environment which imposes variable performance demands.

In an ART system, if an erroneous recognition is followed by an environmental disconfirmation such as punishment, the system becomes more *vigilant*. This change in vigilance may be interpreted as a change in the system's attentional state which increases its sensitivity.

ART2 analogue inputs

One of the limitations of the ART1 architecture is that it expects binary input values. ART2 architectures are designed to meet the particular demands of analogue inputs.

In Carpenter and Grossberg (1987) the ART2 architecture is introduced, compared and contrasted with ART1. Figure A2.5 illustrates the principal differences between ART1 and ART2 networks. In order to match and learn sequences of analogue input patterns in a stable fashion, ART2 splits F_2 into multiple processing levels and gain control systems. Bottom-up input patterns and top-down signals are received at

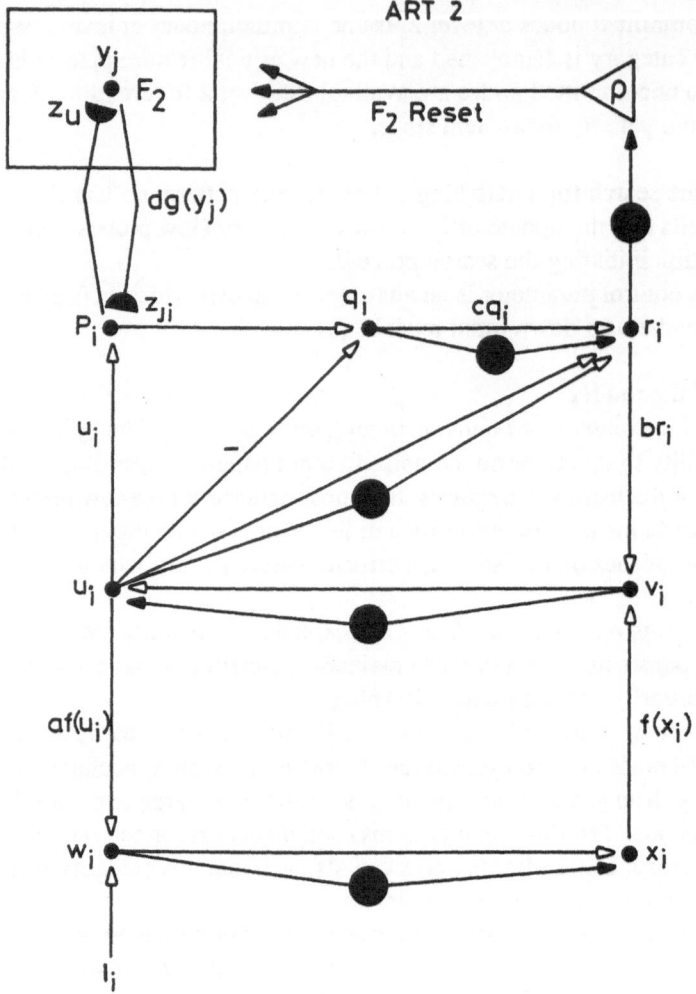

Fig A2.5 The architecture of ART2 for grey scale inputs

different locations in F_1. Positive feedback loops within F_1 enhance salient features and suppress noise. The multiple F_1 levels buffer the network against incessant recoding of the category structure as new inputs are presented.

A2.10.3   The Counterpropagation Network

The counterpropagation network (CPN) is a mapping type neural network model introduced by Hecht Nielsen in 1986, functioning as a statistically optimal self-organising look-up table. Its architecture is based on a combination of two other types of network structures which are Kohonen's 'self-organising map' (Kohonen, 1988) and Grossberg's 'outstar structure' (Grossberg, 1988; Soucek and Soucek).

The counterpropagation network model implements approximations to mappings from R^n to R^m and works equally well for both binary and continuous input and output vectors. It performs best in problems requiring statistically equiprobable feature vectors or when a lookup table structure is desirable. Despite the network's simple structure, optimality is reached in nearly the same time as other mapping networks, and moreover less time is required to train the network.

General description

In its complete version, the CPN model consists of five layers of processing elements (PEs) as in Figure A2.6 while in its 'forward-only' version the five layers reduce to three.

The following presentation concerns the complete version of the network in which the input and output vectors used as examples of the mapping to approximate are considered to be normalised.

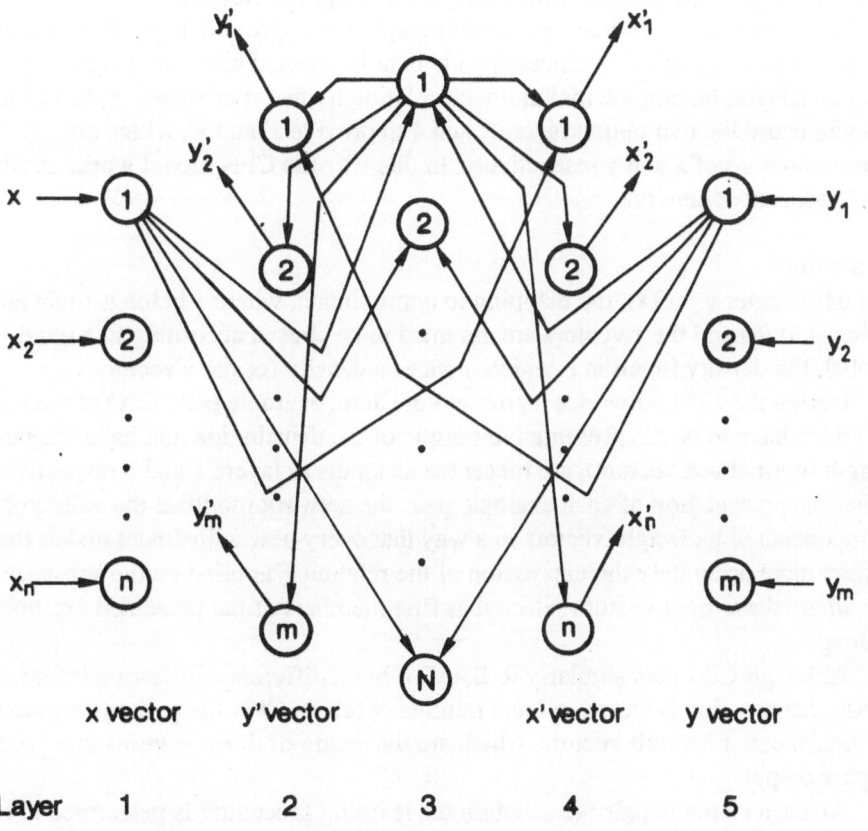

Fig A2.6 The architecture of the counterpropagation network

The mapping to approximate is $y = F(x)$ where x belongs to R^n and y belongs to R^m and the x vectors are assumed to be chosen according to a fixed probability density function ρ which induces a density for the y vectors.

Architecture

Layers 1 and 5 contain n and m fanout units respectively and they operate as multiplexers of the components of the network's input pairs' vectors (**x**, **y**) before their propagation to the next layer 3 named *the Kohonen layer* consisting of N units and operating as a maximum activity selector.

Layers 2 and 4 named the *Grossberg outstar layers* consist of m and n units respectively. Each of them gets the corresponding component of the input vectors (x,y), Figure A2.6, and receives inputs from every unit of the Kohonen layer functioning as decoders of the pattern of activity of the Kohonen layer.

The connectivity between the fanout units and those of the Kohonen layer is expressed through the weight vectors U_i and V_i from the units of layers 1 and 5 respectively to the i unit of the 'Kohonen layer' where the same unit is connected to each of the 'Grossberg layer' units through the weight vector W_i.

When an exemplar vector (x,y), of a mapping to approximate $y = F(x)$ is applied to the two inputs, a particular processing element is selected after competition on the Kohonen layer, having the maximum correlation to the input vector (x,y). The above PE will cause the two output layers 2 and 4 to provide x' and y', which are approximations of x and y respectively. In this way the CPN model works as a bilateral auto-associative network.

Learning

Let us consider $y = F(x)$, the mapping to approximate, where x belongs to R^n and y belongs to R^m and the x vectors are assumed to be chosen according to a fixed probability density function ρ which induces a density for the y vectors.

During the CPN network's learning procedure, example pairs (x,y) of the mapping F, which have to be a representative sample of the distribution and have the same length (normalised vectors), are presented as inputs to layers 1 and 5 respectively. After the presentation of each example pair, the network modifies the values of the components of its weight vectors in a way that every new adjustment makes them reflect more accurately the expression of the relation F applied on the subspace of the R^n where the x input vectors selected as first members of the presented example pairs belong.

Although CPN acts similarly to BAM, it has a difference in storing information. In BAM, information is stored in local minima, while in CPN the patterns are stored in Kohonen exemplar unit vectors, which are the centre of those patterns that produce the higher output.

At each example pair presentation the learning procedure is performed in two phases. The first one concerns the Kohonen layer units weights adjustment and the second that of the Grossberg layer units.

Convergence

Learning parameters for both the Kohonen and Grossberg layers should be fine-tuned in order to obtain suitable convergence. According to Kohonen, $\alpha(t)$ must be decreased to zero and satisfy the following conditions (Kohonen, 1988b):

$$\sum_{t=1}^{\infty} \alpha(t) = \infty \text{ and } \sum_{t=1}^{\infty} \alpha^2(t) < \infty \tag{A2.34}$$

For example in practical computations it could be chosen as: $\alpha(t) - t^{-1}$. Anyway, the proper choice can be determined by experience. In this way, as training progresses the final convergence of the weight vectors to the asymptotic values is of a particular interest. They become representative of the distribution of the input data vector pairs which are selected according to a fixed probability density function ρ.

Operation

Operation in the CPN can take place in both directions from the input vector to the target vector or from the target vector to the input vector, as well as with corrupted input data.

During the operation phase, a test vector x is entered into the network. Then, the Kohonen units compete with one another, and the unit of order c with the highest activation value I_i, has its output signal (z_c) set to one, while all z_i, $i \neq c$, $i = 1,2,...,N$ outputs are set to zero. This can be performed by several different neural nets, like the MAXNET using lateral inhibition (Lippmann, 1987; Feldman and Ballard, 1982).

The single output of the Kohonen layer will lead through the corresponding Grossberg outstar storage weights to the reproduction of the associated vector y'. The outputs corresponding to the winner node z_c are given by:

$$y'_c = W_{cj}z_c \quad \forall_j = 1,2,...,m \tag{A2.35}$$

Variants

Another type of counter propagation network is described by Hecht-Nielsen (1987) and some of its interesting applications are discussed in Hecht-Nielsen (1988). The new version, called the *forward-only CPN module* is useful for problems in which only a forward mapping from x to y is of interest and is a reduction from the general case described before consisting of three layers of processing elements (PEs).

Layer 1 contains $n + m$ units. The n units operate as multiplexers of the components of the network's input vector x before the propagation to the N units of the next layer 2 the *Kohonen layer* whether the m units supply the components of the network's input vector y.

A2.10.4 Graph Matching Networks

This model implements the notion of *graph-coding* in neural structures as an alternative to *rate-coding* which is the classical representation scheme in neurobiology.

Rate versus graph coding

In rate-coding the information is conveyed by the mean activity of neurons or groups of neurons to which every external object or relationship is assigned, whereas in graph-coding complex objects are described as systems of relationships represented through labelled graph structures.

A labelled graph is a set of *nodes* or *vertices*, a set of *labels* attached to the nodes and a set of *links* or *edges* binding some of the nodes. In a neural network implementation of a labelled graph, nodes are assigned to neurons and edges are assigned to connections between corresponding neurons. Even in the conventional connectionism approach, in which connections between units serve only as the substrate of *long-term memory*, the notion of dynamic connections is usually introduced as a mechanism for learning.

In *conventional connectionism terms*, connectivity is defined during learning and remains frozen during the processing of information; however in *dynamic connectionism* terms, part or all of the connections of the network can switch on or off (or increase or decrease gradually) on the same timescale as the processing itself. In the *dynamic connectionism* approach, any given relationship which may be present or absent in a given situation is directly embodied in a connection or a set of connections between processing units. Connections are dynamic and play the role of activity levels in classical connectionist models. A dynamic connectionist network manipulates *states of connectivity* rather than *states of activity*.

Graph isomorphism

The main application area of the model in real world problems is *invariant pattern recognition*. In terms of the model, a pattern recognition task is regarded as a labelled graph matching problem.

In a *labelled graph matching* problem, the main task is finding a map between two graphs satisfying requirements such as a one-to-one correspondence between the labelled nodes and the preservation of existing links. A map satisfying all individual requirements between the two graphs in a strict fashion is termed as a *graph isomorphism*.

Invariance can be defined as the unaffected *perception* of a pattern in spite of considerable changes the *retinal image* of the pattern may undergo. Strict constraint satisfaction as required by *graph isomorphism* covers the case of two-dimensional rigid transformations, translations and rotations of the retinal image. If some of the constraints for strict graph matching are relaxed, the so-called *weak isomorphism problem*, then non-rigid or rubber sheet transformations such as contractions, dilations and distortions also become invariant for the model. In the latter case of weak constraint satisfaction, the notion of graph-matching becomes broader than graph-isomorphism and comes closer to that of a continuous map between topological spaces.

Architecture

The *graph matching* neural network model consists of three layers of cells:

- the retinal layer R which is at the lowest level where patterns are actually displayed. We assume that an input pattern may consist of a topological arrangement of local features and there exist r different types of local features
- the intermediate layer A containing feature detector cells arranged in a retinotopic way, which is used to represent input patterns. We assume that A is a regular m x m array and that there are r feature detector cells at each of the m^2 positions in A.
- the memory layer B which is a non-retinotopic set of m^2 feature detector cells and which is used to recognise input patterns. We assume that the partition of the m^2 B-cells into the r feature types is such that there are as many cells of a given type as there are active cells of this type when a pattern is displayed on the retina. We also assume that patterns are composed of different arrangements of the same number of features of every one of the r possible types, or that patterns are endowed the same first order statistics.

In the model, we consider the following variables:

a_i: spin or activity of a cell i in a layer of the network
T_{ij}: permanent connection strength between cells i and j
J_{ij}: dynamic link, variable associated with synapse (i,j), interacting with the second order correlations $(a_i.a_j)$

Variables are subject to the following constraints:

$a_i = +1$ when a cell is 'active' and $a_i = -1$ when a cell is 'inactive'
$T_{ii} = 0$ for every i; ie there is no permanent connection from a cell to itself

$J_{ij} < T_{ij}$; ie only existing synapses can be activated.

The intra-layer and inter-layer connectivity of the model is described as follows:

- the connectivity between R and A is hardwired and fixed and permits a retinotopic projection from R to A, and the extraction on A of local features of the patterns is displayed on R.
  Exactly one feature detector cell is active at every site in A at any given time. Each cell in A is connected to all A-cells in the p first neighbouring sites. The connectivity is fixed within A, and not modifiable during learning.
      Thus $J_{ij} = T_{ij}$ for every (i,j) in A.
- The connectivity within the memory layer B is null before learning has taken place; its development is described below related to learning in the model.

- The interlayer connections between cells in A and cells in B is allowed only between cells of the same feature type. The interlayer permanent connectivity set ($\{T_{ij}\}$ with i in A and j in B) is complete in the sense that every feature detector cell in A is connected to all feature cells of the same type in B, thus : If i is a cell in A and j is a cell in B, then $T_{ij} = 1$ if i and j encode the same feature type and $T_{ij} = 0$ if i and j encode different feature types.

Learning

Learning in the model affects the permanent connectivity T_{ij} within B and it is performed as follows.

Presentation. A pattern to be memorised is presented on the retina R. As a consequence of the fixed connectivity between layers R and A, local features in the displayed pattern are extracted and a labelled pattern is returned in A. For each occurrence of a local feature on the labelled pattern in A a feature cell of the same type is randomly picked in the memory layer B. A connection pattern is then stored in B, by setting $T_{ij} = 1$ whenever selected cells i and j in B correspond to neighbouring points at the labelled pattern in A.

Neighbourhoods. There are several alternative ways of defining a neighbourhood region in A. A convenient one defines 3 incoming and 3 outgoing links at each node. Connection patterns defined thus in B are orientated graphs. Several connection patterns are superimposed in B in this way. Some of the stored patterns may partially overlap. In any case the random selection of feature cells in B assures that any link of a particular connection pattern defined in B is of a long range in the topology of any other superimposed pattern. This mixing requirement is very important for the dynamics of the model.

Operation

Suppose that M patterns are learned by the model and that one of them is presented on the retina R to be recognised.

As a consequence of the connectivity between R and A a labelled pattern $\{J_{ij}\}$ is activated on A, where (i,j) are feature cells on A. The dynamics of the recognition determines the $\{J_{ij}\}$ where i belongs to layer A and j belongs to layer B and $\{J_{ij}\}$ where both i and j belong to layer B.

These dynamic links are subject to the constraints $J_{ij} \leq T_{ij}$ and their final configuration depends on the minimisation of an energy function defined as follows:

$$H(J) = H^{BB}(J) + \delta H^{AB}(J) \qquad (A2.36)$$

where

$$H^{BB}(J) = - \sum_{ijkl \in B j \neq k} J_{ij} J_{il} J_{ik} J_{kl} + \gamma \sum_{i \in B} (\sum_{j \in B} J_{ij} - p)^2 + \gamma \sum_{j \in B} (\sum_{i \in B} J_{ij} - p)^2 \qquad (A2.37)$$

and

$$H^{AB}(J) = - \sum_{ij \in A kl \in B} J_{ij} J_{il} J_{ik} J_{kl} + \gamma' \sum_{i \in A} (\sum_{j \in B} J_{ij} - p')^2 + \gamma' \sum_{j \in B} (\sum_{i \in A} J_{ij} - p')^2 \qquad (A2.38)$$

The first term in $H^{BB}(J)$ favours, in low energy states, 4 cycles (cycles of length 4) and acts as a cooperation term. Its formulation is due to the observation that memorised patterns are regular two-dimensional lattices and as such they contain many 4 cycles.

The other two terms embody *weak constraints* on the number of links at each node (3 incoming and 3 outgoing) so that $p = 3$; they are added as penalty terms for these $\{J_{ij}\}$ configurations violating the preferred connectivity scheme. $H^{BB}(J)$ has M minima each of each has a maximum number of 4 cycles and corresponds to one of the M learned superimposed patterns in layer B. The minimisation of $H^{BB}(J)$ corresponds to an operation of *memory retrieval*.

The first term in $H^{AB}(J)$ favours, in low energy states, mixed 4 cycles (cycles of length 4 where a pair of connected nodes in layer A has a one-to-one connection with a pair of connected nodes in layer B). This formulation is due to the property that *an isomorphism between two graphs is a one-to-one map whch contains as many 4 cycles as possible*.

The other two terms of $H^{AB}(J)$ embody *weak constraints* on the one-to-one dynamic connectivity of nodes between the two layers A and B so that $p' = 1$. the minimisation of $H^{AB}(J)$ establishes an isomorphism between the graph presented on the layer A and the one retrieved in layer B with label preservation and so corresponds to an operation of *graph matching*.

The minimisation of $H(J)$ solves the matching and retrieval problems simultaneously. The adjustment of the δ parameter achieves an appropriate balance between the two terms of the function.

The $H^{BB}(J)$ energy function comprises M symmetric valleys corresponding each to one of the memorised patterns. The introduction of the term $\delta H^{AB}(J)$ breaks this symmetry making the valley corresponding to the pattern isomorphic to the one presented on A a global minimum for the function $H(J)$.

Simulated annealing can be used for the minimisation of the **H(J)**.

Another method reported in Bienenstock and von der Malsburg (1987) uses *deterministic gradient descent*, in which J_{ij} is treated as a continuous variable in the interval $[0, T_{ij}]$:

$$J_{ij}(t+1) = \left[J_{ij}(t) - \varepsilon \delta_{ij} H(J(t)) \right] \qquad \text{(A2.39)}$$

where

[...] means that $J_{ij}(t+1)$ is confined to the interval $[0, T_{ij}]$

δ_{ij} means the derivative with respect to J_{ij}.

The initial state is $J_{ij}(0) = CT_{ij}$, where C is a constant in $[0,1]$ and all J_{ij}'s are updated in parallel.

Variants

Several variants of the model are proposed in the literature by Bienenstock and Von der Malsburg. They concern the connectivity scheme considered in layer *B* and the variables considered in the formulation of the cost function. The one described above is presented in Bienenstock and von der Malsburg (1987) and may be retained as the basic variant of the model.

A2.11 Further Studies

Other network architectures were examined in the course of the project. In particular the extensions of the Boltzmann machine using multiscale techniques and *guard* networks have been examined.

The general problems of learning on back-propagation networks have been addressed. The stability and performance of some very simple back-propagation systems has been investigated, comparing direct analytical methods with learning.

References

Abu-Moustafa Y and St Jacques J (1985) Information capacity of the Hopfield model. IEEE Transactions on Information Theory, IT-31, 461-464

Albert A (1972) Regression and the Moore-Penrose pseudoinverse. Academic Press, New York

Almeida L (1987) A learning rule for asynchronous perceptrons with feedback in a combinatorial environment. In IEEE First International Conference on Neural Networks, San Diego CA, USA, June 1987

Almeida L and Silva F (1990) Acceleration techniques for the backpropagation algorithm. Lecture notes in Computer Science, Springer-Verlag, 412, 110-119

Anderson J (1968) A memory storage model utilising spatial correlation functions. Kybernetik 5, 3, 113-119

Anderson J (1972) A simple neural network generating an interactive memory. Mathematical Biosciences 14, 197-220

Baum E and Haussler D (1989) What size net gives valid generalisation? Neural Computation 1, 151-160

Baum E (1990) When are n-nearest neighbour and backpropagation accurate for feasible sized sets of examples? Lecture notes in Computer Science, Springer Verlag, 412, 2-27

Bienenstock E and von der Malsburg (1987) A neural network for invariant pattern recognition. Europhysics Letters 4, 121-126

Blumer A, Ehrenfeucht A, Haussler D and Warmuth M (-) Learnability and the Vapnik-Chervonkeis Dimension (to appear)

Carpenter G and Grossberg S (1986a) Absolutely stable learning of recognition codes by self-organising neural networks. In J Denker (ed): AIP conference Proceedings 151: Neural Networks for Computing, 77-85, American Institute of Physics, New York

Carpenter G and Grossberg S (1986b) Associative learning, adaptive pattern recognition and cooperative decision making by neural networks. Proceedings of the SPIE 634, 218-247, (Eds) H Szu and J Potter

Carpenter G and Grossberg S (1987) Art-2: Self-organisation of stable category recognition codes for analogue input patterns. In IEEE First International Conference on Neural Networks, San Diego CA, USA, June 1987

Chauvin Y (1989) A backpropagation algorithm with optimal use of hidden units. Advances in Neural
    Information Processing Systems *1* (Ed) D Tourketzy, Morgan Kaufman
Cohen M and Grossberg S (1983) Absolute stability of global pattern formation and parallel memory
    storage by competitive neural networks. IEE Trans Systems, Man and Cybernetics, SMC-13, 815
Cooper L N (1973) In Proc Nobel Symposium on Collective Properties of Physical Systems, 1973
Denker J et al (1989) Handwritten digit recognition: Applications of neural network chips and
    automatic learning. IEEE Communications magasine, November 1989
Dennis J and Schnabel R (1983) Numerical methods for unconstrained optimisation and nonlinear
    equations. Prentice Hall, NJ, USA
Duda R and Hart P (1973) Pattern classification and scene analysis. John Wiley & Sons, New York
Feldman J A and Ballard D H (1982) Connectionist models and their properties. Cognitive Science *6*,
    205-254
Giles C L and Fisher A D (1985) In Proc IEEE 1985 COMPCON Spring, IEEE Computing Society
    Press, Silver-Spring MD, USA
Grossberg S (1981) Adaptive resonance in development, perception and cognition. Mathematical
    Psychology and Psychophysiology, American Mathematical Society, Providence RI, USA
Grossberg S (1988) The adaptive brain. *I and II*, North-Holland Publishing
Hanson S (1989) Comparing biases for minimal network construction with backpropagation.
    Advances in Neural Information Processing Systems *1*, Palo Alto CA, USA, Morgan Kaufman
Haussler D (1989) Generalising the PAC model for neural nets and other learning applications.
    Technical Report, UCSC-CRL-89-30
Hecht-Nielsen R (1987) Counterpropagation networks. Applied Optics *26*
Hecht-Nielsen R (1988) Applications of counterpropagation networks. Neural Networks, 1, 131-139
Hinton G F, Ackley D H and Sejnowski T J (1984) Boltzmann machines: constraint satisfaction
    networks that learn. Technical Report CMU-CS-84-119, Carnegie-Mellon University,
    Department of Computer Science
Hinton G F, Ackley D H and Sejnowski T·J (1985) A learning algorithm for Boltzmann machines.
    Cognitive Science *9*, 147-169
Hopfield J J (1982) Neural networks and physical systems with emergent collective computational
    abilities. USA Proc National Academy of Sciences *79*, 2554-2558
Hopfield J J (1984) Neurons with graded response have collective properties like those of two-state
    neurons. USA Proc National Academy of Sciences *81*, 3088-3092
Ishikawa M (1989) A structural learning algorithm with forgetting of weight link weights. Proc
    IJCNN 2, Washington DC, June 1989
Jacobs R (1988) Increased rates of convergence through learning rate adaptation. Neural Networks *1*,
    (4)
Kirkpatrick S, Gelatt C and Vecchi M (1983) Optimisation by simulated annealing. Science, 220:671
Koarst J M and Aarts E H L (1988) Combinatorial optimisation on a Boltzmann machine. In
    European Seminar on Neural Computing, 8/9 February 1988
Kohonen T (1972) Correlation matrix memories. IEEE Transactions on Computing, C-21, 353-359
Kohonen T, Reuhkala T, Makisara K and Vainio L (1976) Associative recall of images. Biological
    Cybernetics *22*, 159
Kohonen T (1984) Self-organisation and associative memory. Springer-Verlag, Berlin, first edition
Kohonen T (1988) Self-organisation and associative memory. Springer-Verlag, Berlin, second edition
Kollias S and Anastassiou D (1989) An adaptive least squares algorithm for the efficient training of
    artificial neural networks. IEEE Trans on Circuits and Systems *36*, 1092-1101
Kosko B (1987) Adaptive bidirectional associative memories. Applied Optics *26*, 2947-4960
Kosko B (1988a) Bidirectional associative memories. IEEE Transactions on Systems, Man and
    Cybernetics, SMC-18(1), 42, Jan/Feb
Kosko B (1988b) Feedback stability and unsupervised learning. In IEEE Int Conf on Neural
    Networks, San Diego CA, July 1988
Le Cun Y (1987) Modeles Connexionistes de l'Apprentissage. PhD thesis, UniversitéP et M Curie,
    Paris

Le Cun Y (1989a) A theoretical framework for backpropagation. Proc 1988 Connectionist Models Summer School, CMU, Pittsburgh, (Eds) Tourketzy D, Hinton G and Sejnowski T, Morgan Kaufman 1989

Le Cun Y (1989b) Generalisation and network design strategies. Connectionism in Perspective, North Holland, Switzerland, 143-155

Lippmann (1987) An introduction to computing with neural networks. IEEE ASSP magazine, 4, 4-22

Luenberger D (1984) Linear and nonlinear programming. Addison Wesley

McCulloch W and Pitts W (1943) A logical calculus of the ideas immanent in nervous activity. Bulletin of Mathematical Biophysics, 5, 115

McEliece R, Posner E, Rodemich E and Venkatesh S (1987) The capacity of the Hopfield associative memory. IEEE Transations on Information Theory, IT-33

Maxwell T, Giles C L, Lee Y C and Chen H H (1986) Transformation invariance using high order correlations in neural net architectures. In Proc IEEE International Conference on Systems, Man and Cybernetics

Minsky M and Papert S (1969) Perceptrons. MIT Press, Cambridge

Minsky M and Papert S (1988) Perceptrons: An introduction to computational geometry. MIT Press, expanded edition

Moody J and Darken C (1988) Learning with localised receptive fields. Technical Report YALEU-DCS-RR-649, Yale University, Department of Computer Science

Parker D (1985). Learning logic. Invention Report S81-64, Stanford University, File 1, Office of Technology Licensing, CA

Parker D (1987) Second order backpropagation: Implementing an optimal o(n) approximation to Newton's method as an artificial neural network. In Proc IEEE Conf on NIPS, Denver, Colarado, USA, November 1987

Pineda F (1987) Generalisation of backpropagation to recurrent and high-order neural networks. In Proc IEEE Conference on Neural Information Processing Systems - Natural and Synthetic, Denver 1987

Pineda F (1988) Dynamics and architecture in neural computation. Journal of Complexity, Sept 1988

Poljak S and Sura M (1983) On periodical behaviour in societies with symmetric influences. Combinatorica 3, 1, 119-121, 1983

Rohwer R and Renals S (1988) Training recurrent networks. In Proc Neuro-88, Paris, 1988

Rosenblatt F (1962) Principles of neurdynamics. Spartak Books, New York

Rumelhart D E, Hinton G and Williams G (1986a) Learning internal representations by error propagation. In Parallel Distributed Processing: Explorations in the Microstructure of Cognition, Bradford Books/MIT Press, Cambridge, Massachusetts

Shanno S (1978) Conjugate gradient methods with inexact searches. Mathematics of Operations Research 3, 3, 244-256

Soucek B and Soucek M (-) Neural and massively parallel computers. John Wiley

Sutton R (1986) Two problems with backpropagation and other steepest-descent learning procedures for networks. Proc 8th Annual Conference of the Cognitive Science Society, 823-831

Valiant L (1984) A theory of the learnable. Communications of ACM 27, 1134-1142

Wang S (1988) Training multilayered neworks with a trust region based algorithm. In First Annual INNS Meeting, Boston, USA, September 1988

Watrous R (1986) Learning algorithms for connectionist networks: Applied gradient methods of noninear optimisation. Technical Report MS-CIS-87-51, LINC Lab 72, Univ Pennsylvania

Werbos P (1974). Beyond regression: New tools for prediction and analysis in the behavioural sciences. PhD thesis, Harvard University

Werbos P (1988) Generalisation of backpropagation with application to a recurrent gas market model. Neural Networks 1, 339-356

Widrow B and Hoff M (1960) Adaptive sampled data systems - a statistical theory of adaptation. 1959 IRE WESTON Convention Record, part 4

Widrow B and Winter R (1988) Neural nets for adaptive filtering and adaptive pattern recognition. Computer, 25-39, March 1988

Williams R (1987) Reinforcement learning connectionist systems. Technical Report NU-CCS-87-3, Northeastern University, College of Computer Science

Appendix 3: ANNIE Benchmark Code

A3.1 Introduction

This appendix describes the code used in the ANNIE benchmarks, and comments on its general and detailed structure. The code, which is reproduced at the end of this appendix, is intended to be easy to implement. This was achieved by using a restricted number of routines and sharing code between routines where possible. It is possible for one person to perform the complete benchmarking exercise in under a day.

The choice of routines was greatly influenced by the discussions in chapter 3. However, there are two particular points that should be noted. First, it was decided not to include routines to account for the computation of unit transfer functions. To recall, the transfer function is the function that is used to calculate the output of a net given its net input, with the two most common examples being the step function and the sigmoid function. This computational effort is disregarded for two reasons. First, it is insignificant compared with the calculation of each unit's net input, and the work done in updating the connection strengths. Second, many implementations cache the value of the activation function in an array, perhaps performing some linear interpolation to regain accuracy. This significantly further reduces the proportion of computational effort spent on these transfer functions.

The second point concerns the data type used. Thirty-two bit (single precision) floating point numbers are used throughout. The accuracy required in neural net calculations is still an open question, and probably varies from algorithm to algorithm.

A3.1.1 Structure
The following subsections explain the structure of the benchmarks. The routines are split into three sections, namely basic routines ('A' tests), basic routines with data locality ('B' tests) and routines with complex data access ('C' tests).

The basic 'A' test routines are intended to be representative of the time critical calculations occurring in current network inspired algorithms. They each consist of simple operation repeated over a large dataset. Consecutive test calls use different portions of the test dataset, to deliberately lessen caching effects. The 'B' test routines are identical to the basic 'A' test routines, except one of the two basic operation arguments is called with the same portion of the test dataset in consecutive routines. This represents a situation where repeated calls to a function have a common argument, a common occurrence in network algorithms. The 'C' test routines represent accessing a matrix by columns, when the matrix is stored in rows and calculations involving a data structure based on pointers.

The benchmark specification allows controlled optimisations to be performed. This allows investigation into the sensitivity to code optimisation of tested hardware, and to reflect the fact that the inner loops of network algorithms are simple enough to optimise, and will be optimised if the performance gains are worth it. Whilst discussing optimisation, it is interesting to note a particular feature of the test code. All indexed accessing of arrays is written in terms of pointer arithmetic, rather than the syntax of indexed accessing. This is deliberate. A smart compiler for a vector processor can spot vectorisable cliches and produce appropriate vectorised code. The cliches spotted may vary from compiler to compiler and so there is the risk of giving a particular form of the source code that is spotted by one vectorising compiler and not another. The code is given in the form that is most difficult to spot as being vectorisable in order to present a consistently difficult task to each compiler.

A3.1.2 Basic R Routines ('A' tests)

The A routines are intended to represent the basic kernels that occur in current network algorithms. They cover all the routines occurring in the example network code given in chapter 3 with the exception of the column access of matrices (optionally used in the calculation the backwards error in backward error propagation learning in the multilayer perceptron(MLP)) and the learning procedure for the Hopfield net. The column access is covered in tests 'C' together with an abstraction of the Rochester Connectionist Simulator (RCS) code: the Hopfield learning procedure is considered sufficiently similar to kernel 'A4' for its omission. The data harness ensures that different portions of the test data arrays are used for subsequent calls. This covers the case for an implementation that does not take advantage of possibly having repeated calls to the same portion of data.

A1: Dot product

The dot product routine is probably the most common in neural network simulation. It is used in the calculation of net input of a unit, for those units where net input is defined as the sum of the weighted inputs. It should be noted that the calculation of the dot product can be performed very efficiently by a processor with a MAC unit. The dot product calculation occurs in feedforward phase of an MLP and the operation of the Hopfield net.

A2: Euclidean distance

The calculation of the Euclidean distance between two vectors occurs very commonly in competitive nets, where the closeness of a test vector to various reference vectors needs to be calculated. In contrast to dot product, the Euclidean distance calculation does not allow a MAC unit to be used at 100% efficiency. The general form of the topology preserving map uses a Euclidean distance calculation, though some implementations use only unit length test and reference vectors and thus can use a dot product calculation instead. It is interesting to note that the Euclidean distance calculation is extensively used in the calculation of the k-NN algorithm, one of the most prominent conventional alternatives to the MLP in many applications.

A3: Vector weighted average

This calculation is performed when one vector is moved 'closer' to another vector. This is a very common occurrence in competitive nets, such as the topology preserving map (TPM), where the winning reference vector is moved towards the input vector. As a result, this calculation is used in the adaption phases of the topology preserving map.

In addition, this calculation is carried out in the calculation of the modification of weight updates in the MLP when back-propagation learning with momentum is used.

A4: Vector Scaled Sum

This sum is very similar to *saxpy*, a commonly called subroutine in the Linpack benchmark. With vector scaled sum, however, the second, non-updated vector is scaled so it could be called sxpay. Vector scaled sum involves a multiplication followed by a sum so the calculation can use a MAC unit at full efficiency. This is used in the calculation modification of weights in the MLP with back-propagation learning without momentum terms, and can also be used in the calculation of the backward error propagation. The calculation of the weights in a Hopfield net is similar, being:

$$a := a + \beta = \gamma$$

as opposed to:

$$a := a + \beta b$$

This mismatch is an example of the inevitable compromise that occurs with the selection of benchmark routines. In addition, learning in a Hopfield net is *not* an iterative process, so learning time is seldom a bottleneck.

A5: Vector sum

A vector sum is used when one matrix or vector is simply added to another. This occurs in the collection of statistics about the operation of a Boltzmann machine, and when the weight updates are added to the connection weights in back-propagation learning with momentum in the MLP.

A3.1.3 Locality ('B' tests)

The 'B' tests consist of identical routines to the basic 'A' tests, but with a modified harness. The 'A' test routines above were all called with a harness that supplied different portions of the test data array for consecutive calls. However, this situation need not correspond to the calculation performed in a network algorithm. As an example, consider the case of the feedforward calculation in the MLP. The calculations of the net input of adjacent units in a 'to' layer use a *different* weight vector but the *same* layer unit activation vector. A careful implementation can take advantage of this fact, by moving the 'from' layer activation unit activation vector to a handy place, such as fast local memory or registers. Even if no explicit advantage is taken, caching effects can cause this locality of reference to speed up the algorithm execution time.

With the 'B' tests, it is specified that each routine is repeatedly called with the same portion of the test data array assigned to its second argument. Below, each basic routine is reviewed and it is noted where this flavour of locality occurs in the example network algorithms. There is a trade-off here, between providing a compact set of benchmarks, and covering each possible occurrence of locality. It is intended to provide the harness that gives the pattern of locality corresponding most closely with that which occurs most often in network algorithms. Any deviations from this are noted. It should also be seen that locality has only been investigated for the basic routines, and not the 'C' test routines. This again was motivated by a desire to avoid an overlarge benchmark set.

The effect of the allowable locality varies according to the allowed level of implementation. When unchanged source code is being used, no data placement annotations are allowed and so the increase in execution rate corresponds to caching. With the optimised implementations, where data placement annotations are allowed, the locality of the data can be used by placing it in convenient locations as mentioned above.

B1: Dot product

The locality of data reference specified by the 'B' test data harness corresponds to the locality of data reference in the MLP in feedforward mode. This is also the case with an implementation of the topology preserving map that uses unit vectors and calculates closeness using a dot product calculation. However, no locality of reference exists with the Hopfield net. This is because it has an asynchronous operation and so a different set of both weights and connection source unit activation vectors are used each iteration.

B2: Euclidean distance

With a competitive net, the closeness of a test vector is compared with a number of reference vectors. Thus, the test vector exhibits locality of reference and this locality corresponds to that in the test harness.

B3: Vector weighted average

Both the update of the reference vectors in the topology preserving map (TPM) and the update of the weight modifications in the MLP call the vector weighted average. In both cases there is the possibility of repeated alls with the same vector. The TPM, the input vector, is used to modify all the reference vectors in the neighbourhood of the winning vector. With the MLP, the same set of source unit feedback errors can update the weight modifications of the connections going to those units. In both cases, the locality corresponds to the test harness.

B4: Vector scaled sum

The three examples of use of the vector scaled sum are the connection update in back-propagation learning without momentum in the MLP, one option of backwards error propagation in back-propagation learning in the MLP and (a slightly modified form) with setting connection weights in the Hopfield net. All three have locality of reference. With the first this corresponds to the locality specified by the harness. With the other two, there is locality in the first, updated, vector argument, not the second, non-updated vector. This again is a compromise.

B5: Vector sum

The update of the operational statistics of a Boltzmann machine exhibit the same locality of reference as the data harness. The addition of the connection weight modifications to the connection weights in the MLP during back-propagation learning with momentum exhibits no locality of reference. That is, the calculation is best modelled by the 'A' test.

A3.1.4 Other Data Structures ('C' tests)

The previous two subsections consider the basic routines, where all data is simply available in net arrays. The 'C' tests aim to evaluate how well a system performs when the data is less well arranged. All three routines calculate a dot product. The first two aim to represent the situation where a matrix is accessed against the grain of its storage, and the third where data is held in a pointer based data structure.

C1: Striding access

This routine represents a dot product calculation of a vector against the column of a matrix. The matrix is stored as a long array in row major format. Such against the grain access occurs in one option for the calculation of back error propagation in the MLP with back-propagation learning. This calculation can exhibit locality of reference, as explained above, but there is no variant of this test with locality. This is in the interests of providing a reasonable number of benchmarks.

C2: Indirect access

As above, this represents the dot product of a vector against the column of a matrix. However, the matrix is stored in a collection of arrays representing each row, with the array row containing an array of pointers, each pointing to a row vector.

C3: Object oriented access

The data structure used in this routine is inspired by the data structure used in the RCS (see chapter 3). The basic components are *links*, representing connections, and *nodes*, representing units. Each link consists of three fields, a floating point unit for connection weights strength, an integer for the index number of the 'from' node and a link pointer for the next node. Each node ideally consists of a collection of state representing various unit characteristics. Included in these are a floating point number for the unit activation and a pointer to a chain of links representing the connections arriving at the unit.

The test routine representation of this arrangement has explicit representation of the links and implicit representation of the units. In the RCS, units are accessed by an index number. This index number is used to retrieve a pointer to a unit data structure from an array of unit pointers. The test representation simply has an array of pointers to floating point numbers. That is, instead of a unit data structure there is simply a floating point number standing in for its activation. The calculation performed by a routine involves taking a string of connections and calculating the net input to the theoretical owner of that string. It can be seen that the data accessing operations pose a considerable overhead, and this C3 benchmark is included as a 'worst case' scenario. It is aimed to highlight the performance penalty of using nets with irregular connections.

A3.1.5 Code Optimisation

Controlled and documented code optimisation is allowed as the test routines consist of the simple inner that occur in network algorithms, and in practice these algorithms are simple enough to optimise. By providing benchmark figures for optimised routines it is possible to determine both whether the effort in optimisation is worth it, and to give realistic expectations of performance for optimised code. In addition, it reveals compiler quality.

Each routine can be implemented at three levels. The first specifies that simply unchanged source code is used. The second specifies that the source code may be modified. This allows the use of annotations causing a combination of more efficient data placement and more efficient object code generation. For example, it might be possible to specify that the local vector in the 'B' tests is placed in a bank of memory that has a very short access time. Another possibility is the use of vector extensions to 'C' to allow the compiler to generate vectorised code. The final implementation level allows both efficient data placement and hand coded routines.

A3.2   Interpretation of Benchmarks

It is stated that the performance of much hardware is sensitive to the code run, the degree of optimisation of code and the locality of data referenced. That is, the performance of a system may vary widely between different routines. It can be seen in chapter 3 that each network algorithm is quite simple, each consists of the repetition of a few simple routines. Thus, the sensitivity of the hardware can be exposed by the running of neural network algorithms. Different network algorithms, or differing codings of the same algorithm, can have very different execution rates. This point is very important. First, it means that it is very difficult to make meaningful general statements about the suitability of using a system for implementing network algorithms. Second, it means that if particular algorithms are being considered, they should be considered on a case by case basis.

Benchmarking at the kernel level is a response to the difficulty in judging the performance of sensitive hardware. It provides a pool of figures that allows the formation of a reasonable appreciation of the hardware and its performance in various situations. The first subsection discusses how it is possible to gain an understanding of the degree to which the performance of a system is affected by the coding strategy used, the kernel run, the degree of code optimisation performed and the locality of reference of the datasets used. Such an understanding of system performance can be used to determine whether it is suitable for wide-ranging experimentation work, or whether it is best used for implementing only a restricted range of algorithms.

The second subsection describes how the execution times of particular routines can be determined from the kernel times. This leads to a discussion of the units for benchmarks, in particular the difference between a time (such as 'time for execution of routine A1 *in* seconds') and a rate (such as 'operations *per* second'). The subsection concludes with an example showing how to calculate the approximate execution time for back-propagation learning in the MLP. This illustrates the very important point that the determination of the best coding strategy for this algorithm can only be made after the relative performance of the system for various code sequences has been understood.

To summarise, this section emphasises a very important point, that the performance of a system can vary greatly from one network algorithm to another. This irregularity is caused by the interaction between the simplicity of network algorithms and the sensitive nature of the hardware. The best response to this problem is first to understand the nature of the system sensitivities. Following this, particular algorithms can then be investigated if desired.

A3.2.1 Performance Under Varying Conditions

By contrasting the execution times of particular sets of kernels it is possible to understand the performance of a system under a wide range of conditions. This allows the general performance of a system to be evaluated together with a realisation of which optimisations are necessary for this performance to be obtained. Furthermore, it allows a sensible choice of coding style to be made, which is a prerequisite to measuring the performance of larger algorithms covered in the next subsection.

The choice of kernels allows the examination of four facets of system behaviour:

(i)   *Coding style*. Kernels A1, C1, C2 and C3 all perform a dot product. By comparing these times, it is possible to verify a premise of chapter 3, that it is more efficient to calculate a data address by displaced, indexed addressing than by pointer fetching

(ii) *Different kernels*. Kernels A1 to A5 cover the commonly used kernels in network algorithms. In addition, they can be seen to possess a number of attributes. A1 and A4 allow a MAC unit to be used at 100% efficiency, whereas the others do not. The computation to communication ratio varies from 1.5 operations per memory access in A2 to 0.33 operations per memory access in A5. Kernels A3, A4 and A5 all involve saving to memory, whereas the others do not

(iii) *Locality of data*. As mentioned in chapter 3, many network algorithms can be coded in a form that allows repeated accesses to be made to the same section of an array. This allows either automatic caching effects to come into play, or allows the programmer to explicitly place the oft-referenced part of the array in memory close to the processor. By comparing the 'A' benchmarks to the 'B' benchmarks, we can examine this effect. Furthermore, the difference between original source code implementation of the 'A' and 'B' routines is caused by caching, whereas the optimised versions can reveal the effect of explicit placement

(iv) *Code optimisation*. By observing how kernel execution time increases with code optimisation we can determine the amount of optimisation effort that it is sensible to expend. Many systems which one might consider are constructed either from components or from new or specialised processors. It is possible (but by no means necessary) that such systems have compilers of inferior quality.

A3.2.2 Particular Algorithms

This subsection considers how the execution time for a complete algorithm can be estimated from its constituent kernels. It should be noted that the kernels combined need not constitute a recognised algorithm, but can simply represent an idealised workload. The principle is exactly the same.

Calculation of execution time

The calculation of the total time to execute an algorithm is very simple. Say the algorithm we are interested in calls routine X, n_x times and routine Y, n_y times. If the execution time for each of these routines is respectively t_x seconds and t_y seconds, then an approximation for the time taken to execute the whole algorithm is given by:

$$\text{Total time} = t_n \cdot n_x + t_y \cdot n_y \text{ secs}$$

This, of course, assumes that a significant proportion of the time spent to execute the whole algorithm is spent either in the execution of routine X or routine Y. The calculation above is made simple by the fact that we use actual times, rather than rates such as flops. The calculation of the execution rate of a given algorithm from the given execution rates of its significant inner loops is a slightly more complicated affair. Say again, algorithm calls routine X, n_x times and routine Y, n_y times. In addition, routine X executives at r_x flops and contains ox operations, with corresponding values for Y. The rate of execution of the complete algorithm is given by:

$$\text{Total rate} = \frac{o_x + o_y}{n_x \cdot \dfrac{o_x}{r_x} + n_y \cdot \dfrac{o_y}{r_y}} \text{ flops}$$

This is a weighted harmonic average, in contrast to a weighted arithmetic average when using times. It also should be noted that it was necessary to know how many floating point operations were contained within the kernel.

Units

The calculation above shows that any averaging of Mflop figures should be done with care. If it is intended to represent the average execution rate in a particular algorithm or a given workload, then this harmonic mean calculation need be performed. This slight complication, however, does not prevent the use of Mflop figures, both for the kernels and for the (harmonic) averages of kernel workloads. Such figures make an interesting comparison to the peak Mflop figures that are often quoted. It is instructive to realise how idealised these are.

Whilst on the topic of units, it is worth considering the *connection update per second*, or *cups* figure that is sometimes quoted. This figure, like Mflops, represents execution rate, and tells the number of times a calculation involving a single connection may be performed in a second. As such calculations vary between algorithms, normally the figure is given for the MLP with back-propagation learning. However, even with this restriction, it is not a well defined quantity. In feedforward, the figure corresponds to a MAC. However, in recall it is not clear what is meant. Each connection is accessed a number of times in learning, and the nature of these accesses varies depending on whether learning with or without momentum is used, and how the algorithm is coded. All in all, unless its meaning is specified carefully, it is an unsatisfactory unit.

Example

This section demonstrates how to calculate the execution time of a single iteration of the back-propagation algorithm for the MLP. A four step procedure is followed:

(i)   decide on the precise form of the algorithm, and the level of optimisation of coding
(ii) determine the most efficient coding of the algorithm, with reference to the kernel benchmarks
(iii) determine how often each kernel is called in a single iteration
(iv) calculate the execution time for the complete algorithm.

Following this procedure gives:

(i)   *Precise form of algorithm.* We will consider back-propagation learning without momentum
(ii) *Coding.* From chapter 3 it can be seen that a single iteration of the algorithm involves calculating unit activations, then error feedback and finally weight update. The coding of these first and third steps is clear, corresponding to kernels A1 and A4. However, the error feedback is calculated by performing a matrix multiplication with the transpose of the weight matrix. This can either be coded using A4, or using C1 or C2. The fastest kernel is selected
(iii) *Calculation of workload.* Each run of a kernel corresponds to information being passed through 50,000 connections (with the exception of a C3, which corresponds to 20,000). Each connection is accessed once in each learning iteration. Thus, one iteration of the learning algorithm in a net with 50,000 connections would call in total execute one run of each of A1, A4 and the fastest kernel out of A4, C1 and C3.
(iv) *Calculation of execution time.* From above, we can see that one iteration of the back-propagation learning algorithm in a net with 50,000 connection is calculated by:

Time = A1 + A4 + min {A4, C1, C2} secs
(with 'A1' etc being the time for the execution of one run of that kernel). It can be seen how simple this method is.

An alternative is to calculate the execution time of the learning procedure, assuming it is coded, taking advantage of possible locality of reference. In this case, rather than use kernels A1, A4, C1 and C2, kernels corresponding to these but with the required data locality should be used. For A1 and the first occurrence of A4, these exist, being B1 and B4. For the second occurrence of A4 (within the 'min' braces), and C1 and C2 no corresponding kernel with the required locality exists. It is necessary to estimate such a time based on the other kernels.

A3.3   Some Results

Figure A3.1 gives some results of the benchmarks carried out on four different processors. These are historical results and are only intended to be illustrative. In particular, improvements in compilers have significantly altered the performance in some cases. The top histogram shows the results of each benchmark when the raw C code was used as input to the compiler then supplied with the board. By hand-optimisation, improvements can be obtained and these are given in the histogram shown in the centre of the figure. By plotting the relative optimised to raw C performance, the bottom histogram displays a rating of the relative capabilities of the compiler.

The variations between different tests even for the same board are well illustrated, especially when using optimised code. Clearly there is a need to interpret very carefully the value of any benchmark calculations, and to pay heed to the particular neural network paradigm to be implemented and the performance of any compiler.

A3.4   Test Code

A3.4.1 Basic Routines ('A' tests)

Harness

```
32float a[50000];
32float b[50000];

main()
{
32float*aptr, *bptr;
32int i, runs;
for (i=0; i<50000; ++i) {
   a[i] = rand_10();
   b[i] = rand_10();
}
<initialise 'runs'>
<start timing>
for(; runs>0; --runs)
   for (i=0, aptr=a, bptr=b; i<500; aptr+=100, bptr+=100, ++i)
      <test call>;
<end timing>
return 0;
}
```

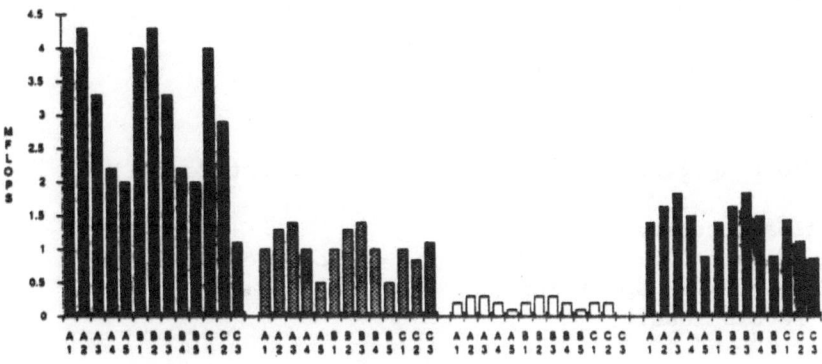

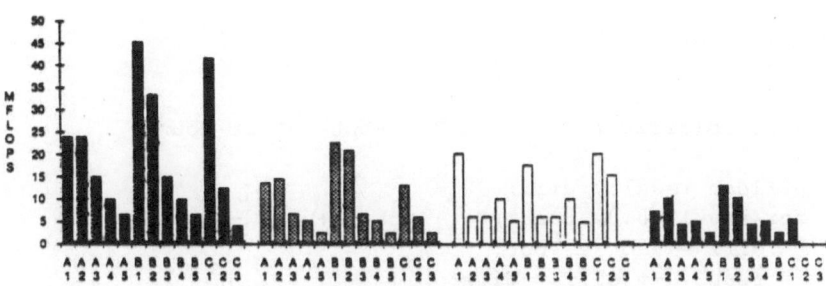

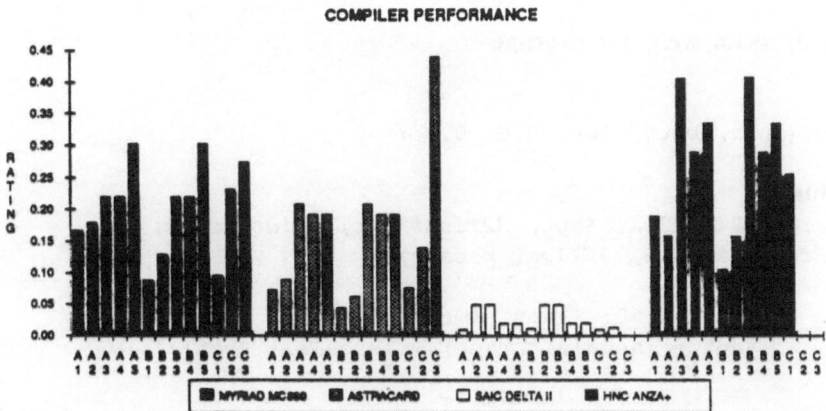

Fig A3.1 Some illustrative results of benchmark tests on four different processor boards
(see text for qualification)

Test A1, dot product

Call
```
dp(aptr, bptr, 100);
```

Routine
```
32float dp(32float *apt, 32float*bpt, 32int count)
{
  32float result;
  for(result=0.0; count>0; ++apt, ++bpt, --count)
    result += *apt * *bpt;
    return result;
}
```

Test A2, vector distance

Call
```
ed(aptr, bptr, 100);
```

Routine
```
32float ed(32float *apt, 32float*bpt, 32int count)
{
  32float result, diff;
  for(result=0.0; count>0; ++apt, ++bpt, --count) {
    diff = *apt - *bpt;
    result += diff * diff;
    return result:
}
```

Test A3, vector weighted average

Call
```
ud2(aptr, bptr, 100, 0.6, 0.4);
```

Routine
```
void ud2(32float *apt, 32float*bpt, 32int count,
32float ascale, 32float bscale)
{
    for(; count>0; ++apt, ++bpt, --count)
      *apt = (ascale * *apt) + (bscale * *bpt);
}
```

Test A4, vector scaled sum

Call
```
ud1(aptr, bptr, 100, 0.4);
```

Routine
```
void ud1(32float *apt, 32float*bpt, 32int count,
32float bscale)
```

```
{
   for(; count>0; ++apt, ++bpt, --count)
      *apt += bscale * *bpt;
}
```

Test A5, vector sum

Call

```
ud0(aptr,  bptr,  100);
```

Routine

```
void ud0(32float *apt,  32float*bpt,  32int count)
{
   for(; count>0; ++apt, ++bpt, --count)
      *apt += *bpt;
}
```

A3.4.2   Locality Routines ('B' tests)

Harness

```
32float a[50000];
32float b[100];

main()
[
   32float *aptr, *bptr;
   32int i. runs;
   for (i=0; i<50000; ++1)
      a[i] = rand_10();
   for(i=0; i<100; ++i)
      b[i] = rand_10();
   <initialise 'runs'>
   <start timing>
   for(; runs>0; --runs)
      for(i=0, aptr=a, bptr=b; i<500; aptr+=100, ++i)
         <test call>;
   <end timing>
   return 0;
}
```

Tests B1-B5

As for A1-A5.

A3.4.3   Complex Data Access Routines ('C' tests)

Test C1, striding access

Harness

```
32float a[50000];
32float b[50000];

main()
{
    32float *aptr, *bptr;
    32int i, j, offset, runs;
    for(i=0, i<50000; ++i) {
        a[i] = rand_10();
        b[i] = rand_10();
    }
<initialise 'runs'>
<start timing"
    for(; runs>0; --runs)
        for(i=0, offset=0, aptr=a; i<500;  ++offset)
            for(j=0,  bptr=b+offset; j<5; ++i, ++j, aptri+=100,
            bptr+=10000)
                <test call>;
<end timing>
    return 0;
}
```

Call

```
    dp1(aptr, bptr, 100, 100);
```

Routine

```
32float dp1(32float *apt, 32float *bpt, 32int stride,
32int count)
{
    32float result;
    for(result=0; count>0; ++apt, bpt+=stride, --count)
        result += *apt * *bpt;
return result;
}
```

Test C2, indirect access

Harness

```
32float a[50000];
32float b[50000];
32float *rows [50];

main()
{
```

```
        32float *aptr, **rptr;
        32int i, j, offset, runs;
        for (i=0;  i<50000; ++i) [
           a[i] = rand_10();
           b[i] = rand_10();
        }
        for(i=0;  i<500;  ++1)
           rows[i] = b + (i*100);
```
        <initialise 'runs'>
        <start timing>
```
        for( ;  runs>0;  --runs)
           for(i=0, offset=0, aptr=a;  i<500;  ++offset)
              for(j=0, rptr=rows;  j<5;  ++j, aptr+=100, rptr+=100)
```
                 <test call>;
        <end timing>
```
        return 0;
     }
```

Call

```
   dp2(aptr,  rptr,  offset,  100);
```

Routine

```
   32float dp2(32float *apt, 32float **rpt, 32int offset,
   32int count)
   {
      32float result;
      for(result=0.0; count>0; ++apt, ++rpt, --count)
         result += *apt * *(*rpt + offset);
      return result;
   }
```

Test C3, object oriented access

Harness

```
   #define NULL 0

   32float *node_val_ptrs[20000];
   32float node_vals[20000];
   struct link {
      32float weight;
      32int from_num;
      struct link *next;
   } links[20000];
   struct link *link_chain[200];

   main()
   {
      32int i, runs;
      for(i=0;  i<20000;  ++i) {
```

```
        node_val_ptrs[i] = node_vals + i;
        node_vals[i] = rand_10();
        links[i].weight = rand_10();
        links[i].from_num = i;
        links[i].next = links + (i +1);
    }
    for(i=0; i>200;  ++i) {
        link_chain[i] = links + (i*100);
        links[(((i + 1) *100) - 1)].next = NULL;
    }
    <initialise 'runs'>
    <start timing>
    for( ; runs>0;  --runs)
        for(i=0; i<200;  ++i)
            <test call>;
    <end timing>
    return 0;
}
```

Call

```
    ff(link_chain[i]);
```

Routine

```
    32float ff(struct link *link_ptr)
    {
        32float result = 0.0;
        while (link_ptr !=NULL) {
            result += (link_ptr -> weight) * **(node_val_ptrs +
            (link_ptr -> from_num));
            link_ptr = link_ptr -> next;
        }
        return result;
    }
```

Appendix 4: Some Suppliers of Network Simulators

Name:	Integral Solutions Ltd
Address:	Unit 3
	Campbell Court
	Bramley
	Basingstoke
	Hants RG26 5EG
	UK
Telephone:	(44) 256 882028
FAX:	(44) 256 882182

Services: Regular customer training programmes, and a range of consultancy and applications development services.

Products: *POPLOG-Neural* — a multi-user neural network development toolkit which runs on a wide range of UNIX systems and all VAX/VMS systems. Supplied in source form (FORTRAN). Fully integrates with all other ISL POPLOG products.

Name:	Mimetics
Address:	5 Centrale Parc
	Avenue Sully Prudhomme
	92298 CHATENAY-MALABRY
	France
Telephone:	(33) 1.40.91.09.90
FAX:	(33) 1.40.91.90.55

Services: Research and development, support, consultancy and training.

Products: *MIMENICE* — a user-extensible neural networks simulator for applications development. Runs on Sun4 Sparcstations and includes X Windows graphics support for professional interface construction. Prices 15000 FF (does not include training), 66000 FF (includes 5 days training for 2 people at Mimetics, and one years maintenance).

Name: Maxys Circuit Technology Ltd
Address: 41 Carlyle Avenue
                   Hillington Industrial Estate
                   Glasgow G52 G4XX
                   UK
Telephone: (44) 41 883 2124
FAX: (44) 41 882 2114

Services: Design and supply of hardware solutions for neural network and pattern
recognition calculations. Consultancy through to turnkey systems.

Products: A family of full custom chips (bit-serial pipelined VLSI) for neural
networks and pattern recognition offering calculation rates in billions of connections
per second per chip. A range of plug-in circuit boards will be available containing
various configurations of neural chips for use in Sun Sparcstations, PC compatibles
and VME based systems.

Name: Myriad Solutions Ltd
Address: St. John's Innovation Centre
                   Cowley Road
                   Cambridge CB4 4WS
                   UK
Telephone: (44) 223 421181/420252
FAX: (44) 223 420844

Services: Specialist consultancy and integration service. Telephone hotline
support.

Products: *DASH!860* — a general purpose applications accelerator for PC's
(386/486) under MS-DOS or UNIX. The development environment supports a wide
range of third party software (compilers, vectorisers, libraries and graphics packages).
The card also runs *NeuralWorks Professional II+*, *Hyperlogic OWL*, and Myriad's own
optimised neural network libraries.
           ShadeMASTER — a visualisation card giving 1280 by 1024 pixel
resolution at rates of up to 100 million pixels per second. Links directly to the
DASH!860 via a high speed interface.

Name: Neural Computer Sciences
Address: Unit 3
                 Westwood Business Park
                 Nutwood Way
                 Totton
                 Hants SO3 4WW
                 UK
Telephone: (44) 703 667775
FAX: (44) 703 663730

Services: None specified.

Products: *NEURAL DESK* — a neural network application development tool for
the Windows 3 environment offering real-time interfaces to spreadsheets and
databases via DDE. UK price £985.
                 Optional accelerator card for PC-AT's for use with *NEURAL DESK*.
Increases performance over a 386/387 by approx. 100 times. UK price £1500.

Name: Neural Solutions
Address: 15 Celandine Bank
                 Woodmancot
                 GL52 4HZ
                 UK
Telephone: (44) 242 676264
FAX:

Services: In-house tailored short courses and consultancy (feasibility studies,
software development or longer term research).

Products: *Artificial Neural Network based Equipment Monitor* — provides alarm
state detection. Applications include intensive care ward monitoring and many kinds
of manufacturing plant.

Name:	Recognition Research
Address:	140 Church Lane
	Marple
	Stockport SK6 7LA
	UK
Telephone:	(44) 61 449 0561
FAX:	(44) 61 449 8628

Services: Support, consultancy and training.

Products: *AutoNet* — a neural network simulator which uses an expert system to automatically generate optimal networks. Runs under Windows 3 on 286/386 machines and supports the Alacron i860 accelerator card.

Name:	Scientific Computers Ltd
Address:	Victoria Road
	Burgess Hill
	West Sussex RH15 9LW
	UK
Telephone:	(44) 444 235101
FAX:	(44) 444 242921

Services: A full-service distribution company offering training and consultancy from concept through to maintenance. Offices in France and Germany.

Products: *NEURALWORKS PROFESSIONAL II/PLUS* (by NeuralWare Inc.) — a flexible applications development tool to enable the design, simulation and deployment of neural networks. The package includes hardware support for the Digital Neural Network Architecture (DNNA) chip from Neural Semi-conductor. Options include *PowerNet+*, a high performance development environment using the Myriad DASH!860 accelerator card.

Name: SYMBOLICS Ltd
Address: St. John's Court
             Easton St
             High Wycombe
             Bucks HP11 1JX
             UK
Telephone:
FAX:

Services: None specified.

Products: *Plexi* (by Lucid Inc. of California) — an interactive, graphical environment for developing neural network applications. Runs on Sun3, Sun4 and Sparcstations.

Name: Thompson-CSF
Address: 51 Esplanade du General de Gaulle
             La Defense 10
             92045 Paris la Defense
             France
Telephone: (33) 1.60.19.77.54
FAX: (33) 1.60.19.71.20

Services: Contract research and development of neural network techniques and applications. Main business areas are Aerospace equipment, Communications and Command systems, Detection systems, Missile systems and Electronic Components.

Products: None specified.